D1261537

SETTLEMENT AND METAMORPHOSIS OF MARINE INVERTEBRATE LARVAE

SETTLEMENT AND METAMORPHOSIS OF MARINE INVERTEBRATE LARVAE

Proceedings of the Symposium on Settlement and Metamorphosis of Marine
Invertebrate Larvae, American Zoological Society Meeting, Toronto, Ontario, Canada,
December 27-28, 1977

Editors:

Fu-Shiang Chia
Department of Zoology, University of Alberta, Edmonton, Alberta, Canada

Mary E. Rice
Department of Invertebrate Zoology, National Museum of Natural History,
Smithsonian Institution, Washington, D.C., U.S.A.

ELSEVIER · NEW YORK
NEW YORK · OXFORD

© 1978 by Elsevier/North-Holland Biomedical Press

Published by:
Elsevier North-Holland, Inc.
52 Vanderbilt Avenue
New York, New York 10017

Sole distributors outside the U.S.A. and Canada:
Thomond Books
(A Division of Elsevier/North-Holland Scientific Publishers, Ltd.)
P.O. Box 85
Limerick, Ireland

Library of Congress Cataloging in Publication Data

Symposium on Settlement and Metamorphosis of Marine
    Invertebrate Larvae, Toronto, Ont., 1977.
    Settlement and metamorphosis of marine invertebrate larvae.

    Bibliography: p.
    Includes index.
    1.   Marine invertebrates—Metamorphosis—
    Congresses.   2.   Larvae—Invertebrates—Congresses.
    I.   Chia, Fu-Shiang, 1931-   II.   Rice, Mary E.
    III.   Title.
QL363.5.S95   1977   592'.03'34   78-15934
ISBN 0-444-00277-4

Manufactured in the United States of America

Professor Robert L. Fernald, to whom this Symposium is dedicated.

Robert L. Fernald received his Ph.D. in 1941 from the University of California, Berkeley. He taught in the Department of Zoology, University of Washington, Seattle, from 1946 to 1975 and was the Director of the Friday Harbor Laboratories for 14 years (1958–1972). He retired from the Zoology Department in 1975 and is now living in his house on Griffin Bay, San Juan Island. He continues to teach the course in Comparative Invertebrate Embryology at the Friday Harbor Laboratories.

CONTENTS

PREFACE

During recent years considerable attention has been directed to studies of reproduction and development of marine invertebrates. This interest has been evidenced by the publication of numerous research papers, review articles (*e.g.*, Chia,[1] Mileikovsky,[2,3] Scheltema,[4] Stearns[5]), textbooks (Kume and Dan,[6] Reverberi[7]), reference books (Anderson,[8] Jägersten[9]), a multi-volume treatise (Giese and Pearse[10,11,12,13]), and various symposia. The American Society of Zoologists has sponsored the following series of symposia dealing with invertebrate development: Developmental Biology of Cnidaria in 1972,[14] Developmental Biology of Echinoderms in 1973,[15] Developmental Biology of Spiralia in 1974,[16] and Effects of Environmental Pollutants on Developing Invertebrates in 1976. Other recent sysmposia which have been devoted entirely or in part to invertebrate larvae are the Fourth European Marine Biology Symposium,[17] the Conference on Marine Invertebrate Larvae,[18] held in Rovinj, Yugoslavia, and the Symposium on Reproductive Ecology of Marine Invertebrates,[19] sponsored by the Belle W. Baruch Institute for Marine Biology and Coastal Research, University of South Carolina. None of these symposia has emphasized settlement and metamorphosis of marine larvae.

In view of the fact that there are at least 100,000 species of marine benthic invertebrates which have pelagic larval stages of two to several weeks'duration,[20] it is not surprising that problems of larval settlement and metamorphosis are diverse and complicated. In terms of larval settlement some very exciting discoveries, especially in relation to induction of settlement by specific substrata, have been made in recent years. This literature has been reviewed by Crisp[21,22] and Scheltema.[4] At present the literature on metamorphosis of marine larvae is rather scattered and remains largely at the level of descriptions of morphological changes. For most marine organisms our information can hardly compare with what is known about metamorphosis of insects and amphibians.[23]

It is our judgment that many more biologists in the near future will turn their attention to the mechanisms of induction of both settlement and metamorphosis, and it is expected that results of such investigations will contribute significantly to the general fields of developmental biology and marine ecology. It seemed, therefore, both timely and important to bring together persons working in the field of invertebrate development for an exchange of ideas and synthesis of existing information. In view of these considerations, the Symposium on Settlement and Metamorphosis of Marine Invertebrate Larvae, the proceedings of which are reported in this volume, was conceived and organized.

The Symposium was sponsored by the American Society of Zoologists at its annual meeting in Toronto in December, 1977. Divided into three sessions, it extended over a period of one and one-half days (December 27, 28) and included 19 papers, consisting of reviews as well as reports of original research on settlement and metamorphosis in a total of 11 phyla. This publication of the proceedings of the Symposium, with reviews of the literature on settlement and metamorphosis, is intended to serve as a general reference for biologists and also as a reference for graduate students and investigators who are working on problems of invertebrate development. It is our hope that the contents of this volume will stimulate further research on these subjects and will point out new directions for future studies.

Support from several organizations is gratefully acknowledged. The Symposium was supported in part by the Office of Naval Research, Microbiology Program and Oceanic Biology Program, Naval Biology Project, under contract #N00014-77-G-0067, NR 205-027/11-22-77. Other support was provided by the National Science Foundation, Division of Physiology, Cellular, Molecular Biology, Grant No. PCM77-09751. In addition, the American Society of Zoologists provided assistance from its Symposium Fund as well as through the Division of Invertebrate Zoology. Preparation of the publication was funded, in part, by the Fort Pierce Bureau of the Smithsonian Institution. Numerous individuals contributed their time, effort, and devotion to make the Symposium a success and this publication a reality. We wish to thank those persons who acted as chairmen of the three sessions of the Symposium: Richard M. Eakin, Eugene N. Kozloff and Arthur H. Whiteley. We express our gratitude to Mary A. Wiley, Business Manager of the American Society of Zoologists, for her cheerful efficiency and able assistance in every phase of our endeavors. We are grateful to the editors of Elsevier North Holland, J. G. Hillier and Margaret Quinlin, for their facilitation of our publication efforts. Special thanks go to Margaret Quinlin for her many helpful suggestions during the preparation of this volume and for her warm personal concern for our publication. And we are indeed indebted to Venka V. Macintyre who, with unusual dedication, diligence and expertise, provided invaluable editorial assistance and typed the entire volume for camera-ready copy.

This volume is dedicated to Robert L. Fernald, Professor of Zoology at the University of Washington, Seattle on the occasion of his retirement from the Department of Zoology and in recognition of his innumerable contributions over the past three decades to the field of comparative invertebrate embryology. Both as Director of the Friday Harbor Laboratories and as a teacher of embryology, Dr. Fernald has guided and inspired many students toward professions in teaching and research in invertebrate embryology and development. The fact that more invertebrate embryology classes are being taught today than ever before in so many places by his students and students' students is indeed a tribute to his achievements as an educator and a scholar. Each of the contributors to this Symposium has been influenced either directly or indirectly by his scholarship. We dedicate this book to him with appreciation and fond affection.

Fu-Shiang Chia
Mary E. Rice
May 1, 1978

REFERENCES

1. Chia, F. S. (1976) Thalassia Jugosl., 10, 121-130.
2. Mileikovsky, S. A. (1971) Mar. Biol., 10(3), 193-213.
3. Mileikovsky, S. A. (1976) Thalassia Jugosl. 10, 171-180.
4. Scheltema, R. S. (1976) Thalassia Jugosl., 10, 263-278.
5. Stearns, S. C. (1976) Quart. Rev. of Biol. 51, 3-47.
6. Kume,M. and Dan K. (1968) Invertebrate Embryology, NOLIT Publ. House, Belgrade (TT 67-5805D, clearing house for Fed. Sci. and Tech. Info., Springfield, Virginia), 546 pp.
7. Reverberi, G. (1971) Experimental Embryology of Marine and Freshwater Invertebrates, American Elsevier Publ. Co., Inc., New York, 587 pp.
8. Anderson, D. T. (1973) Embryology and Phylogeny in Annelids and Arthropods, Pergamon Press, Oxford, 495 pp.
9. Jägersten, G. (1972) Evolution of the Metazoan Life Cycle, Academic Press, New York, 282 pp.
10. Giese, A. C. and Pearse, J. S. (1974) Reproduction of Marine Invertebrates, Vol. I, Acoelomate and Pseudocoelomate Metazoans, Academic Press, New York, 546 pp.
11. Giese, A. C. and Pearse, J. S. (1975) Reproduction of Marine Invertebrates, Vol. II, Entoprocts and Lesser Coelomates, Academic Press, New Yor, 344 pp.
12. Giese, A. C. and Pearse, J. S. (1975) Reproduction of Marine Invertebrates, Vol. III, Annelids and Echiurans, Academic Press, New York, 343 pp.
13. Giese, A. C. and Pearse, J. S. (1977) Reproduction of Marine Invertebrates, Vol. IV, Molluscs: Gastropods and Cephalopods, Academic Press, New York, 369 pp.
14. American Society of Zoologists: Symposium (1974). The Developmental Biology of the Cnidaria, Amer. Zool., 14(2), 440-866.

15. American Society of Zoologists: Sympsoium (1975) Developmental Biology of Echinoderms, Amer. Zool., 15(3), 485-775.

16. American Society of Zoologists: Symposium (1976) Spiralian Development, Amer. Zool., 16(3), 277-626.

17. Crisp, D. J. (1971) Fourth European Marine Biological Symposium, Cambridge University Press, 599 pp.

18. Proceedings of the Conference on Marine Invertebrate Larvae (1974) Thalassia Jugosl., 10(1/2), pp. 1-424.

19. Stancyk, C. E. (in press) Reproductive Ecology of Marine Invertebrates, The Belle W. Baruch Library in Marine Science, No. 9, Univ. South Carolina Press.

20. Thorson, G. (1971) Life in the Sea, World University Library, Weidenfeld and Nicolson, London, 256 pp.

21. Crisp, D. J. (1974) in Chemoreception in Marine Organisms, Grant P. T. and Mackie, A. M. eds. Academic Press, New York, pp. 177-265.

22. Crisp, D. J. (1976) Thalassia Jugosl., 10, 103-120.

23. Etkin, W. and Gilbert, L. I. (1968) Metamorphosis, A Problem in Developmental Biology, Appleton-Century-Crofts, New York 459 pp.

MECHANISMS OF LARVAL ATTACHMENT AND THE INDUCTION OF SETTLEMENT AND METAMORPHOSIS IN COELENTERATES: A REVIEW

Fu-Shiang Chia and Louise R. Bickell

Department of Zoology, University of Alberta, Edmonton, Alberta, Canada T6G 2E9

Literature is reviewed concerning behavior, mechanisms of attachment, induction of settlement and metamorphosis, and intercellular communication and sensory receptors in planulae. Behavioral responses of planulae, either taxic or kinesic, are shown to increase the opportunity for settlement on favorable substrates. Evidence is reviewed demonstrating that larval attachment is made either by glandular secretion or by firing of cnidocytes. In some species, organic surface films have been implicated as metamorphic triggers, either by providing an excitatory chemical cue, or by providing an attractive surface texture. In some symbiotic hydroids, components of the host species induce or enhance settlement and metamorphosis. Nervous and neuroid conduction, and diffusion of hormone-like chemicals are discussed as possible mechanisms for intercellular coordination during settlement and metamorphosis.

INTRODUCTION

The life cycle of most coelenterates includes a planula larval stage. Planulae are usually elliptical and are constructed of two body layers, ectoderm and endoderm, separated by a thin mesoglea. The endoderm of most planulae forms a solid core; only in a few species of actinarian planulae does the endoderm form a functional gastrovascular cavity. Relative to endoderm, the ectoderm displays a considerable degree of cellular differentiation. Characteristic ectodermal cell types include the sustentacular cells (usually ciliated), various gland cells, cnidocytes, and putative sensory cells.[1,2,3,4] In some actinarian planulae, ectodermal cells at the aboral end form a specialized structure known as the apical organ. This structure is suspected to perform a sensory role, either during feeding,[1,5] or during substratum selection.[6,7]

Planula motility, either swimming or crawling, is usually mediated by the activity of cilia. These organelles are characteristically distributed over the entire surface of the larva. In most coelenterates, metamorphosis transforms the motile planula into a sessile polyp. As a result, the planulae of many species, especially those which are symbiotic, possess the ability to select settlement substrates which enhance post-metamorphic survival.

In this review, we shall attempt to examine the literature concerning settlement and metamorphosis of planula larvae. By doing so, we hope to reveal those areas where information is inadequate or lacking, and where future research might be most beneficial.

PLANULA BEHAVIOR AND SITE SELECTION

Many species of planula possess a limited repertoire of behaviors which are directed by various types of environmental cues. These behaviors, defined as taxic or kinesic responses, influence the eventual site of settlement.

In the laboratory, free-swimming planulae often exhibit two major behavioral phases during larval life. The initial period is characterized by active swimming near the water surface. This is followed by a period in which the planulae descend to the bottom of the culture bowl where swimming becomes sluggish or is halted altogether. This behavior has been reported in the planulae of *Stomphia didemon*,[8] *Siderastrea radians*[9] and *Pocillopora damicornis*.[10] Some species such as *Coryne uchidai*[11] and *Favia fragum*,[12] commence crawling during the latter phase.

1

Harrigan[10] and Lewis[12] have shown that this behavioral transition in both *Pocillopora dami-cornis* and *Favia fragum*, results from a gradual reversal of the positive phototaxis exhibited by young planulae. In these cases the reversal is apparently dictated by endogenous programming. Whatever the causative mechanism of this behavior pattern, it would seem to facilitate an initial period of dispersal (none of the species mentioned possess a medusoid morph in the life cycle), followed by a period of contact with a settling substrate.

Some planulae change their reaction to light only after reception of a specific chemical cue. Williams[13] documented this phenomenon in the crawling planula of *Clava squamata,* a hydroid epiphytic on fronds of the algae, *Ascophyllum nodosum.* Planulae of this species are normally positively phototactic. However, if older planulae are exposed to directional light in the presence of *Ascophyllum,* they become negatively phototactic. The effect of this algal-induced photonegativity, and the tendency of the planulae to settle in pits and grooves on the surface of the plant thallus, promote settlement in sites protected from desiccation and extreme water currents, respectively.[13]

A second example of this type of behavior is exhibited by the planulae of *Hydractinia echinata,* a colonial hydroid characteristically found on the shells of hermit crabs. Schijfsma[14] observed both positive phototaxis and photokinetic responses by the crawling planulae of this hydroid. However, the planulae became indifferent to light when a hermit crab was present in the bowl. *Hydractinia* planulae are not directly attracted to hermit crabs;[14] their eventual transfer to the shell of the crab depends upon the chance of the crab dragging its shell past the planulae.[15,16] This evidence suggests that the light reactions of *Hydractinia* planulae facilitate dispersal, but a mechanism exists to inhibit these dispersive activities in the proximity of a hermit crab.

In its natural habitat, the scyphistoma of *Cyanea capillata* is typically found hanging upside down from the lower face of overhanging objects. During an investigation of this scyphozoan, Brewer[17] noticed that the planulae increase their activity and swim upward in sea water with a high concentration of dissolved carbon dioxide. Such conditions may develop in calm water at the surface of sediments. Brewer[17,18] has suggested that upward swimming by planulae which encounter such conditions will enhance the possibility of contact with the underside of objects.

Prior to settlement, some species of planulae are capable of responding to the texture and contour of the substrate. This is particularly true of crawling planulae. Crawling rate by planulae of *Clava squamata* increases on glossy surfaces and decreases on rough surfaces.[13] In addition, the planulae of this species, as well as the crawling planulae of *Sertularella muirensis* and *Favia fragum,* will settle preferentially in pits and grooves on the surface of the substratum.[13,19,12]

Williams[20] has described congregating behavior by the crawling planulae of the hydroids, *Nemertesia antenina* and *N. ramosa.* He attributed the apparent gregariousness of these species to positive thigmotaxis.

MECHANISMS OF PLANULA ATTACHMENT

Two types of substrate attachments by planulae are recognized. The first, which is weak and transient, may be called temporary attachment. For example, the swimming planulae of both *Favia fragum* and *Tealia crassicornis* occasionally attach to objects but subsequently resume swimming.[12,6] The second type, which is much stronger and sometimes permanent, is the settlement attachment. Settlement attachments are always accompanied by metamorphosis.

Two types of ectoderm cells, gland cells and cnidocytes, may potentially establish both temporary and settlement attachments.

Gland Cells

When the planulae of the sea pen, *Ptilosarcus gurneyi,* are offered suitable substrate, they become covered with a mucous secretion.[21] In a comparative fine structural study of the pre- and post-metamorphic stages of *P. gurneyi,* Chia and Crawford[4] identified two types of gland cells in the planula which were not present in the polyp. Assuming these are not merely developing stages of polyp secretory cells, their products must be functional only during larval life—presumably to facilitate settlement attachment.

Vandermeulen[3,22] compared the epidermal ultrastructure of the planula and polyp of the coral *Pocillopora damicornis.* The aboral epidermis of the planula contained four morphologically distinct secretory cells. However, in the six-hour settled polyp, only one secretory cell, the calicoblast cell, was found in the aboral epidermis. Vandermeulen[3] has suggested that the secretions of at least some of the planula gland cells facilitate settlement attachment.

The planulae of both *Phialidium gregarium* and *Hydractinia echinata* secrete a mass of viscous, mucoid material at the onset of metamorphosis.[23,24] Bonner[23] demonstrated that the settlement process in the planula of *Phialidium gregarium* is coincident with the disappearance of PAS-positive gland cells from the ectoderm. Prior to settlement, these intensely staining cells are more abundant at the aboral end of the planula, the end which fixes to the substrate. Similarly, Nyholm[25] found mucous cells at the aboral end of the planula of *Protanthea simplex.* On this evidence, he also suggested that mucus was responsible for substrate attachment during the settlement process.

Finally, it is known that crawling planulae employ mucus for adherence to the surfaces over which they creep.[13]

Cnidocytes

The temporary attachments exhibited by some planulae result from the discharge of mechanosensitive nematocysts. If an object, such as a coverslip or mollusc shell, is dragged past the planula of *Hydractinia echinata,* the planula will attach and transfer to this object by discharging atrichous isorhiza nematocysts.[15,16] Müller *et al.*[16] have stated that "zeta potentials produced by the movement of the object, in addition to the impulse of collision, constitutes the condition that stimulates the larva to fire its nematocysts. . . ." However, this nematocyst-mediated attachment is not followed by metamorphosis unless further, specific triggers are provided (see section on induction of settlement). It is interesting to recall that tentacle atrichous isorhizas are used by hydra for adherence to the substrate during somersaulting locomotion.[26]

A second example is provided by the actinula larva of *Tubularia larynx* (the planula stage of this hydrozoan is passed within the parent body). This larva possesses a number of stiff tentacles which radiate outward from the body proper. The tentacle tips, which are swollen with large numbers of cnidocytes, temporarily attach to surfaces which they contact.[27] Attachments were observed even on clean glass surfaces. Haws[28] has shown that discharge of tentacle nematocysts is responsible for this adhesion, and the readiness of the nematocysts to respond to mechanical contact increases with the age of the actinula.

Cnidocytes which are sensitive to specific chemical or contact-chemical stimuli may also facilitate attachments between the planula and a substrate. Donaldson[29] has made a detailed study of such a phenomenon in the planula of a symbiotic hydroid, *Proboscidactyla flavicirrata.* Members of this genus live only on the tube rims of certain sabellid polychaetes.[30] Campbell[31] showed that the free-swimming planula of *Proboscidactyla flavicirrata* readily attaches to the tentacular pinnules of the host sabellid worm by discharging nematocysts. The

3

experiments of Donaldson[29] indicate that contact-chemoreception of some factor from sabellid body surfaces provides the exclusive stimulus for the discharge of microbasic eurytele nematocysts. The nematocysts do not respond to tissues from other organisms or to simple mechanical stimulation applied in the presence of host mucus or tissue extracts. Following temporary attachment to the pinnules, the planulae soon elongate and the cnidocytes lose their sensitivity to sabellid tentacles. Instead, the effective stimulus for the cnidocytes becomes the tube of the worm. If the planulae contact the worm tube, nematocysts will discharge and the planulae will transfer to the tube and metamorphose. Because the orientation of the planulae is unimportant for attachment to the tentacles and the tube, Donaldson[29] has suggested that either two functional populations of cnidocytes are intermingled or only one type exists whose specificity can be altered.

Microbasic euryteles are often found in other hydrozoan, as well as scyphozoan planulae.[1] In addition, all anthozoan planulae are provided with spirocysts, whose ultrastructure strongly suggests an adhesive function.[32] In an examination of the planulae of *Anthopleura elegantissima,* we have found that the aboral half of the planula contains five times more spirocysts than the oral half.

INDUCTION OF LARVAL SETTLEMENT AND METAMORPHOSIS

Species of coelenterate planulae vary in their degree of dependence on environmental factors for the induction of settlement and metamorphosis. As in many other groups of invertebrates, the planulae of generalist species exhibit little or no dependence on external inducing factors, while specialist species (in habitat or diet) are likely to be obligatory to environmental induction.

The planulae of *Phialidium gregarium* will settle on a variety of substrates, including glass, wood, plastic and algae.[33] Similarly, many other hydrozoan planulae, including *Aequorea aequorea,* and *Sarsia tubulosa* will settle in glass bowls without requiring apparent physical or chemical inducing factors (personal observations). However, these observations should not be considered conclusive proof against the existence of subtle discriminatory abilities by the planula. An instructive example is provided by *Hydractinia echinata.* When Schijfsma[14] cultured the crawling planula of this hydroid in bowls containing sand and provided it with continuous circulation of fresh sea water, the planulae began to settle and metamorphose two days after fertilization. He concluded that these planulae metamorphose according to an endogenous time table. However, Müller and his associates[34,35,24,36,37,38] reinvestigated settlement by *H. echinata* planulae and found that if planulae were cultured under sterile conditions, they would not metamorphose. These investigators found that certain marine, gram-negative bacteria emit a product at the end of their exponential growth phase which induces metamorphosis in planulae of *H. echinata.* The effective bacteria are regularly found in natural sea water, on marine substrates, and on hermit crab shells, and they can be cultured under aerobic conditions in solutions of meat extract. Two observations indicate that the effective bacteria do not exert their inductive capacity by conferring an attractive surface texture on substrates: (a) if the bacterial cells are removed by filtration, the supernatant will effectively induce metamorphosis; (b) when planulae are suspended in solutions of the bacterial inducer, they may metamorphose without attachment to a substrate.

If the bacterial culture is subjected to osmotic shock prior to filtration (a procedure which is known to enhance release of surface-bound enzymes and secretion of leakage products[24]), the supernatant, but not the bacterial cells, has inductive capacity.

Müller and associates have also attempted to determine the physiological mechanism by which the stimulus activates metamorphosis. If various concentrations of the inductive bacteria, or the isolated product, were plotted against the percent metamorphosis effected by each dose, the resultant dosage-response curve exhibits Michaelis-Menten saturation kinetics[38,35,24] (Fig. 1).

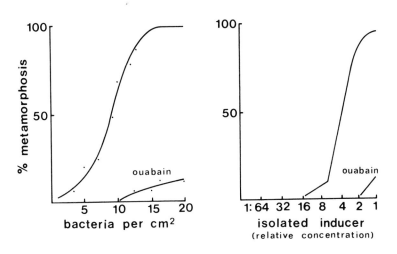

Fig. 1. Induction of metamorphosis in *Hydractinia echinata* by bacteria and the isolated inducer. The effect of ouabain is shown in each case (from Müller, 1973).

This evidence suggested that an enzyme or carrier system was involved in the larval response.

It was further discovered that bacterial induced metamorphosis was inhibited by ouabain, but that ouabain inhibition could be abolished at high concentrations of the isolated bacterial product. Therefore, Müller[35,24] suggested that the inducer may operate by stimulating the Na^+/K-ATPase of larval cell membranes.

Supporting evidence for this hypothesis was subsequently derived from experiments using various monovalent cations. It was found that the effect of the bacterial inductor could be imitated by Cs^+, Li^+, Rb^+, and K^+.[38,36,24,37] These experiments indicate that the planula receptor which receives the inductive stimulus possesses membrane-bound binding sites for cations. Experiments using various combined concentrations of K^+ and Ca^{++} revealed that the percent metamorphosis was optimal when the relative ratio of these ions was K^+/Ca^{++}=40. This is the Gibbs-Donnan ratio. Ouabain was found to antagonize the inductive capacity of all the monovalent cations except potassium (Fig. 2). From this and other results, it was concluded that potassium induction was based on a passive event controlled by the Gibbs-Donan principle, but Cs^+ induction and induction by the bacterial product was an active event—apparently stimulating the activity of the carrier system. These observations are similar to experimental results obtained with the Na^+/K^+-ATPase of other systems.[24,36]

Müller et al.[16] have recently shown that the distinctive shell vibrations produced by a hermit crab provide a second type of stimulus for metamorphosis in *H. echinata* planulae. Metamorphosis takes place on inhabited shells even when these have been previously sterilized. The mechanism of this induction has not been determined.

The planula of the athecate hydroid, *Coryne uchidai*, provides another example of a settlement response elicited by a chemical stimulus. *Coryne uchidai* is an epiphytic hydroid associated with certain Sargasso-algae.[39] In laboratory experiments, planulae which were offered a choice of a number of algal species displayed greater settlement on Sargasso-algae.[40] In a subsequent experiment, the behavior of planulae was compared both in the presence and in the absence of solutions of *Sargassum* extract.[11] Planulae in the control group initially displayed swimming activity, but gradually they began crawling on the bottom and side of the culture bowls. None of the planulae in this group metamorphosed before three days after their release. However, in the presence of *Sargassum* extract, the swimming and crawling activity of planulae was markedly attenuated. In this group, metamorphosed polyps began to appear within one day

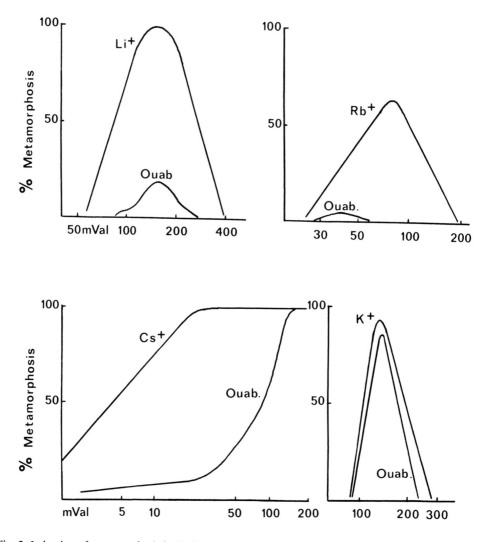

Fig. 2. Induction of metamorphosis in *Hydractinia echinata* by monovalent cations. The effect of ouabain is shown in each case (from Müller, 1973).

after release of the planulae. Nishihira[11] has therefore suggested that some chemical factor from Sargasso-algae enhances settlement and metamorphosis by causing ciliary arrest in the planulae.

Kato *et al.*[41] have succeeded in extracting and purifying the chemical component of *Sargassum tortile* which stimulates settlement and metamorphosis of *Coryne uchidai.* The most potent molecule was found to be the epoxide of δ-tocotrienol ($C_{27}H_{40}O_3$).

Nishihira[19,42] also examined substrate selection by a second epiphytic hydroid of Sargasso-algae, *Sertularella muirensis.* Field experiments showed that these planulae preferentially settle on certain species of *Sargassum,* even when ecological factors other than the algae were controlled. However, the mechanism of selection was not investigated.

The chemical induction of settlement and metamorphosis of *Proboscidactyla flavicirrata* by the mucus of the host polychaete has already been reviewed in the previous section on planula attachment.

Settlement and metamorhposis of some species of planula are enhanced by the presence of a thin organic surface film. Such films may provide an excitatory chemical cue, as in *H. echinata,* or they may influence settlement by conferring an attractive surface texture on the substrate. A possible example of the latter is found in the planula *Pocillopora damicornis*. The optimal settlement substrate for this species is a thin organic film comprised of green filamentous algae, diatoms, and bacteria.[10] However, they will also settle favorably on such chemically diverse substrates as coralline algae, brown algae, and films of bacteria and diatoms alone. Substrates avoided by the planulae include clean sand, clean inorganic surfaces, red algae, and encrusting animals such as ascidians.

Nishihira[43] examined the distribution of newly settled polyps of several epiphytic hydroids of marine grasses. *Clytia edwardsia* and *Tubularia radiata* are characteristically found on eel grass (*Zostera marina*) while *Plumularia undulata* grows on surf grass (*Phyllospadix iwatensis*). In every case, settlement densities were a function of the age of the blade, the older portions of the plant receiving the heaviest settlement. Nishihira[43] suggested that this phenomenon is related to the diatom and microbial film present on older parts of the host plants. If so, the attractiveness of the film may be either its chemical nature or its surface texture.

The planula of *Ptilosarcus gurneyi* is competent to metamorphose at six days post-fertilization. However, metamorphosis will be delayed considerably unless specific substrate qualifications are fulfilled.[21] It was found that planulae preferentially metamorphose on sand samples from sea pen beds and further experimentation proved that the organic component of the sand actively induced settlement and metamorphosis.

In experiments performed by Williams,[20] crawling planulae of *Clava squamata* and *Kirchenpaueria pinnata* were observed to preferentially settle around attached polyps of their respective species. In addition, more planulae of *K. pinnata* settled on surfaces previously coated with planula mucus, than on clean surfaces or on surfaces covered with a bacterial mucous film. Williams[20] concluded that in these species planula settlement was enhanced by an intraspecific factor present in the mucous secretions of both the pre- and post-metamorphic stages.

Siebert[8] has reported that adults of the anemone *Stomphia didemon* can be found on many types of hard substrates (rocks, bottles, old shells). The planulae of this species will begin to settle on coarse sand and fine gravel at eight days, but settlement on broken scallop shells and on plain glass bowls requires almost twice as long. No explanation was given for the differential attractiveness of these substrates.

Although larva of *Tealia crassicornis* will eventually settle on wax and plain glass dishes, metamorphosis occurs 15 days earlier if tubes of the polychaete *Phyllochaetopterus* are present.[6] Siebert and Spaulding[44] have reported that *Phyllochaetopterus* tubes also enhance metamorphosis of *Cribrinopsis fernaldi* planulae. Paradoxically, however, the habitat of these anemones is quite different from that of *Phyllochaetopterus*. The polychaete is an inhabitant of sand and mud flats, while *Tealia* and *Cribrinopsis* are found on rocky substrates.

Finally, Theodor[45] has reported that the planulae of *Eunicella stricta* preferentially settle on coral debris or coral skeletons with attached pieces of coralline algae, bryozoan zooecia, and worm tubes; few planulae settle on clam shells or broken pieces of glass or plastic.

INTERCELLULAR COMMUNICATION AND SENSORY RECEPTORS

Depending on the species of planula, the metamorphic trigger may be either simple contact with a surface, or a specific type of physical or chemical cue. Regardless of the nature of the trigger, settlement and metamorphosis involve many types of cellular processes: ciliary arrest, nematocyst discharge, glandular secretion, cellular differentiation, and bodily shape changes. Coordination between the many cells seems necessary and some form of intercellular communication and integration must be involved. Although the actual mechanism of communication remains speculative at the present time, there are three possibilities: nervous conduction, neuroid conduction, or diffusion of hormone-like chemicals.

Histological studies have reported the presence of a "fibrillar layer" located immediately below the ectodermal epithelium.[1,2,3,4] Widersten,[1] Lyons,[2] and Chia and Crawford[4] have suggested that this layer is composed of nerve cell processes, while Vandermeulen[3] believes the layer is formed of elongate basal processes of the overlying ectodermal cells. Whatever their identity, the fibers exhibit ultrastructural characteristics of axons , including microtubules and possible neurosecretory vesicles.

Although neuroid conduction has been shown in many types of coelenterate epithelia,[46] no report is available demonstrating such activity in the ectodermal epithelium of planulae. Fine structural studies have failed to show gap junctions between planula ectoderm cells.[2,3,4]

Circumstantial evidence for nervous or neuroid conduction comes from the studies of Müller's group[38,24,36,37] on metamorphosis of *Hydractinia echinata* planulae. It will be recalled that the bacterial inducer may act via stimulation of the Na^+/K^+-ATPase. This system is responsible for maintaining the elctrochemical potential across cell membranes. Conceivably, increased activity of the Na^+/K^+-ATPase may transduce the chemical stimulus of the bacterial metabolite into an ionic generator potential. Unfortunately, electrophysiological experiments on metamorphosis in *H. echinata* planulae have not been performed.

At present, evidence for intercellular communication via hormone-like chemicals is only conjectural. When planulae of *Proboscidactyla flavicirrata* attach to tentacle pinnules of the host sabellid, the cnidocytes require 30 seconds to become desensitized to sabellid tissue and subsequently require several minutes to acquire sensitivity to the worm tube (respecification).[29] It has also been necessary to maintain both desensitization and respecification. Donaldson[29] has pointed out the discrepancy between the actual time required for conduction of the information, and the time which might be expected if nervous or neuroid conduction was responsible. Based on this evidence he proposed that a metabolite, which diffuses from the planula-pinnule attachment site, is responsible for altering the sensitivity of the planula's cnidocytes. Müller *et al.*[16] have reported that conduction of "pacemaker" impulses in metamorphosing *H. echinata* planulae is a slow process. In this case, however, the actual rate of conduction was not specified.

There is considerable controversy concerning the identity of the cells which might receive the settlement trigger. In some cases the receptive field seems to be distributed over the entire surface of the planula. For example, if planulae of *H. echinata* are cut transversely into several pieces and each piece exposed to a sufficiently large concentration of the bacterial inducer, each piece, except the tail, will metamorphose into the corresponding part of the polyp. Donaldson[29] similarly found that all portions of the planula were receptive to both the sabellid tentacle stimulus and the worm tube stimulus. In this case, the extent of the sensitivity is correlated with the uniform distribution of the cnidocytes, which appear to be the receptor cells initiating attachment and possibly metamorphosis.

Due to the fact that planulae swim and crawl with the aboral end foremost, and settlement attachment takes place with the aboral end, investigators have often searched for sensory cells in this region. The most obvious aboral specialization is the apical organ, present only in some actinarian planulae. We have made a preliminary fine structural study of this organ in the planula of *Anthopleura elegantissima*. It is composed of a number of very tall columnar cells, each giving rise to a long cilium (Fig. 3). In life, the cilia tend to adhere together in the form of a tuft which is capable of some bending movement.[47,1,17] However, the cilia of the tuft are not fused, as has been stated by Siebert.[7] The ciliary rootlets of these cells are exceedingly long; they extend to the base of the columnar cells. Widersten[1] described uptake of both methylene blue and silver stain by the apical organ cells in the planula of *Metridium senile.* In *A. elegantissima* we have found nerve-like processes concentrated just below the columnar cells of the apical organ (Fig. 3). These processes contain both clear and dense-cored vesicles (Fig. 4). The settlement qualifications of *A. elegantissima* are still unknown. Siebert[7] was unsuccessful in his

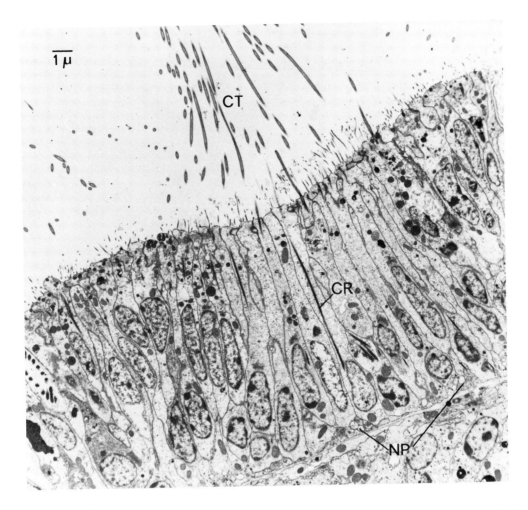

Fig. 3. Aboral ectoderm of the planula of *A. elegantissima* showing the tall, columnar cells of the apical organ. The layer of nerve-like processes is evident at the base of the ectodermal epithelium. CR, ciliary rootlet; CT, ciliary tuft; NP, nerve-like processes.

attempts to induce metamorphosis in this species. However, other species which possess an apical organ, for example *Tealia crassicornis* and *Cribrinopsis fernaldi,* are sensitive to *Phyllochaetopterus* tubes as settling substrate.[6,44]

Ultrastructural studies of planulae which are not equipped with an apical organ have failed to show identifiable sense cells in the ectoderm.[2,3,4] However, these authors have pointed out the marked structural similarity between one type of the ciliated sustentacular cells of the planula, and the ciliated sense cells of post-metamorphic polyps. This similarity has prompted Vandermeulen[3] and Chia and Crawford[4] to suggest a possible involvement by these cells in substrate selection. Both *Ptilosarcus gurneyi* and *Pocillopora damicornis* are selective in their choice of settlement substrates.

CONCLUDING REMARKS

It is apparent from this review that considerable progress has been made concerning the mechanisms of settlement and metamorphosis of planula larvae, particularly as a result of the studies on *Hydractinia, Coryne, Proboscidactyla,* and *Clava.* These hydrozoans are ideal subjects

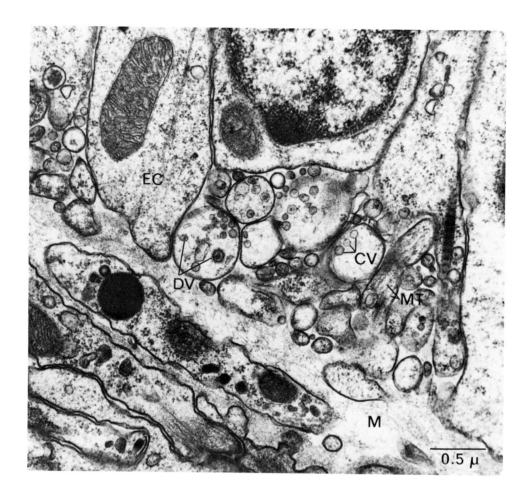

Fig. 4. Section through the nerve-like processes located between the ectoderm and the mesoglea. Clusters of vesicles are evident within the processes. CV, clear vesicles; DV, dense-cored vesicle; EC, ectoderm cell; M, mesoglea; MT, microtubule.

for settlement studies because they are associated with very specific host species. In our judgment, additional research on these and other symbiotic species promises to yield exciting results. However, the settlement phenomenon in planulae of anthozoans and scyphozoans must also receive additional attention for a more complete understanding of the processes involved.

Despite the extensive research which has been performed on the settlement behavior of hydrozoan planulae, nothing is yet known of their ultrastructure. Electron microscopy should also help to clarify the identity of the fibrillar layer of planulae. Use of radioactive precursors of transmitter substances might elucidate the content of the so-called neurosecretory vesicles within the processes of this layer. In addition, cell dissociation techniques, such as those employed on hydra,[48] may enable identification of nerve cells.

Further research on the sensory physiology of planulae is required. Pertinent questions include: What cells are receptive to environmental factors? Are the receptor cells also the effectors, or must the sensory information be communicated to other cells? What is the role of the apical organ? With respect to the latter problem, it might be interesting to perform the experiment of Müller et al.,[16] in which planulae of H. echinata were cut transversely and each piece subsequently induced, on planulae provided with an apical organ. This may denote whether the apical organ is necessary for initiating settlement and metamorphosis.

Preliminary biochemical studies of planulae, both before and during metamorphosis, have begun to yield interesting results.[49,50] Such studies involve the use of autoradiography and inhibitors of cellular processes to determine mechanisms of cell differentiation. It might be instructive to compare biochemical events between planulae which metamorphose spontaneously and those which require a specific type of metamorphic trigger.

ACKNOWLEDGMENTS

Preparation of this paper and the original research reported here were supported by a grant to F. S. Chia from the National Research Council of Canada. We thank Mr. Ron Koss for technical assistance.

REFERENCES

1. Widersten, B. (1968) Zool. Bidr. Upps., 37, 139-182.
2. Lyons, K.M. (1973) Z. Zellforsch., 145, 57-74.
3. Vandermeulen, J.H. (1974) Marine Biol., 27, 239-249.
4. Chia, F.S. and Crawford, B.J. (1977) J. Morph., 151,131-158.
5. Widersten, B. (1973) Zool. Scripta, 2, 119-124.
6. Chia, F.S. and Spaulding, J-G. (1972) Biol. Bull., 142, 206-218.
7. Siebert, A.E. (1974) Can. J. Zool., 52, 1383-1388.
8. Siebert, A.E. (1973) Pac. Sci., 27, 363-376.
9. Duerden, J.E. (1902) Am. Nat., 36, 461-471.
10. Harrigan, J.E. (1972) Am. Zool., 12, 723.
11. Nishihira, M. (1968) Bull. mar. biol. Sta. Asamushi, 13, 91-101.
12. Lewis, J.B. (1974) J. exp. mar. Biol. Ecol., 15, 165-172.
13. Williams, G.B. (1965) J. mar. biol. Ass. U.K., 45, 257-273.
14. Schijfsma, K. (1935) Arch. neerl. Zool.,1, 262-314.
15. Teitelbaum, M. (1966) Biol. Bull., 131, 410-411.
16. Müller, W.A., Wieker, F. and Eiben, R. (1976) in Coelenterate Ecology and Behavior, Mackie, G.O. ed., Plenum Press, New York, pp. 339-346.
17. Brewer, R.H. (1976) Biol. Bull., 150, 183-199.
18. Brewer, R.H. (1976) in Coelenterate Ecology and Behavior, Mackie, G.O. ed., Plenum Press, New York, pp. 347-354.
19. Nishihira, M. (1967) Bull. mar. biol. Sta. Asamushi, 13, 35-48.
20. Williams, G.B. (1976) Ophelia, 15, 57-64.
21. Chia, F.S. and Crawford, B.J. (1973) Marine Biol., 23, 73-82.
22. Vandermeulen, J.H. (1975) Marine Biol., 31, 69-77.
23. Bonner, J.T. (1955) Biol. Bull., 108, 18-20.
24. Müller, W.A. (1973) in Recent Trends in Research in Coelenterate Biology, Tokioka, T., and Nishimura, S. eds., Publ. Seto Mar. Biol. Lab., 20, 195-208.
25. Nyholm, K.G. (1959) Zool. Bidr. Upps., 33, 69-77.
26. Ewer, R.F. (1947) Proc. zool. Soc., London, 117, 365-376.
27. Pyefinch, K.A. and Downing, F.S. (1949) J. mar. biol. Ass., U.K., 28, 21-44.
28. Hawes, F.B. (1958) Ann. Mag. Nat. Hist., 13,147-155.
29. Donaldson, S. (1974) Biol. Bull., 147, 573-585.
30. Hand, C. (1954) Pac. Sci., 8, 51-67.
31. Campbell, R.D. (1968) Pac. Sci., 22, 336-339.
32. Mariscal, R.N. (1974) in Coelenterate Biology: Reviews and New Perspectives, Muscatine, L., and Lenhoff, H.M. eds., Academic Press, New York, pp. 129-178.
33. Roosen-Runge, F.C. (1970) Biol. Bull., 139, 203-221.
34. Müller, W.A. (1969) Zool. Jb. Abt. Anat. Ontog., 86, 84-95.
35. Müller, W.A. (1973) Wilelm Roux' Arch.,173, 107-121.
36. Müller, W.A. and Buchal, G. (1973) Wilhelm Roux' Arch., 173, 122-135.
37. May, G. and Müller, W.A. (1975) Wilhelm Roux' Arch., 177, 235-254.
38. Spindler, K.D. and Muller, W.A. (1972) Wilhelm Roux' Arch., 169, 271-280.
39. Nishihira, M. (1965) Bull. mar. biol. Sta. Asamushi, 12, 75-92.

40. Nishihira, M. (1968) Bull. mar. biol. Sta. Asamushi, 13, 83-89.
41. Kato, T., Kumanireng, A.S., Ichinose, I., Kitahara, Y., Kakinuma, Y., Nishihira, M., and Kato, M. (1975) Experientia, 31, 433-434.
42. Nishihira, M. (1967) Bull. mar. biol. Sta. Asamushi, 13, 49-56.
43. Nishihira, M. (1968) Bull. mar. biol. Sta. Asamushi, 13, 125-138.
44. Siebert, A.E. and Spaulding, J.G. (1976) Biol. Bull., 150, 128-138.
45. Theodor, J. (1967) Vie Milieu, Ser. A, 18, 291-301.
46. Spencer, A.N. (1974) Am. Zool., 14, 917-929.
47. Gemmill, J.F. (1920) Phil. Trans. roy. Soc., London, 209(B), 351-375.
48. David, C.N. (1973) Wilhelm Roux' Arch., 171, 259-268.
49. Müller, W.A. and Spindler, K.D. (1972) Wilhelm Roux' Arch., 170, 152-164.
50. Lesh-Laurie, G.E. (1976) in Coelenterate Ecology and Behavior, Mackie, G.O. ed., Plenum Press, New York, pp. 365-375.

THE SETTLEMENT OF *HALICLYSTUS* PLANULAE

Joann J. Otto

Department of Biology, University of Pennsylvania, Philadelphia, Pennsylvania 19104
and Friday Harbor Laboratories, Friday Harbor, Washington 98250

The planula of *Haliclystus salpinx* moves by a series of extensions and retractions. The extension phase of movement is caused by microfilaments which lie circumferentially around the periphery of the endoderm whereas the retraction phase is caused by longitudinal microfilaments in the ectoderm. When the planula settles, these movements cease and the microfilaments responsible for them disappear. In addition, the endodermal cells, linearly arranged in the planula, rearrange to surround a minute gastric cavity. When the planulae settle, they sometimes aggregate. This aggregation may be due to the presence of a chemotactic or settling factor. Further development of the settled planulae was observed in only one culture. Most cultures fail to continue development perhaps because the settled planulae are an overwintering stage. The settled planulae produce a unique coating composed of plaques of hexagonally packed subunits which is unlike those described for other scyphozoans.

INTRODUCTION

The planula larva of Stauromedusae differs from that of other coelenterates in that it is non-ciliated and creeps rather than swims.[1,2,3,4,5] The planula usually has a constant number of endodermal cells; this is also unusual because most coelenterates do not exhibit cell constancy. After a relatively short motile existence, the planulae settle in aggregates of various sizes. Sometime after settlement, the endodermal cells, which are arranged in a stack in the planula, rearrange to surround a small central gastric cavity.[2,3] In only one case has development beyond this stage been reported, probably because of the rarity of Stauromedusae and the difficulties of obtaining spawning and development. Wietryzykowski[2,3] observed that settled planulae fused and then budded off new planulae which in turn settled and grew. Both the original, settled aggregate and newly budded planulae then developed into juvenile polyps. Similar sexually immature juveniles have been collected from the field.[5,6,7] Aside from these observations on development and field collections, little is known about the life cycle of Stauromedusae.

In this paper I examine the morphological and ultrastructural changes which occur during settlement of the planula of a stauromedusan, *Haliclystus salpinx,* and analyze the aggregation of the planulae. I also suggest that the reason it is so difficult to obtain the development of the settled planulae to juvenile polyps might be that the settled planulae represent an overwintering stage or cyst and that the proper stimuli for further development have not yet been duplicated.

MATERIALS AND METHODS

Haliclystus salpinx (Scyphozoa, Stauromedusae) was collected from blades of *Ulva lactuca* and *Laminaria saccharina* throughout the summer of 1977 from Smallpox Bay, San Juan Island, Washington. The animals were kept for one week in culture dishes at 13° to 16°C without feeding.

During this week, a natural light-dark cycle allowed spawning both in the morning and evening. The embryos were removed from the adults after fertilization, as evidenced by the elevation of a fertilization membrane, and were placed either in glass dishes or Falcon tissue culture plastic petri dishes. The embryos were kept in either filtered or unfiltered sea water at 13° to 16°C. The cultures with unfiltered sea water had available a variety of small crustaceans as potential food. Pieces of *Ulva* or *Laminaria* were added to some cultures during or after planula settlement.

The planulae were observed at various stages of settlement with brightfield and Nomarski differential interference microscopy in order to determine their morphological changes.

For observations on the aggregation of the planulae during settlement, varying numbers of planulae were placed in different 35 mm petri dishes and allowed to settle. A grid (0.5 cm x 0.5 cm square) placed under the petri dish was used to determine the number of larvae per aggregate and the number of larvae per square was counted under a dissecting microscope. At least 3.5 cm^2 area was observed for each concentration of larvae.

For histological examination, the planulae were placed in petri dishes and fixed at various developmental stages in 1% glutaraldehyde in filtered sea water (pH 7.2 to 7.5) for 0.5 hours and postfixed in 0.5% OsO_4 in 0.4M sodium acetate (pH 6.0) for 0.5 hours. Thick and thin sections were cut on a Porter Blum MT-1 ultramicrotome with glass knives. Thick sections were stained with Richardson's stain,[8] and thin sections were stained with uranyl acetate and lead citrate. Thin sections were observed on a Phillips 200 electron microscope.

RESULTS

Planula Behavior

The stauromedusan planula is composed of a stack of highly vacuolate, coin-shaped endodermal cells, covered by a layer of ectodermal cells. The planula of *Haliclystus salpinx* has 16 endodermal cells, and, like other stauromedusan larvae, is quite small, measuring approximately 20 x 100 μm in its relaxed state. The planula moves by a series of extensions and retractions (Fig. 1). Elongation of the endodermal cells in an anterior to posterior sequence causes the entire planula to lengthen. Then, as the endodermal cells shorten and regain their coin-like shape, also in an anterior to posterior sequence, the planula shortens. One cycle of elongation and retraction takes about 2 minutes. These movements result in net forward locomotion when the planula interacts with the substrate. The planula secretes a sheath[3,5] which probably acts as the effective substrate for movement as planulae are often observed crawling through it in the water column as well as on the bottom of the dish. The cell shape changes which occur during movement have been analyzed previously.[5]

The planula moves in this manner for about 1 day. During this time there is no endodermal cell division. The endodermal cell number remains constant at 16. It is not clear if there is ectodermal cell division because the cell nuclei cannot be seen in living specimens with brightfield, phase, or Nomarski optics.

Two days after fertilization (and 1 day after the planulae form) most planulae of *H. salpinx* begin to settle. The anterior end of the planula attaches to the substrate or to an already settled embryo. The ectoderm in the anterior portion of the embryo (except for that at the very tip) becomes thicker as the planula gradually shortens. Endodermal cell movements continue although they appear to be reduced in the anterior part of the embryo. Finally, endodermal cell

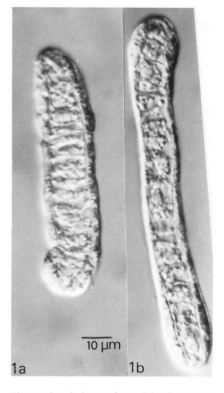

10 μm

1a   1b

Fig. 1. Planula larva of *H. salpinx* in retracted (a) and extended (b) state.

14

movements cease. At this time the endodermal cells appear much less vacuolate,
perhaps due to rearrangements within the
cell as it increases in diameter and shortens
in height. The shape of the anterior portion of the embryo (now the basal end) is
indeterminate and simply conforms to
that of the adjacent embryos or structures.
Figure 2 shows different stages in planula
settling. The settlement process, from the
time a planula makes contact with either
the substrate or another planula until
movement ceases, takes approximately 30
minutes. Soon after settlement, the endodermal cells rearrange to surround a minute gastric cavity (Fig. 3).

After the planulae settled, they were
not observed to move again or to reattach
if dislodged. One half of a dish of settled
embryos was gently scraped. The embryos did not reattach although they appeared to remain alive and to maintain a normal morphology for at least 6 days. During
this time the settled embryos in the original dish did not move to the unoccupied
side.

As the planulae settle, they often
aggregate (Fig. 2). The mechanism of
aggregation could involve either an
adhesion of the planulae to one another
or some form of chemotaxis, and/or the

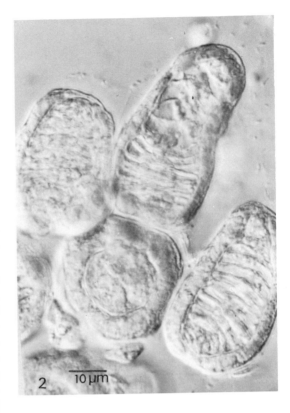

Fig. 2. Settlement of planulae. The four planulae are in
ferent stages of settlement. The central one has stopped
moving, the upper left one is in the final stages of settlement, the anterior cells in the upper right one have stopped
moving, and the lower right one is still moving.

presence of a factor which stimulates settling. In order to test which of these alternatives was
the likely cause of aggregation, the effect of density on aggregation behavior was observed. Two
experiments were performed in which increasing numbers of planulae were placed in different
dishes. In one experiment, shown in Figure 4a, the number of planulae per aggregate was counted 3 days after fertilization (early in the settling process). Among those planulae settled, there
appeared to be no difference in the size of the aggregates. At low densities, a larger fraction of
the planulae had not yet settled compared to higher densities. In another similar experiment,
the number of larvae was counted about 3 weeks after fertilization (Fig. 4b). By this time all
the planulae had settled (the longest time a planula has been observed to move before settling is
13 days). For all densities, the majority of the planulae settled in groups containing 1 to 8 planulae even at high densities where there were aggregates of large size. This experiment was done
twice with similar results.

A number of treatments of the settled planulae were tried in order to induce further development. These included: placing a piece of *Ulva lactuca* or *Laminaria saccharine* in a dish;
keeping the culture (beginning approximately 1 week postfertilization) at 4°C for 2 weeks and
then returning it to 13° to 16°C; using the previous treatment and then leaving the culture at
room temperature (*ca.* 24°C) for 3 to 6 hours daily for a week with the remainder of the time
at 13° to 16°C; putting the cultures with or without *Ulva* at room temperature daily for a week

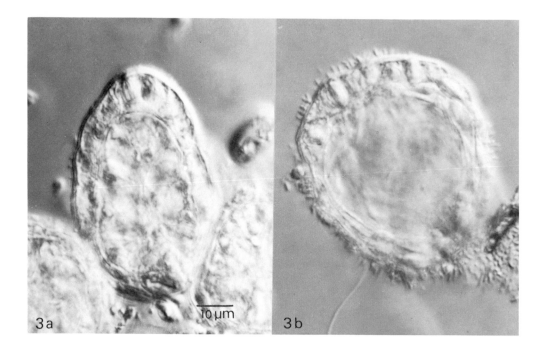

Fig. 3. Settled planulae of *H. salpinx*. (a) Approximately 1 week old. (b) With gastric cavity.

without previous chilling; keeping the cultures at 13° to 16°C as long as 10 weeks; or allowing the embryos to settle on either *Laminaria* or *Ulva* instead of glass or plastic. Further development of the settled planulae was observed in only one culture. The embryos in this dish were fertilized on 15 July 1977. On 17 July, a piece of *Ulva* was added. By 25 July, some of the settled planulae had undergone further cell division and developed large gastric cavities (Fig. 3b). This particular culture degenerated over the next week. Other cultures were similarly treated but failed to develop further. This failure to obtain further development may be because the settled planulae are an overwintering stage and I have not yet been able to reproduce the cues necessary to stimulate further development. Field observations also suggest that these larvae do not develop immediately after settling. Only adults are found in the field during the time of greatest spawning activity (July and August) whereas juveniles have been found in the spring (Dr. R. Fernald, pers. comm.) and early summer.

Morphology

During the planula stage the most striking behavior exhibited by the planula is its movement. The mechanical basis of planula elongation apparently resides in microfilaments (approximately 60 Å in diameter) which encircle the periphery of the endodermal cells (Fig. 5). As these microfilaments contract, the endodermal cells would elongate from a short to a longer cylinder with a smaller diameter. The retraction and bending movements of the planula are probably caused by microfilaments which run longitudinally in the ectoderm (Fig. 6). The longitudinal microfilaments occur in strips and are not continuous laterally (not shown in Fig. 6). In addition to the contraction of longitudinal ectodermal microfilaments, the relaxation of the endodermal microfilaments plus elastic elements in the planula, such as mesogleal components,[9] may contribute to retraction.

16

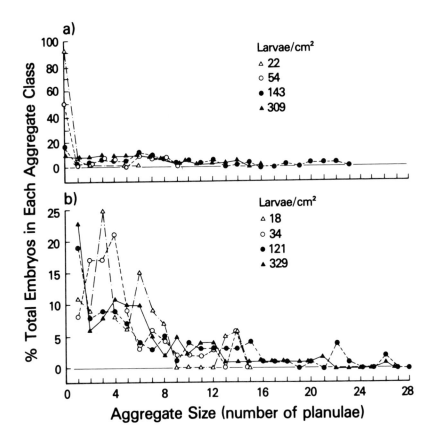

Fig. 4. The effect of density on planula settlement and aggregation. The percentage of total planulae (ordinate) making up each size aggregate (abscissa) is plotted for 4 different planula densities (shown in upper right). (a) Fertilized 12 July 1977 and counted on 15 July. (b) Fertilized 30 July 1977 and counted 20 August 1977. "O" represents unsettled planulae.

The ectodermal epithelium in the planula is composed of highly active synthetic cells. The cytoplasm is packed with ribosomes, rough endoplasmic reticulum, and Golgi complexes. Vesicles near the surface may contribute to the amorphous sheath which surrounds the planula. Adjacent cells are connected by septate desmosomes (Fig. 7). Interstitial cells are also present in the planula's ectoderm. The endodermal cells are highly vacuolate. The nucleus and most of the cytoplasm of endoderm cells lie near either the anterior or posterior cell membrane, and, as mentioned above, the thin peripheral cytoplasm contains microfilaments. Yolk granules occur in both the ectoderm and endoderm. No gastric cavity is present at this stage.

Soon after the planula settles, movement stops and the endodermal cells become rearranged from a linear stack to an arrangement having 2 to 3 cells per cross section. I rarely observed planulae which had not settled but had endodermal cells which had slipped passed one another to form arrangements such as diagrammed in Figure 8.

Following settlement a noticeable change occurs in the extracellular material. The planula is surrounded by an amorphous sheath whereas plaques of nearly hexagonally packed subunits enclose the settled planula (Fig. 9). These plaques are visible not only on the surface of the settled planulae, but also in vesicles in the ectodermal cytoplasm where they are presumably being made and transported to the exterior. The microfilaments which are probably responsible for the movement of the planula are no longer present. Following endodermal cell rearrangement, no other major histological changes were observed in the settled planulae.

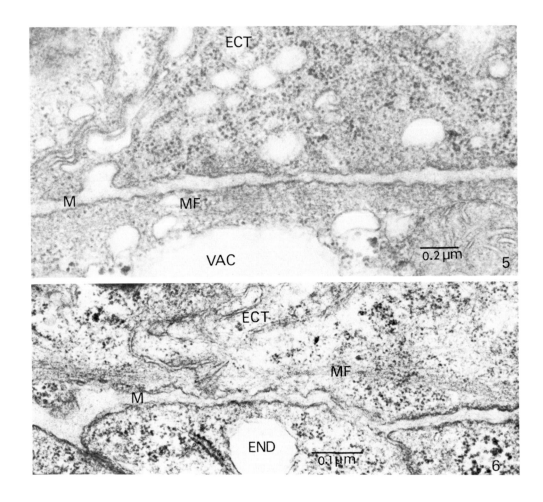

Fig. 5. Cross section of planula. Microfilaments (MF) lie around the periphery of the endoderm adjacent to the mesoglea (M). VAC, endodermal vacuole; ECT, ectoderm.

Fig. 6. Longitudinal section of planula. Microfilaments (MF) run longitudinally in the ectoderm near the mesoglea (M). ECT, ectoderm; END, endoderm.

In coelenterates, interstitial cells, besides reproducing themselves, are considered to be multipotent stem cells which give rise to nematocytes, nerves, gametes, and perhaps gland cells. Cytologically, interstitial cells usually have a denser cytoplasm than epithelial cells; they are filled with ribosomes and rough endoplasmic reticulum and Golgi complexes, and they lie nestled between epithelial cells. In *H. salpinx* interstitial cells are present in the ectoderm in the posterior end of the planula at the time of hatching (Fig. 10). Whether interstitial cells are present in gastrulae is hard to assess because all the cells are quite dense and compact. During the planula stage, some of the interstitial cells begin to differentiate into nematocytes. Five days after fertilization, fully differentiated nematocysts are mounted in both the settled (Fig. 3) and unsettled planulae in the original posterior end. The ectoderm is generally thickened where the nematocysts are mounted (see Fig. 3). In the settled, 1 week old, planula, there are usually 5 to 7 nematocysts mounted and 2 to 4 unmounted along the sides or at the base (original anterior). This number does not increase with time (up to 8 weeks).

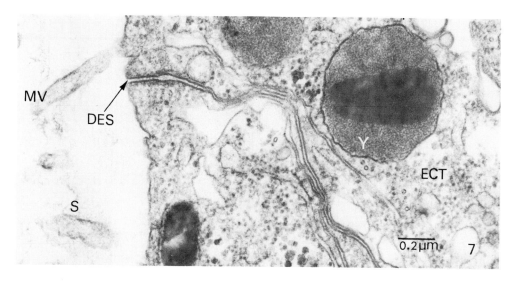

Fig. 7. Longitudinal section of *H. salpinx* near planula surface. Microvilli (MV) protrude into an extracellular sheath (S). Septate desmosomes (DES) connect adjacent ectodermal cells (ECT) which contain yolk granules (Y) as well as rough endoplasmic reticulum, microtubules, and numerous ribosomes.

The nematocyst of the planula is different from that of the adult. Figure 11 shows the three nematocyst types found in the tentacles, exumbrellar ectoderm, and nematocyst sacs of the adult and the one nematocyst type found in the settled planula. The adult nematocysts have been identified as microbasic euryteles (Fig. 11b, c) and atrichous isorhizas (Fig. 11a).[10,11] The nematocyst of the settled planulae is probably also a microbasic eurytele but of slightly different dimensions (Fig. 11d). As fully developed nematocysts such as these do not change size, these represent different developmental populations of nematocysts which are specific to settled planulae or adults.

DISCUSSION

These changes in endodermal cell shape during planula movement and the resulting movement have been analyzed previously.[2,3,4,5] The mechanical basis for the elongation phase is probably the contraction of microfilaments which run circumferentially around the periphery of the endodermal cells, as shown in this paper. Contraction of longitudinal microfilaments in the ectoderm probably causes planula retraction. The longitudinal microfilaments in the ectoderm are also probably responsible for planula shortening during the settlement phase. They appear to be more abundant in the ectoderm at this time although specific counts have not been made. After planula movement stops, these organelles disappear from their location during

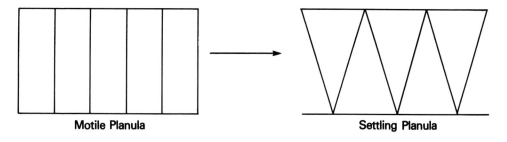

Fig. 8. Diagram of possible endodermal cell rearrangements. Only the endodermal cells are indicated.

19

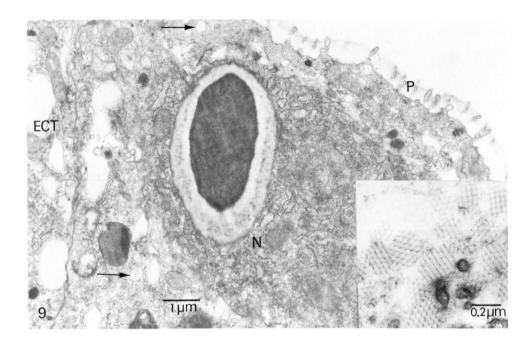

Fig. 9. Ectoderm of settled planula. (a) Ectodermal epithelial cells (ECT) contain vacuoles with forming surface plaques (arrows) as well as extracellular plaques (P). A nematocyte (N) with a developing nematocyst lies between the epithelial cells. Inset: Surface plaques showing subunits.

movement. Microfilaments (60 Å in diameter) have been described in other contracting systems such as the contractile ring in dividing cell[12] and in ascidian tail resorption.[13] In several systems, microfilaments have been shown to bind heavy meromyosin and, therefore, probably contain actin (for review, see reference 14).

At the end of planula movement, many of the planulae aggregate. Wietryzykowski observed that the aggregated settled planulae fused to form individuals which then budded off new planulae of various sizes.[2,3] These new planulae then developed into small polyps. In the one culture which developed gastric cavities, the size of the settled planulae was at most twice as large as those without gastric cavity, and I did not observe planula fusion. In culture, relatively large numbers of planulae do settle alone. Wietryzykowski does not state whether planulae which settle individually can develop further.

The mechanism of aggregation may involve simple adhesiveness of the planulae to one another or some form of chemotaxis. My experiments on aggregation do not rule out adhesiveness as the reason for the clumps of planulae, but they do suggest that chemotaxis or a settling factor may be involved. As expected for all the mechanisms, with increasing planula density, the size of the aggregates increases. However, at high planula densities, relatively more planulae settled alone (Fig. 4b). If there are point sources for some chemotactic factor (perhaps from the initially settled planulae around which the others aggregate), then at high densities, a gradient which exists around the source may be obliterated by the high concentration of factor. Lacking a concentration gradient in which to orient, the planulae may settle alone rather than aggregate. Alternatively, if a settling factor is produced by the initially settled planulae, then at high planula concentrations, the concentration of factor would be high throughout the dish and not just around settled planulae. Planulae would then be stimulated to settle wherever they were in the dish. In addition, at low planula densities, settlement is delayed (Fig. 4a); this also supports the concept of the lack of a settling factor.

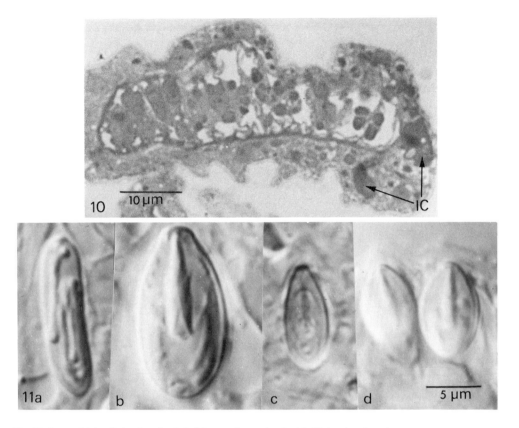

Fig. 10. Interstitial cells in planula. A 0.5 μm section stained with Richardson's stain.

Fig. 11. Nematocysts found in the adults (a-c) and embryo (d) of *H. salpinx*.

The ultrastructural changes which occur at, or just after, settling mainly involve endodermal cell rearrangement. The endodermal cells in the planula are arranged in a stack with their anterior and posterior membranes touching adjacent endodermal cells (except for the most anterior and posterior cells) and their side membranes are in contact with the mesoglea. Following endodermal cell rearrangement, the cells become pyrimidal with one side in contact with the mesoglea and the original anterior and posterior cell membranes still in contact with adjacent endodermal cells. The new tip of the cell protrudes into the minute gastric cavity. How these rearrangements occur is not known, but they must involve changes in adhesiveness both for the mesoglea and for adjacent cells, as well as perhaps more active changes in cell shape.

Nematocyst types are known to differ during the life cycle of coelenterates. For example, Spangenberg[15] and Calder[16] have shown that at least one nematocyst is different between the scyphistoma and medusa in *Aurelia* and *Chrysaora*. In *H. salpinx*, the larval nematocyst (a microbasic eurytele) differs from the adult euryteles in size and shape. The control of the formation of nematocyst types during the life cycle by the multipotent interstitial cells has been studied best in *Hydra*. Zumstein and Tardent[17] have shown that a feedback mechanism may affect the number of interstitial cells committed to forming stenoteles. Bode and Smith[18] have also shown some positional dependence on the type of nematocyte formed. An additional level of control must exist for animals having different nematocysts at different stages of the life cycle.

Field observations suggest that the settled planulae of *H. salpinx* may be the overwintering phase of the life cycle on San Juan Island. Young, sexually immature polyps are found only in the spring and early summer and mature individuals in the late summer and fall. A similar situation probably exists in *H. stejnegeri*, a related species of the same area. Young polyps of *H. stejnegeri* are found from May through mid-July. This species becomes sexually mature beginning only in mid-July (unpub. obs.).

If the settled embryos are the overwintering phase, then their encasement is quite different from other coelenterate cysts. The plaques which cover the settled planulae appear quite porous in the electron microscope. The cysts of other scyphozoans, *Chrysaora*[19] and *Aurelia*,[20] are covered with chitin arranged in fibrous lamellae. The plaques also differ from the filamentous material which coats the basal disk of the adult *Haliclystus*. The adult coating, at least on the basal disk, is a filamentous adhesive material.[21,22] The composition of the plaque will thus be interesting to study because its structure appears to differ from other extracellular coatings of coelenterates and because it apparently self assembles in vesicles before it is released from the cell.

ACKNOWLEDGMENTS

I thank Dr. Robert Fernald for many helpful discussions, Dr. Tom Schroeder for advice on electron microscopy and for the use of his microscopes, and Drs. Joseph Bryan, Richard Campbell, and Sally Zigmond for critically reading the manuscript. Much of this study was carried out at the Friday Harbor Laboratories. The use of their facilities is gratefully acknowledged.

REFERENCES

1. Kowalevsky, A. (1884) Zoologischer Anzeiger, 7, 712-717.
2. Wietryzykowski, W. (1910) Archives de Zoologie Expérimentale et Générale, 5, x-xxvii.
3. Wietryzykowski, W. (1912) Archives de Zoologie Expérimentale et Générale, 5e Serie, 10, 1-95.
4. Hanaoka, K. (1934) Proc. Imp. Acad. Japan, 20, 117-120.
5. Otto, J. (1976) in Coelenterate Ecology and Behavior, G. Mackie ed.., Plenum Press, New York, pp. 319-329.
6. Uchida, T. (1929) Jap. J. Zool., 2,103-192.
7. Berrill, M. (1962) Can. J. Zool. , 40, 1249-1262.
8. Richardson, K., Jaret, L. and Finke, E. (1960) Stain Technol., 35, 313-323.
9. Wainright, S. and Koehl, M. (1976) in Coelenterate Ecology and Behavior, G. Mackie ed., Plenum Press, New York, pp. 5-21.
10. Weill, R. (1934) Trav. Sta. Zool. Wimereux, 10 & 11, 1-701.
11. Mariscal, R. (1974) in Coelenterate Biology: Reviews and New Perspectives, Muscatine, L. and Lenhoff, H. eds., Academic Press, New York, pp. 129-178.
12. Schroeder, T. (1968) Exp. Cell Res., 53, 272-276.
13. Cloney, R. (1966) J. Ultrastruc. Res., 14, 300-328.
14. Pollard, T. and Weihing, R. (1974) CRC Crit. Rev. Biochem., 2, 1-65.
15. Spangenberg, D. (1965) J. Exp. Zool., 160, 1-9.
16. Calder, D. (1971) Trans. Amer. Micros. Soc., 90, 269-274.
17. Zumstein, A. and Tardent, P. (1971) Revue suisse Zool., 78, 705-714.
18. Bode, H. and Smith, G. (1977) Wilhelm Roux's Archives, 181, 203-213.
19. Blanquet, R. (1972) Biol. Bull., 142, 1-10.
20. Chapman, D. (1968) J. mar. biol. Ass. U.K., 48, 187-208.
21. Westfall, J. (1968) Am. Zool., 8, 803-804.
22. Singla, C. (1976) in Coelenterate Ecology and Behavior, G. Mackie ed., Plenum Press, New York, pp. 533-540.

THE COMPARATIVE STRUCTURE
OF THE PREORAL HOOD COELOM IN PHORONIDA
AND THE FATE OF THIS CAVITY DURING AND AFTER METAMORPHOSIS

Russel L. Zimmer

*Department of Biological Sciences, University of Southern California, Los Angeles,
California 90007*

Until recently, there has been general agreement that the preoral hood of the larval actino-troch did not possess a discrete coelom (protocoel) and that the entire hood was lost at meta-morphosis. Consequently, the epistome of the adult was considered to be formed *de novo* after metamorphosis. The cavity of this adult structure was usually regarded to be coelomic, but, as it is in communication with the lophophoral coelom, it has been interpreted as derived from the latter.

In 1964, the existence of a protocoel in advanced larvae of *Phoronopsis harmeri* was docu-mented and in 1974, the existence of a larval protocoel and its retention through metamorpho-sis as the coelomic cavity of the juvenile's epistome was demonstrated for *Phoronis muelleri*. In the present paper, eight of nine species of actinotrochs examined are revealed to have proto-coels; in the ninth species, a transitory protocoel is found in early developmental stages. Addi-tionally, the retention through metamorphosis of the protocoels of three species is followed.

Collectively, these several studies demonstrate that (1) a protocoel is characteristically pres-ent in the preoral hood of larval phoronids, (2) the larval preoral hood and adult epistome are ontogenetically linked as they share (consecutively) the same coelomic cavity, and consequent-ly, (3) these preoral body regions of larval and adult phoronids are equivalent to the protosome of tri- or oligomerous deuterostomes.

INTRODUCTION

Metamorphosis of the planktonic actinotroch larva to the benthic tubiculous adult phoronid is one of the most dramatic transformations in nature. Within a period of about 15 minutes, the free-swimming larva undergoes a thorough reorganization of its body parts, literally turning its posterior half inside out, develops functional adult systems from "unassembled" components preformed during the larval periods, and loses significant portions of its body as transitory lar-val structures.

The rapidity and complexity of the transformation, combined with the fact that one is deal-ing with a small but highly differentiated organism, have made difficult the resolution of a num-ber of important features of the metamorphosis. Perhaps the most persistent and complex controversy concerns the structure of the preoral hood of the larva, the fate of this hood at metamorphosis, and consequently, the relationship (if any) of the hood to the epistome of the juvenile. Interrelated are questions concerning the structure of the epistome and the possible in-terpretation of the hood or the epistome, or both, as a preoral region (*i.e.*, a protosome) of an oligo- or trimerous animal.

Synopses or enumerations of the earlier literature concerning these subjects are available from several sources, *e.g.*, [1,2,3,4] but several critical articles are reviewed below to introduce the current findings.

The possibility that the preoral hood of the larval phoronid contains its own coelomic cavity was first advanced by Masterman[5] who described an unpaired protocoel filling the interior of the hood of *Phoronis muelleri* larvae. During the subsequent six decades, studies of various other actinotrochs resulted in the nearly unanimous conclusion that such a coelom was absent, although an imperfect septum (the preoral septum) separating the cavities of the hood and collar was observed in some species and a transitory, mesodermally lined vesicle within the hood, in others. Hyman[6] acknowledged that a hood coelom might exist on the basis of the presence of the incomplete septum in some species, but doubted its correspondence to a protocoel as it is not paired in origin and lacks communication to the exterior. In 1964, Zimmer[1] demonstrated that an extensive protocoel fills the preoral hood during early developmental stages of *Phoronis vancouverensis,* but that it could not be detected by the completion of the lecithotrophic development or in free-swimming larvae. He observed a similar coelom to form and seemingly disappear during the early development of *Phoronopsis harmeri,* but found an apparent remnant of the cavity in the later larvae. Subsequently a protocoel was described from the early embryos of *Phoronis psammophila,*[7] *P. ijimai,*[4] and *P. muelleri,*[8] and advanced larvae ("Actinotrocha branchiata") of *P. muelleri.*[4]

Most research papers prior to the 1960's, including those by Selys Longchamps[9] and Ikeda,[10] concluded that the entire preoral hood is cast off (and usually swallowed) at metamorphosis. Veillet's[11] diagrams of the dehiscence of the entire hood as well as most of the collar in *Phoronis psammophila* seemingly provide convincing evidence for this interpretation. In review articles,[6,12] it was stated that the epistome of the juvenile is formed *de novo* after metamorphosis. Recognizing the morphological and functional similarities and the positional identities of the larval hood and adult epistome, Beklemischev[13] suggested that the epistome represented a regenerated hood.

In 1964, Zimmer[1] expressed doubts that the entire hood was a transitory larval structure and suggested that the just "rediscovered" larval protocoel as well as the overlying epidermis, including the larval ganglion, were retained through metamorphosis as the primordium of the epistome and the adult ganglion, respectively, in *Phoronopsis harmeri.* Subsequently, Zimmer[14] stated without documentation that this pattern of retention had been observed in a total of four species. As will be demonstrated later in this paper, his opinion that the larval brain serves as the primordium of the adult ganglion was incorrect. Quite recently, through the use of photomicrographs of serial sections, Siewing[4] provided the first incontrovertible evidence for the fate of the preoral hood. During metamorphosis of *Phoronis muelleri,* one part (but not all) of the extensive protocoel, the entire exumbrellar epidermis of the hood (including the larval brain), and all but a small preoral portion of the subumbrellar epidermis are cast off. The retained portion of the protocoel is the proximal recess which was situated between the larval brain and esophagus in the larva. This remnant is enfoded between the esophagus, to which it has remained attached, and the preoral epidermis which is reflected back over it. This composite fold is called the "epistomal fold" by Siewing in recognition of its function as the primordium of the epistome of the juvenile.

The question of whether the phoronid larvae or adults, or both, are oligo- or trimerous in the same context as echinoderms and hemichordates has been vehemently debated since Masterman's inflammatory speculations.[5,15] Selys Longchamps,[9] refuting many of Masterman's arguments, denied the possibility of oligomery; Cori[12] subsequently supported it, Hyman[6] again denied it, and recently Zimmer,[1,14] Emig,[3] and Siewing[4] (among others) have defended it. This history suggests that one should not be overly confident of current interpretations but the recent findings that the larva does possess a discrete protocoel[1] and that this is retained through metamorphosis as the cavity of the adult epistome[4] suggest that a definitive answer in favor of oligomery is at hand. Because these critical findings were based on only one or two species, additional comparative studies on the structure of the larval protocoel and its fate at metamorphosis are essential.

This paper attempts to obtain such comparative information on the protocoel and its fate.

MATERIALS AND METHODS

Living larvae were isolated from the plankton samples taken with a ¼-m, 253-micron mesh plankton net towed vertically through the upper 15 m of the water column, or through the en-

tire water column in more shallow areas. For the present study, larvae of *Phoronis vancouverensis**
and *Phoronopsis harmeri* are from Friday Harbor, San Juan Island, Washington; "actinotroch D"
and larvae of *Phoronopsis viridis*† from Morro Bay, California; larvae of *Phoronopsis californica*
from Big Fisherman Cove, Santa Catalina Island, California; and "actinotroch C" and larvae of
Phoronis architecta, *P. pallida* and *P. psammophhilla* from the Los Angeles Bight, California.

Advanced larvae were maintained at 15° to 18°C and observed periodically for fortuitous
metamorphosis. Warming the cultures to room temperature after chilling to about 5°C and
addition of a solution of Janus Green B to give a perceptible coloration to the medium were
sometimes useful in inducing metamorphosis. ‡

Larvae of all ages were narcotized for a fraction of to several minutes using 1 part of isotonic
magnesium chloride to 3 to 5 parts sea water. Metamorphosing individuals were usually narcotized briefly to prevent excessive contraction, but occasionally were fixed directly.

One specimen of *Phoronis architecta* in the process of metamorphosis was fixed in Stockard's
fluid and embedded in paraffin; sections were cut at 7 μm and stained with Harris' hematoxylin
and counterstained with eosin. For all other materials, primary fixation was for 1 hour at room
temperature or 0° in phosphate buffered 2.5% glutaraldehyde or bicarbonate buffered 2% osmium tetroxide. Fixation in glutaraldehyde was followed by postfixation in phosphate buffered 1%
osmium tetroxide. Embedment in Epon followed dehydration through a graded alcohol series
and exchange in propylene oxide. One-μm sections were stained with Richardson's solution.
Photographs were made on Panatomic X film using a Zeilss WL research microscope equipped with
planchromatic and planapochromatic objectives.

To avoid confusion, the following description of the hood and definition of terms are given.
When developed and carried in its normal position (not elevated above the mouth so that its
long axis parallels that of the larva), the hood joins the rest of the larval body at the collar in a
frontal, not a transverse plane. It extends from this insertion ventrally and laterally as well as
posteriorly with reference to the larval body. The terms proximal (rather than posterior) and
distal (rather than anterior) refer to the insertion and free edge of the hood, respectively, and
the concave and convex surfaces of the hood and are called the subumbrella and exumbrella, respectively. The terms anterior and posterior are used only in reference to the larval axis. Thus
the larval brain is at the anterior pole of the larva, but is part of the exumbrellar epithelium near
the proximal border of the hood. The coelomic cavity of the preoral hood is called the protocoel, although this may have undesirable phylogenetic connotations for some readers. When
the protocoel is fully developed the coelomic cavity of the hood is separated from the blastocoelic cavity of the collar region by part of the lining of the protocoel, the so-called preoral septum.[10] Whether the septum is really preoral is questionable as it intercepts the esophagus some
distance from the mouth, but the term is used here as the septum is part of a coelom that is
largely preoral. The position of the preoral septum marks the proximal limit of the preoral
hood at its junction with the collar. That portion of the protocoel localized beneath the larval
brain and delimited laterally and proximally by the median part of the preoral septum is distinguished as the proximal recess of the protocoel.

COMPARATIVE MORPHOLOGY
OF THE PROTOCOEL IN ADVANCED ACTINOTROCHS

The protocoels in the nine species of actinotrochs studied for this paper exhibit a considerable range of morphological variation. One species lacks a discrete protocoel and the remaining

Phoronis vancouverensis has been synonymized with *P. hippocrepia*[16] and *P. ijimai*,[17] but is regarded by
the author as a distinctive species.

†*Phoronopsis viridis* has been synonymized with *P. harmeri*,[16,17] but is regarded by the author as a distinctive species.

‡Recently, Hermann has demonstrated that various bacteria and certain ions will trigger the metamorphosis
of *Phoronis muelleri*[18] and *P. psammophila*[19] but these stimuli were not tested.

eight species, on the basis of conformation of the protocoel, can be assigned to one of three basic patterns.

Pattern 1: *Phoronis architecta*

The definitive configuration of the protocoel in *P. architecta* can most easily be described if one first imagines that the protocoel originally occupied the entire cavity of the hood,* but that secondarily the distal three-fourths or four-fifths have been modified by (1) replacement of the colelomic fluid with a gelatinous matrix traversed by mesenchyme cells, and (2) disappearance of the peritoneal lining. This "acoelomic" portion may have been formed by the dispersion of the component cells of the coelomothelium and their subsequent differentiation into cells that secrete or traverse the gel or form the intrinsic musculature of the hood. The fact that this highly modified region abuts directly against and is continuous with the proximal region that is typically coelomic suggests that it was developed from part of a once more expansive protocoel.

Although "open" distally, the proximal portion of the protocoel is elsewhere lined by a typical mesothelium. This coelomic lining sheathes the ex- and subumbrellar epidermis and forms the preoral septum. The septum consists of a single squamous epithelium which lacks support other than that provided by its attachments to the ex- and subumbrellar surfaces of the hood and by the fluid of the collar blastocoel proximally and that of the protocoel distally. Not surprisingly, the septum is commonly damaged during preparation of sections.

In the transparent larva, the preoral septum is seen to course on each side from the dorso-lateral junction of the free edge of the hood with the collar, to near the disto- or ventrolateral

*This is the case in early embryos of *Phonoris muelleri*,[8] a closely related species.

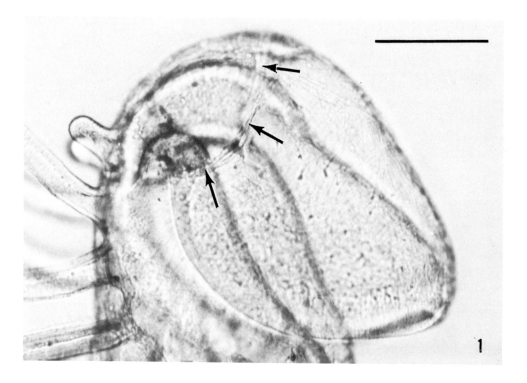

Fig. 1. *Phoronis architecta*. Right lateral view of 20-tentacled actinotroch. The lateral portion of the preoral septum (arrows) is seen through the translucent exumbrellar epidermis of the hood. Bar = 100 μm.

corner of the brain (Fig. 1). In living specimens, these lateral parts of the septum first arch one way and then the other, apparently in response to pressure differences between the coelomic fluid bathing their apical face and the collar blastocoelic fluid bathing their basal surfaces. The septum, in the region of the brain, is continuous with its lateral parts, but is evaginated proximally, forming a pocket that is interposed between the brain and the esophagus. This was called the posterior recess by Ikeda[10] but here is called the proximal recess in keeping with the terminology indicated in **Materials and Methods**. The proximal recess of the protocoel has lateral and proximal walls formed by the space-spanning septum itself, a "ceiling" attached to the brain, and a "floor" attached to the esophagus; distally, the proximal recess is in open communication with the rest of the protocoel.

Figures 2 and 3 are from a near median sagittal section through the proximal portion of the hood (and the proximal recess) of an advanced actinotroch. The specimen had been treated for several minutes with a dilute solution of Janus Green B in an attempt to induce metamorphosis. Although the extensive morphogenetic movements that characterize metamorphosis were not triggered by this dye, certain tissues have undergone degenerative changes that are normal features of metamorphosis, but that usually follow the morphogenetic rearrangement. All affected tissues (the collar epidermis and parts of the epidermis of the hood) are transitory larval ones, but the larval brain, which is also transitory (see below), has retained its integrity. Note that the lining of the protocoel becomes tenuous distally as it approaches the region of transition to the "acoelomic" condition.

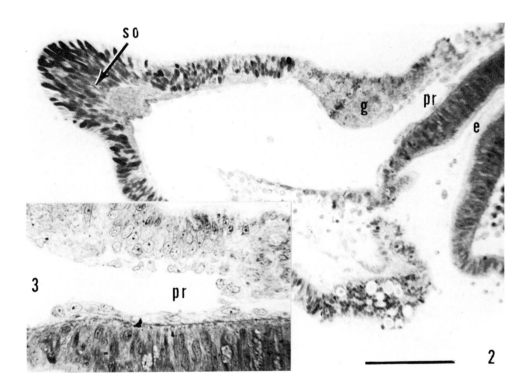

Fig. 2. *Phoronis architecta*. Median sagittal section through part of the hood of an advanced actinotroch. The proximal recess (pr) of the protocoel lies between the larval ganglion (g) and esophagus (e). Note sensory organ (so). The epidermis of the hood (except for the ganglion) and collar are undergoing autolysis as the larva was treated with Janus Green B in an attempt to induce metamorphosis. Bar = 50 μm.

Fig. 3. *Phoronis architecta*. Detail from Fig. 2 of the proximal recess (pr).

**Pattern 2: *Phoronis pallida, P. psammophila,*
"Actinotroch C," and "Actinotroch D"**

The protocoels in these four species are basically identical in construction so that only that of "actinotroch D" is described and figured. The protocoel occupies the entire cavity of the hood except that the exumbrellar epidermis is separated from its coelomic lining in the distal (but not the proximal) portion of the coelom. Thus in median sagittal section (Fig. 4), the coelomic lining of the exumbrellar surface appears closely associated with the brain throughout the proximal recess (in this specimen some separation has occurred), but more distally is separated from that epidermis by a space which is technically blastocoelic. In this region, the exumbrellar portion of the coelomothelium (partly disrupted in this specimen) is anchored to both the ex- and subumbrella by short processes and therefore appears as a zig-zag line in sections. This coelomic membrane can be seen suspended within the preoral hood cavity in the translucent larvae of *P. psammophila,* actinotroch C, and actinotroch D, but not in the opaque larvae of *P. pallida.*

From the study of serial sections the protocoel is found to extend throughout the hood distally as well as laterally and the preoral septum to be complete and to follow a path similar to that described for *P. architecta.*

In *P. psammophila,* the coelomic lining of the protocoel is absent from the distal part of the subumrellar surface (it may have contributed to the conspicuous circular musculature of the subumbrella), but typically developed elsewhere. This does not indicate that a protocoel is absent in this species, but only that the coelomothelium can be transitory.

**Pattern 3: *Phoronopsis californica,*
P. harmeri, and *P. viridis***

In all three west coast species of the genus *Phoronopsis,* the configuration of the definitive protocoel in young as well as advanced actinotrochs is that of a small vesicle situated between the brain and esophagus near the proximal limit of the hood. In essence, the coelom in these larvae corresponds only to the proximal recess, which is here provided with a continuous lining owing to the addition of a distal wall. The origin of this vesicle is not known for any of these species, but it is known that a voluminous coelomic vesicle virtually fills the cavity of the preoral hood during the lecithotrophic stage of ontogeny in *Phoronopsis harmeri.*[1] By the planktotrophic stage, however, only the small vesicle persists. The distal and lateral regions of the hood are in open communication with the blastocoelic space of the collar because the preoral septum is found only near the median line as the lateral and proximal walls of the proximal recess. The non-coelomic regions of the hood cavity contain a gelatinous material.

The diminutive protocoel of *Phoronopsis harmeri* is shown in midsagittal section in Figure 5. Note the complexity of the larval ganglion just anterior to the protocoel (above in the figure) and the presence of the distal wall (left in the figure) of the coelomic vesicle (compare with Figs. 2, 3, and 4, in which no distal wall is present). An arrow to the right of the approximate site of the mouth marks an area in which the subumbrellar epithelium rapidly thins; the epidermis to the right of the arrow will be cast off at metamorphosis, that to the left retained. Serial sections in cross, frontal and sagittal planes confirm the continuity of the protocoelic lining.

The larval protocoel of *Phoronopsis viridis* is almost identical to that of the previous species and can be seen in Figure 8.

Figure 6 is a photograph from the right of an advanced 24-tentacled actinotroch of *Phoronopsis californica.* The exceptional separation of the ex- and subumbrellar surfaces of the hood (and of the epidermis and somatic coelomothelium of the larval trunk) are characteristic of this larva; the resulting spaces are filled with a semi-rigid gel and, in consequence, the hood of this species lacks the flexibility possessed by other species. In correlation with the separation of the surfaces of the hood, the larval protocoel is an elongate, tubular cylinder rather than a box-like

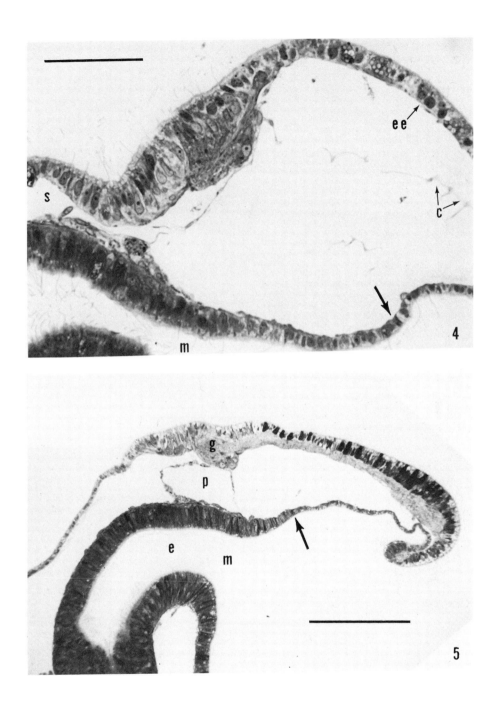

Fig. 4. Actinotroch D. Midsagittal section through preoral hood and its extensive protocoel. Note the preoral septum (s) and the separation between the exumbrellar epidermis (ee) and its coelomothelium (c) except in the region of the ganglion. The arrow to the right of the mouth (m) indicates the approximate site on the subumbrella from which the transitory portions of the hood dehisce. Bar = 100 μm.

Fig. 5. *Phoronopsis harmeri*. Midsagittal section through the proximal portion of the preoral hood. The protocoel (p) is limited to the area between the larval ganglion (g) and esophagus (e). The arrow to the right of the mouth (m) indicates the approximate site on the subumbrella from which the transitory portions of the hood dehisce. Bar = 100 μm.

structure as seen in the previous species. The topological relationships and basic morphology are identical in all three species, however.

Pattern 4: *Phoronis vancouverensis*

In early embryology an extensive protocoel occupies the entire cavity of the hood, but it is virtually eliminated before completion of the lecithotrophic phase of development.[1] In advanced actinotrochs, the intimate apposition of the epidermis of the subumbrella and that of the exumbrella is evident; not only is the protocoel lacking but also there is no preoral hood cavity which it could occupy (Fig. 7). Minor concentrations of mesodermal elements can be detected between the larval ganglion and esophagus. These could represent remnants of the lining of the proximal recess of the earlier voluminous protocoel. There is no evidence that these elements are organized as an epithelium, indicating that not even microscopic remnants of the protocoelic lumen could be present.

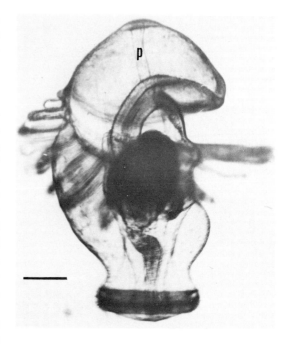

Fig. 6. *Phoronopsis californica*. Right lateral view of 24-tentacled larva near metamorphosis. The protocoel (p) is a tall cylinder between the widely separated brain and esophagus. Bar = 100 μm.

THE FATE OF THE PREORAL HOOD AND PROTOCOEL

In the following section, the fate of the preoral hood and its contained protocoel is followed through metamorphosis in three species, each representing a different pattern of morphology of the definitive protocoel.

Phoronopsis viridis

The general pattern of metamorphosis of *Phoronopsis viridis* is identical with that of *P. harmeri* described earlier by Zimmer.[1] The initial step of metamorphosis is eversion of the metasomal sac caused by contraction of the trunk retractor muscles and peristaltic action of the metasomal sac itself (Fig. 8). Degradation of the integrity of the hood epidermis soon follows, but is obscured by the unique set of tentacles which have rotated about 90º so that they become parallel to the axis of the newly formed trunk, not perpendicular to the larval body. As the digestive tract is drawn out into the everted metasomal sac, the larval trunk collapses into a short cyliner. The pretentacular portion of the larval collar is strongly constricted, drawing the mouth which had been several hundred microns above the level of the tentacle bases to or near that level. With this constriction (because of it?), the tentacle ring becomes indented at

Phoronis architecta

The metamorphosis of *Phoronis architecta* is like that of *Phoronis muelleri*, as described by Siewing[4] and Herrmann.[18]

Figures 13 and 14 are midsagittal sections of a metamorphosing individual in which the metasomal sac is fully everted, but the larval body has not been strongly collapsed. The distal portion of the hood shows strong evidence of autolysis (*e.g.*, pyknotic nuclei) and incipient

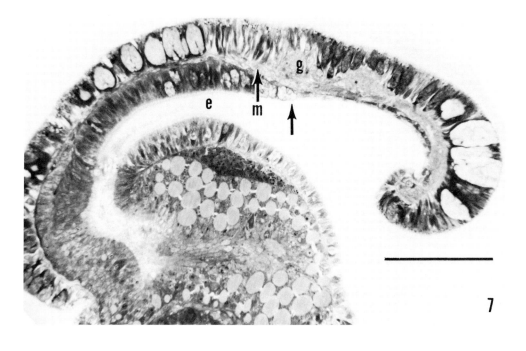

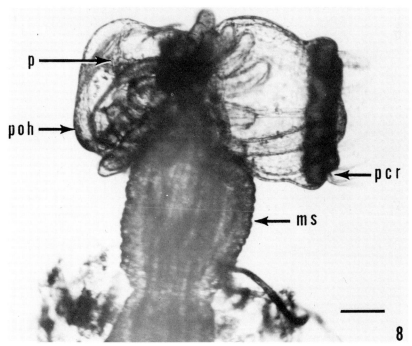

Fig. 7. *Phoronis vancouverensis.* Midsaggital section of advanced, 12-tentacled actinotroch competent to metamorphose. Note the absence of any space between the sub- and exumbrella. A few mesodermal cells (m) are present between the larval ganglion (g) and esophagus (e). The arrow indicates the approximate site on the subumbrella from which the transitory tissues of the hood dehisce. Bar = 100 μm

Fig. 8. *Phoronopsis viridis.* Left lateral view of metamorphosing individual. The metasomal sac (ms) has everted, but the larval body is only slightly contracted and the hood has not begun autolysis. p, protocoel; pcr, perianal ciliated ring; poh, preoral hood.

what was the dorsal midline of the larva. The collapsed larval trunk occupies this indentation. By the time these events are completed, the transitory tissues of the larval hood have been cast off. The precise extent of lysis is difficult to ascertain because of the consolidation of the pretentacular region of the collar, the position of the tentacles, the indentation of the tentacle ring, and the degradation and release of tissue from the preoral hood itself. The details can only be deduced from sections.

Figure 8 illustrates an individual which has just initiated metamorphosis. The everted metasomal sac (only partly shown) extends at right angles from the partly contracted larval body and is seen to contain the gut. The anterior end of the larva bears the slightly constricted preoral hood. Within the preoral hood, the outline of the protocoel is detectable. The posterior end of the larval body bears the anus surrounded by the conspicuous perianal ciliated ring. This same relative orientation is used in all remaining illustrations, but the virtual obliteration of the larval body by contraction, the upward rotation of the mouth and anus, and the lysis of transitory larval tissues obscure the correspondence.

Figures 9 and 10 are from a series of sagittal sections of a metamorphosing individual in which the major morphogenetic movements (e.g., eversion of the metasomal sac, collapse of the larval body, and re-orientation of the tentacles) have been completed. Degradation of the transitory tissues of the larval hood is almost complete as evidenced by the absence of all but a proximal "stump." Figure 9 is a near midsagittal section in which a remnant of the preoral hood containing the collapsed protocoel is present to the right of the mouth. The epidermis on the aboral side of the hood remnant is that part of the exumbrella differentiated as the larval ganglion. This epidermis appears reasonably normal at the light microscope level in this particular section; the finely punctate appearance of its basal cytoplasm is reminiscent of neurofibrillar masses seen in the larval ganglion at the same site (compare with Fig. 5). However, in a section only 16 μm more laterally (Fig. 10) it is evident that this part of the subumbrellar epidermis containing the larval brain is also transitory. The supposed mat of neurofibrillae may be such, but it may also be a by-product of the autolysis of the basal cytoplasm of these epidermal cells. The section clearly indicates that the nuclei and perinuclear cytoplasm of cells in this area are being forced to the exterior through rents in the apical surface of the epithelium. Recognizing the precise association of the larval ganglion with the protocoel, one must conclude that the larval brain (and the entire exumbrellar epidermis of the hood) are dehisced. Because of the difficulty of assigning a precise location to the mouth and of defining the proximal limit of the preoral hood subumbrella, and thus a distal limit to the esophagus, whether one states that the entire subumbrellar surface or all but a small preoral part is lost is a matter of definition. I have accepted the latter interpretation.

A slightly later stage of metamorphosis is represented in midsagittal section in Figure 11. The protocoel is seen to remain in direct contact with the esophagus, but there is no trace of the exumbrellar epidermis which once sheathed its opposite side. Although at this time there is no suggestion of the epistome that shortly will occupy the site of (and contain) the protocoel, the retention of the larval preoral hood coelom through metamorphosis and the degeneration of the larval ganglion are apparent.

Figure 12 is of a median sagittal section through an 8-day old juvenile. The epistome is fully developed as a fold overhanging the mouth. The epistomal coelom, derived directly from the larval protocoel, occupies the narrow space within the epistome. Developed on the aboral face of the epistome is a concentration of nervous tissue, presumably a part of the adult ganglion. The ontogeny of this neural tissue has not been followed in detail and whether the aboral epidermis of the epistome is contributed by a preoral remnant of the larval subumbrellar surface or from collar tissue of the larva remains unclear.

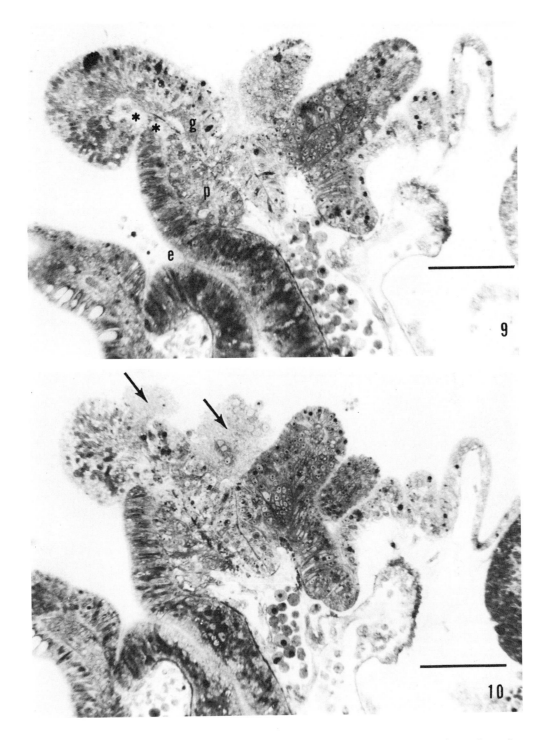

Fig. 9. *Phoronopsis viridis.* Near midsagittal section of individual in early metamorphosis. The persistent (but somewhat collapsed) protocoel (p) retains its position between the esophagus (e) and larval ganglion (g). The distal parts of the preoral hood have already been separated off (at asterisks). Bar = 20 μm.

Fig. 10. *Phoronopsis viridis.* Section 16 μm to one side of that in Fig. 9. Nuclei and perinuclear cytoplasm of cells of the larval brain have spewed into the surrounding medium (arrows). Compare with Fig. 9. Bar = 20 μm.

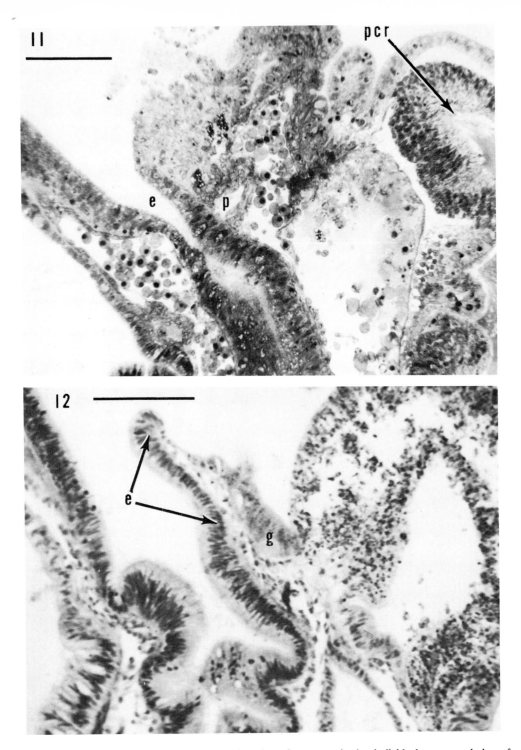

Fig. 11. *Phoronopsis viridis*. Near median sagittal section of metamorphosing individual near completion of the autolysis and dehiscence of the epidermal tissues of the hood. The protocoel (p) retains its association with the esophagus (e). Note the collapsed remnant of the larval trunk with its conspicuous perianal ciliated ring (pcr). Bar = 20 μm.

Fig. 12. *Phoronopsis viridis*. Midsagittal section of 8-day old juvenile. The retained larval protocoel occupies the lumen of the flap-like epistome (e). The apparent ganglionic concentration (g?) on the aboral surface of the epistome possibly represents the adult ganglion. Bar = 100 μm.

dehiscence. In contrast, the proximal part of the hood, including the proximal recess of the protocoel, the preoral septum and the larval ganglion appear normal (compare with Fig. 9).

Figure 15 is also a midsagittal section of a metamorphosing individual in which the larval collar region has undergone extensive construction and in which the tentacles have assumed the adult orientation. Examination of the preoral hood reveals that the subumbrellar surface is now disconnected from the oral region and is in advanced stages of degeneration. The exumbrellar surface of the hood is similarly degraded but has not broken its connection with the collar epidermis. Note particularly that the proximal recess of the protocoel retains its association with the esophagus, but is now separated from the exumbrellar epidermis. The latter has undergone so much autolysis that the larval ganglion, which earlier was in contact with the proximal recess, cannot be recognized. This figure indicates that the proximal recess of the protocoel is the only significant part of the larval preoral hood to be retained through early metamorphosis. That this remnant is indeed retained in later stages is seen in Figure 16. This juvenile, about 24 hours after metamorphosis, is not yet fully differentiated, but the elimination of the larval tissues of the preoral hood and collar is essentially complete. The proximal recess of the protocoel remains attached to the esophagus and the preoral remnant of the subumbrellar epidermis of the hood has folded back over this protocoelic remnant and fused with the thickened epithelium from which the "adult" tentacles arise. The collar epidermis anterior to the "adult" tentacles has been entirely eliminated by this stage. The remnant of the protocoel and its epidermal covering will soon differentiate into the definitive form of the adult epistome. Differentiation of the "epistomal fold" parallels that documented for *Phoronis muelleri* by Siewing[4] and so is not redescribed here.

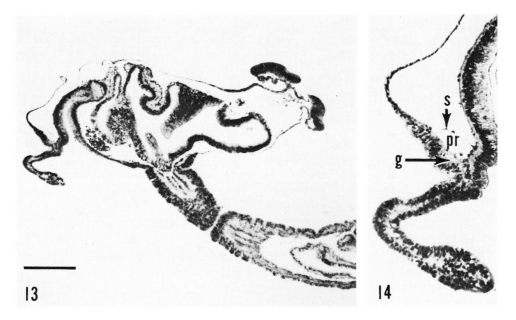

Fig. 13. *Phoronis architecta*. Midsagittal section of individual in first phases of metamorphosis. The everted metasomal sac has carried with it the gut. The distal part of the preoral hood appears degenerate. Bar = 100 μm.

Fig. 14. *Phoronis architecta*. Detail of Fig. 13. Note that the proximal portion of the larval hood, including the larval ganglion (g), proximal recess (pr) of the protocoel, and preoral septum (s), appear normal, but that the epidermis of the distal part of the hood has pyknotic nuclei and is degenerating.

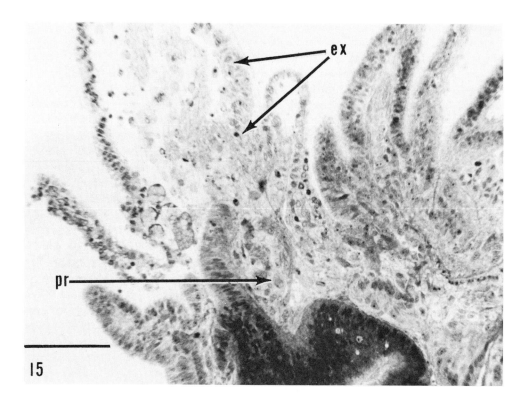

Fig. 15. *Phoronis architecta*. Midsagittal section of metamorphosing individual in which autolysis of the hood, including the proximal epidermis is well advanced. The proximal recess (pr) of the protocoel is retained against the esophagus but the larval brain is detached and cannot be identified within the degenerating exumbrellar epidermis (ex). Bar = 20 μm.

It should be noted that in the metamorphosis of *P. architecta*, only the proximal recess of the protocoel is retained. The proximolateral and all distal parts of the preoral coelom are lost.

Phoronis psammophila

The metamorphosis of *P. psammophila* has been described by Selys Longchamps[9] and (as *Phoronis sabatieri*[17]) by Veillet.[11] Only a single stage of metamorphosis of *P. psammophila* was available for sectioning, but it provided convincing evidence that the proximal recess (but not other parts) of the protocoel is retained through metamorphosis. On superficial examination, this specimen seems to be in an early stage of metamorphosis, for although the metasomal sac is completely everted, the larval trunk has not undergone collapse. In sections, however, lysis of the transitory tissues of the hood and collar is seen to be nearly complete. In Figure 17, the larval portions of the curiously composite tentacles are in advanced autolysis and apparently almost dehisced. A short preoral remnant of the subumbrella has folded at a right angle with the esophagus and abuts against the thickened epidermis at the base of the "adult" tentacles. As described by Veillet,[11] all the collar and hood epidermis have been lost anterior to the level of the tentacle out-pocketings. The proximal recess is the only remnant of the voluminous larval protocoel and it is seen in its original association with the esophagus, but has been covered over by reflection of the preoral remnant of the subumbrella. It would be misleading to state that an "epistomal *fold*" exists at this stage, but a comparison of Figures 16 and 17 reveals the coelomic and epidermal parts of the epistomal primordium of *P. architecta* have precise counterparts in *P. psammophila*.

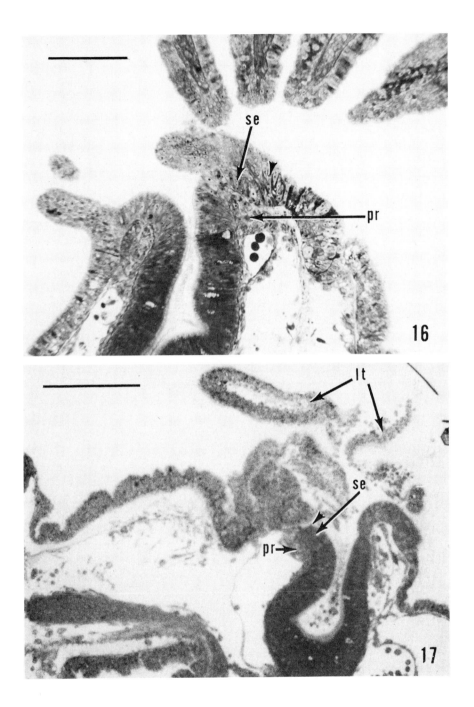

Fig. 16. *Phoronis architecta*. Midsaggital section of individual in which the dehiscence of the epidermal parts of the hood is complete. The proximal recess (pr) of the protocoel remains in contact with the esophagus and is more or less enclosed by the preoral remnant of the subumbrellar epidermis (se). The approximate junction of this epidermis with that of the collar or lophophore is indicated by an arrowhead. Bar = 20 μm.

Fig. 17. *Phoronis psammophila*. Near median sagittal section of metamorphosing individual. The larval trunk is only partly collapsed, but dehiscence of the transitory portion of the larval tentacles (lt) is all but complete and only the proximal recess (pr) of the protocoel persists. The abutment of the preoral remnant of the subumbrellar epidermis (se) against the lophophoral epidermis is indicated by an arrowhead. Bar = 20 μm.

DISCUSSION

Structure of the Larval Protocoel

As the existence of a coelomic compartment in the preoral hood of the larva has usually been denied, even in certain species examined during the course of this paper, it is appropriate to make a brief analysis of the literature relative to the current findings.

The conformation of the protocoel described here for advanced larvae for *Phoronis architecta* is essentially identical to that of the closely related species *P. muelleri* as recently resdescribed by Siewing[4] (see below). In the distal portion of the hood of advanced actinotrochs of *P. muelleri*, the mesothelium is absent as in *P. architecta*, but this region is not filled with gel and fibers. Significantly, the protocoel of young larvae of *P. muelleri* fills the hood and has a complete lining.[8] *P. architecta* has been previously studied by Brooks and Cowles.[20] These two workers actually studied two species of actinotrochs and came to the incorrect conclusions as to the identity of *P. architecta*. This error is irrelevant because they observed the preoral septum in both types of larvae, but, believing that it was incomplete, concluded that neither species had a coelomic cavity in the preoral hood. Similar reasoning led Selys Longchamps[9] to conclude that larvae of *P. muelleri* also lack such a coelom. Damage to the delicate preoral septum probably resulted in the incorrect conclusions of these careful workers.

The pattern of the protocoel recorded for advanced larvae of *Phoronis pallida* and *P. psammophila* and for the two unidentified larvae "actinotroch C" and "actinotroch D" has not been recorded previously. Emig[7] demonstrated that a spacious protocoel fills the hood cavity in early stages of *P. psammophila*, but he did not examine advanced actinotrochs. Such larvae have been studied by Shearer,[21] Selys Longchamps,[9] and Veillet[11] but none recognized the existence of a protocoel. Again, the delicacy of the preoral septum probably resulted in misinterpretation.

The diminutive protocoel characteristic of *Phoronopsis californica, P. viridis*, and *P. harmeri* was first accurately described and figured in 1964 for the latter species,[1] but had in fact been recognized much earlier.

At the turn of the century, both Masterman[5] and Menon[22] observed minute chambers beneath the brains of actinotrochs they were studying, but neither appears to have interpreted these structures correctly. Both workers reported that the preoral hood and collar cavities were filled by an unpaired and paired coeloms, respectively. Masterman described that a "subneural sinus" comparable to the vertebrate structure of the same name was left as a blastocoelic space beneath the larval brain by the incomplete approximation of the hood and collar coeloms. Menon claimed that a similarly positioned cavity in one (or more?) of his three species had a complete lining of its own and therefore represented a coelomic "heart vesicle" (apparently comparable to the lesser developed of the paired protocoels of hemichordates). In retrospect, both workers probably were studying species with protocoels like those described above for *Phoronopsis* species.*

The obliteration of the cavity and protocoel of the preoral hood in larvae of *Phoronis vancouverensis* by the approximation of the sub- and exumbrellar epidermis of the hood is otherwise unknown. This species has been synonymized with *P. hippocrepia*[16] and with *P. ijimai*.[17] The internal organizaition of the larvae of the former has not been studied. Ikeda[10] believed his Species A belonged to *P. ijimai*, but in none of the four larval types he studied were the layers of the hood apposed.

It is profoundly significant that, although an unmodified protocoel filling the preoral hood cavity is found in none of the advanced actinotrochs yet studied (accepting that Masterman[5] and Menon[22] were in error), such a coelom is known in the early ontogeny of one species

---

*In fact, Menon probably was working with *Phoronopsis californica*. Menon did not figure any of his actinotrochs, but Selys Longchamps[9] provided two illustrations of "*Actinotrocha Menoni* X." It is probable that two species, not two different stages of the same species are figured because the smaller individual is remarkably similar to the distinctive larva of *Phoronopsis californica* (compare his figure 7, plate 12 with my Fig. 6). Significantly, the adults of this species are now known to occur in the Indian Ocean.[17]

representing each of the four known definitive patterns: *Phoronis muelleri*,[4] Pattern 1; *Phoronis psammophila*,[7] Pattern 2; *Phoronopsis harmeri*,[1] Pattern 3; and *Phoronis vancouverensis*,[1] Pattern 4. Thus the modest (*Phoronis psammophila*, Pattern 2) or extensive (Pattern 1) loss of the coelomic lining and the modest (Pattern 2), extensive (Pattern 3) or complete (Pattern 4) reduction of the volume of the protocoel are secondary changes. The ontogenetic changes in the protocoel have been studied in detail only in *Phoronis vancouverensis*.[23]

The Fate of the Preoral Hood and its Cavity

Siewing's[4] observation that the larval protocoel is retained (in part) through the metamorphosis as the cavity of the incipient epistome of the juvenile in *Phoronis muelleri* has been confirmed in three additional species—*P. architecta*, actinotroch D, and *Phoronopsis viridis*. As indicated above, if a protocoel is present in advanced larvae, its morphology has one of three patterns. Each of these patterns is represented in the species now studied (*Phoronis architecta* and *P. muelleri*, Pattern 1; actinotroch D, Pattern 2; and *Phoronopsis viridis*, Pattern 3). It is noteworthy that one part, and only one part of the protocoel—the proximal recess— is retained through metamorphosis regardless of the lateral and distal extent of the coelom in premetamorphic larvae *and* that this is the only part of the protocoel common to all advanced actinotrochs with protocoels. The observed fate of the protocoel (and origin of the epistomal coelom) probably is characteristic for the phylum. An obvious exception is *P. vancouverensis* in which advanced larvae have no protocoel. In this species, the putative mesodermal remnants of the protocoelic lining (at the site of the proximal recess of typical larvae) may serve as the primordium of the epistomal cavity.

As concluded in almost all earlier studies, the epidermal components of the hood are indeed transitory with the exception of a small preoral portion of the subumbrellar epithelium (actually part of the larval esophagus?). Specifically, the larval brain is lost, in contrast to the interpretation of Zimmer.[1,14] Observations of (1) early stages of metamorphosis in which the brain had not yet lost its integrity although other transitory tissues had been degraded and cast off, and (2) early imagoes in which a conspicuous aggregation of neural elements is present on the epistome in a position corresponding to that of the larval brain on the preoral hood led Zimmer to the erroneous conclusion that the larval brain was retained as the primordium of the adult one.

Are Phoronids Oligomerous?

These collective findings provide conclusive evidence for the identity (homology) of the larval preoral hood and adult epistome. I share the opinion of others[e.g. 3,4] that the hood and epistome are equivalent to the protosome of tri- or oligomerous animals.* This correspondence has been criticized by Hyman,[6] among others, on the basis that, in contrast to the situation in echinoderms and hemichordates, the phoronid protocoel is unpaired, that it lacks communication with the exterior and that it is in communication (in the adult) with the lophophoral coelom. These criticisms have been considered previously,[e.g.1] but the presence of a communication between the cavities of the epistome and lophophore in the adult requires further attention. This confluence has been interpreted to indicate that the epistome and its cavities are derivatives of the lophophore. It is now evident that this is not the case. The coeloms of both the epistome and lophophore are preformed in the larva, but they *are not in communication* in either the larva or just formed juveniles. That the two cavities are in communication in the adult was first observed at the turn of the century[9] and has recently been confirmed.[24] When, then, does such communication develop? An obvious possibility is that the union is formed shortly after metamorphosis when the two cavities become closely approximated (in contrast to their posi-

---

*Although I regard the lophophorates to be related to the echinoderms and hemichordates and consider the similarity in their regionation an important shared feature, I prefer the neutral terms oligo- or trimery to archimery or achimetamery which now appear frequently in the literature.[e.g. 3,4]

communicate, but that fusion in the adult is a consequence of regeneration of the anterior end after its autotomy. Emig,[25] who reviewed the early literature on this subject, recently confirmed that such communication is formed during regeneration.

The fact that autotomy followed by regeneration is a frequent and characteristic event during the adult life cycle may fully account for the observed connection between the anterior coeloms of the phoronid. Thus Selys Longchamps'[9] argument that the communication of the cavities of the epistome and lophophore indicates that the former is derived from the latter may indeed be true only if the animals have undergone regeneration; it is false if one considers only ontogeny.

Finally, I should like to point out the remarkable parallel that exists between the fates of the preoral lobes of phoronids and asteroids. The lobe is inconspicuous or absent in the adults, but it is a prominent structure provided with a well-developed protocoel in the larvae. In both groups, the epidermal components of the lobe are almost or completely eliminated at metamorphosis and the proximal but not the distal parts of the protocoel are retained and utilized in the adults (in some phoronids the distal parts of the protocoel are lost before metamorphosis). A difference in the functional significance of the parts does occur in the two forms, but such differences also occur between echinoderms and hemichordates. I do not imply a direct relationship here, but point out that the pattern of transformation of the preoral hood and its cavity during metamorphosis in phoronids is not without parallel.

ACKNOWLEDGMENTS

I would like to thank Dr. Bernard C. Abbott, Chairman of the Department of Biological Sciences, the University of Southern California, for providing ship time and facilities. Special thanks go to Dr. Robert Fernald for his long-time friendship and encouragement.

REFERENCES

1. Zimmer, R. L. (1964) Reproductive Biology and Development of Phoronida, Ph.D. Dissertation, Univ. Washington, Seattle, pp. 1-416.
2. Emig, C. C. (1973) Z. Morph. Tiere, 75, 329-350.
3. Emig, C. C. (1976) Z. zool. Syst. Evolut.-forsch., 14, 10-24.
4. Siewing, R. (1974) Zool. Jb. Anat., 92, 275-318.
5. Masterman, A. T. (1897) Quart. J. Microsc. Sci., 40, 281-339.
6. Hyman, L. H. (1959) The Invertebrates. V. Smaller Coelomate Groups, McGraw Hill, New York, pp. 228-274.
7. Emig, C. C. (1977) Amer. Zool., 17, 21-37.
8. Siewing, R. (1975) Verh. Dtsch. Zool. Ges., 1974, 116-121.
9. Selys Longchamps, M. de (1907) Fauna u. Flora Golfes Neapel, 30, 1-280.
10. Ikeda, I. (1901) J. Coll. Imp. Univ. Tokyo, 13, 507-592.
11. Veillet, A. (1941) Bull. Inst. oceanogr. Monaco, 810, 1-11.
12. Cori, C. J. (1937) Phoronidae. Handbuch der Zoologie, Kükenthal und Krumbach eds., 3(2), 71-135. De Gruyter, Berlin.
13. Beklemishev, W. N. (1969) Principles of Comparative Anatomy of Invertebrates, Vol. 1, Promorphology, Engl. transl. by J. M. MacLennan; Kabata, Z. ed., Univ. Chicago Press, Chicago.
14. Zimmer, R. L. (1973) in Living and Fossil Bryozoa, Larwood, G. P. ed., Academic Press, London.
15. Masterman, A. T. (1898) Prov. roy. Soc. Edinb., 22, 270-310.
16. Marsden, J. C. (1959) Canad. J. Zool., 37, 87-111.
17. Emig, C. C. (1971) Bull. Mus. Hist. nat. Paris (Zool.) 8, 469-568.
18. Herrmann, K. (1976) Zool. Jb. Anat., 95, 354-376.
19. Herrmann, K., personal communication.
20. Brooks, W. K. and Cowles, P. (1905) Mem. Nat. Sci. Washington, 10(5), 75-111.
21. Shearer, C. (1906) Mitt. Zool. Stn. Neapel, 17, 487-514.
22. Menon, K. R. (1902) Quart. J. microsc. Sci., 45, 473-484.
23. Zimmer, R. L. Unpublished studies.
24. Emig, C. C. and Siewing, R. (1975) Zool. Anz. 194, 47-54.
25. Emig, C. C. (1972) Z. Morph. Tiere, 73, 117-144.

LARVAL MORPHOLOGY AND SETTLEMENT
OF THE BRYOZOAN, *BOWERBANKIA GRACILIS* (VESICULARIOIDEA,
CTENOSTOMATA): STRUCTURE AND EVERSION OF THE INTERNAL SAC

Christopher G. Reed

Department of Zoology, University of Washington, Seattle, Washington 98195

During the metamorphosis of marine bryozoans, an invagination of the oral epithelium called the internal sac is typically everted to attach the larva to the substratum. Some authors have claimed that larvae of the superfamily Vesicularioidea lack an internal sac, and that the corona or pyriform organ makes the initial attachment at metamorphosis. In the larva of *Bowerbankia gracilis* an internal sac everts at the onset of metamorphosis and secretes an adhesive that cements the settling organism to the substratum. The internal sac and a subjacent layer of undifferentiated cells subsequently move into the center of the metamorphosing larva as the corona involutes.

INTRODUCTION

Metamorphosis in marine bryozoans consists of a series of rapid morphogenetic movements followed by the histolysis of evanescent larval tissues and the differentiation of ancestrular anlagen. Attachment is essential for the transformation of the natant larva into the sessile ancestrula, and is typically effected by the eversion of an invaginated part of the oral epithelium called the internal sac (metasomal sac). Secretions from the everted sac cement the metamorphosing organism to the substratum.[1,2] The size and complexity of the internal sac vary among genera, but the sac always functions as an adhesive organ at the onset of metamorphosis.[3,4]

Some authors have reported that an internal sac is absent in larvae of the superfamily Vesicularioidea. Barrois[5] thought that there was only a remnant of an internal sac in the larva of *Amathia lendigera* (as *Serialaria lendigera*) and that the corona was the attachment organ. According to Hondt,[6,7,8] an invagination of the oral epithelium in the larva of *Bowerbankia imbricata* is not homologous to the internal sacs of other bryozoan larvae and does not evert at metamorphosis. He inferred that secretions from the pyriform organ attach the larva to the substratum in the initial stages of metamorphosis and that the final attachment is made by the pallial sinus epithelium. In contrast, Ostroumoff[9] described the eversion of an internal sac in the larva of *Bowerbankia stationis* (as *Vesicularia stationis*), and internal sacs have been identified in the larvae of *Bowerbankia pustulosa, Amathia semiconvoluta,*[10] and *Zoobotryon verticillatum.*[11] Eiben[12] identified an attachment organ in the larva of *Bowerbankia gracilis,* but did not describe its structure or origin.

In this paper the structure and eversion of an internal sac in the larva of *Bowerbankia gracilis* are described.

MATERIALS AND METHODS

Colonies of *Bowerbankia gracilis* were collected from the undersides of floats in the San Juan Archipelago and maintained in aquaria with running sea water at the Friday Harbor Laboratories of the University of Washington. Positively phototactic larvae were collected from the lighted side of an aquarium and either pipetted directly into the fixative or into dishes where they metamorphosed on pieces of the inner shell membrane of chicken eggs.[13] The methods of fixation and the microscopic procedure used in this study are described in a previous publication.[14]

RESULTS

Larval Morphology

Although the larva of *Bowerbankia gracilis* is anenteric, it is convenient to use the terms aboral and oral to designate the former animal and vegetal poles of the embryo. The larva is elongated in the aboral-oral axis and swims with the aboral pole foremost. Most of the surface is covered by densely ciliated coronal cells that extend from aboral to oral pole, except where the pyriform organ is located in the anterior midline (Fig. 1). The pyriform organ consists of a superior glandular field with a glandular pit, a papilla that bears a ciliary tuft, an inferior glandular field, and a ciliated cleft. At the aboral pole, the apical disc consists of peripheral columnar cells with immobile cilia around a central non-ciliated neural plate. The apical disc is surrounded by a pallial sinus lined with a simple cuboidal epithelium. The pallial sinus forms a shallow furrow anterior to the apical disc, but deepens to two-thirds of the length of the larva on the lateral and posterior sides (Figs. 1, 2, 3). The pallial sinus epithelium consists of cuboidal secretory cells with apical vesicles and basal arrays of parallel cisternae of rough endoplasmic reticulum. An equatorial neuro-muscular ring lies beneath the perikarya of the corona cells.

A glandular region of the oral epithelium is infolded to form a small internal sac with an inconspicuous lumen. The infolded glandular epithelium consists of a single layer of columnar

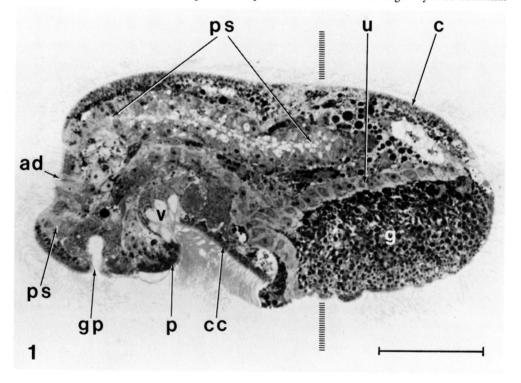

Fig. 1. Median section of larva. The glandular cells of the internal sac (g) are filled with secretory granules and basal lipid droplets. The glandular epithelium is bordered basally by a layer of undifferentiated cells (u) that continues aborally as a median band. The pyriform organ consists of the superior glandular field with a glandular pit (gp), the papilla (p), the inferior glandular field with vacuolated secretory cells (v), and the ciliated cleft (cc). The apical disc (ad) is surrounded by the invaginated pallial sinus epithelium (ps). The pallial sinus is about 130 μm deep on the posterior and lateral sides of the larva, but only 20 μm deep on the anterior side. The corona (c) comprises most of the larval surface. The dashed line indicates the level of section in Fig. 3. Scale bar = 50 μm.

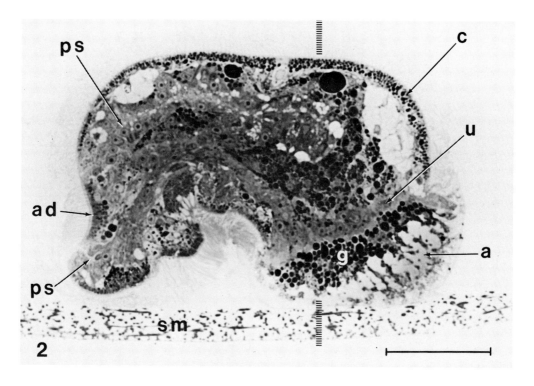

Fig. 2. Median section of a settling larva with the internal sac everted. The glandular cells of the internal sac (g) are in contact with the shell membrane (sm) and have secreted a viscous adhesive (a). Lipid droplets remain in their basal cytoplasm. The dashed line indicates the level of section in Fig. 4. Scale bar = 50 μm.

cells virtually filled with large bipartite secretory granules (Figs. 1, 3). Each cell has a basal nucleus, supranuclear lipid droplets, and scattered tubules of rough endoplasmic reticulum. The cells are joined to each other and to the superficial ciliated oral epithelium by subapical zonulae adhaerentes and septate desmosomes. The internal sac opens as a narrow slit in the median plane of the larva (Fig. 3).

A layer of undifferentiated cells lies immediately beneath the glandular epithelium of the internal sac and extends aborally as a median band between the glandular cells of the pyriform organ and the pallial sinus epithelium (Figs. 1, 3, 5). The polygonal cells of this layer have neither junctional complexes nor parallel axes of polarity characteristic of epithelia. Each cell contains a central nucleus with a large nucleolus, perinuclear mitochondria, and multitudinous free ribosomes and microtubules.

Settlement

Prior to settlement, the larva creeps slowly over the substratum with its pyriform organ against the surface. Secretions from the pyriform organ may temporarily anchor the larva during this exploratory phase. The larva then constricts in the region of the equatorial neuro-muscular ring and the internal sac is everted. The glandular cells contact the substratum and secrete a viscous adhesive that is wafted slowly over the settling larva by the coronal cilia (Figs. 2, 4). The adhesive forms a protective pellicle that encases the metamorphosing larva and anchors it to the substratum (Figs. 5, 6). During eversion of the internal sac, the subjacent concave layer of undifferentiated cells is flattened (Figs. 2, 4). As the polygonal cells are compressed laterally, apparently by the contraction of the muscles of the equatorial neuro-muscular ring, they elong-

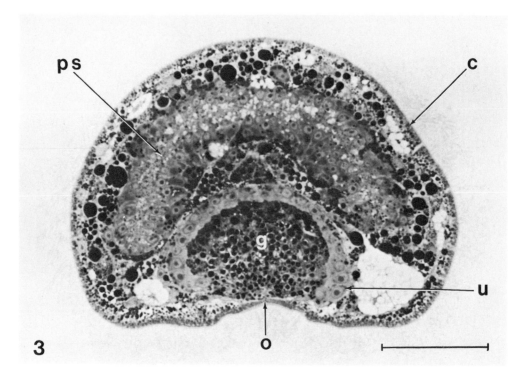

Fig. 3. Transverse section of a larva at the level indicated in Fig. 1. The infolded granular epithelium of the internal sac (g) joins the ciliated oral epithelium at the orifice (o). The subjacent layer of undifferentiated cells (u) follows the contour of the internal sac, but does not extend to the surface of the larva. The lateral and posterior regions of the pallial sinus epithelium (ps) are visible at this level. Scale bar = 50 μm.

ate and bulge into the basal surface of the internal sac. This layer may facilitate the eversion of the internal sac by transmitting the force of contraction of the muscles to the glandular epithelium.

After the glandular cells of the internal sac have released most of their contents, the apical disc is pulled in and the corona begins to inroll at the oral pole, exposing the pallial sinus epithelium (Fig. 5). The internal sac loses contact with the substratum and precedes the corona into the interior of the metamorphosing larva, accompanied by the layer of undifferentiated cells (Figs. 6, 7).

DISCUSSION

An internal sac has been described in the larvae of every marine bryozoan studied except for a few species in the superfamily Vesicularioidea. The larva of *Bowerbankia gracilis* (Vesicularioidea) also everts an internal sac at the onset of metamorphosis, in contrast to the findings of Barrois[5] and Hondt[6,7,8] on closely related species. Internal sacs have been identified in other vesicularioid larvae,[9,10] and it is likely that they are typical of the larvae of this group.

Various authors have interpreted the structure of the internal sacs in vesicularioid larvae differently. The "sacco interno" described by Zirpolo[11] in the larva of *Zoobotryon verticillatum* is actually the pallial sinus; he referred to the true internal sac as a "massa granulare vitellina." According to Calvet,[10] the oral glandular cells in *Bowerbankia pustulosa* and *Amathia semiconvoluta* (his "tampon du sac interne") are a degenerate product of metamorphosis. The same region in the larva of *Amathia lendigera* was identified by Barrois[5] as a vestige of an

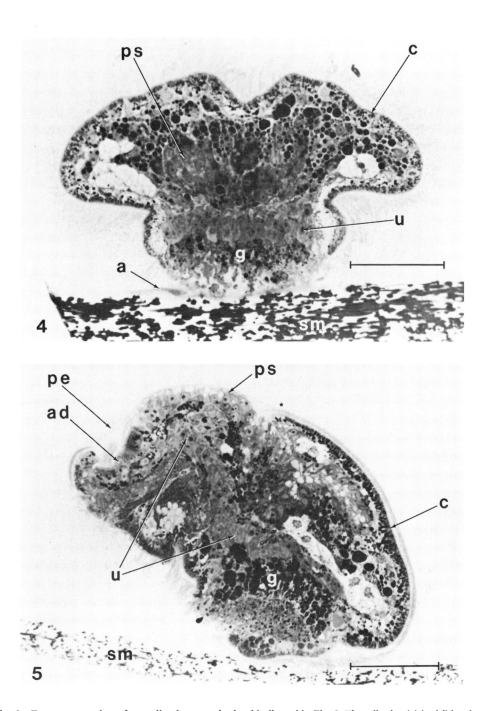

Fig. 4. Transverse section of a settling larva at the level indicated in Fig. 2. The adhesive (a) is visible where the glandular cells (g) of the everted sac contact the shell membrane (sm). The larva is constricted in the region of the neuro-muscular ring. The undifferentiated cells (u) have become columnar and form a flat layer subjacent to the everted glandular epithelium. Scale bar = 50 μm.

Fig. 5. Median section of a metamorphosing larva after pellicle formation. The corona (c) has started to inroll at the oral pole, exposing the pallial sinus epithelium (ps). The pellicle (pe) is visible over the retracted apical disc (ad) and at the aboral margin of the corona (asterisk). The glandular cells of the everted sac (g) are no longer in contact with the shell membrane (sm). The undifferentiated cells (u) form a conspicuous layer that extends from the internal sac to the vicinity of the apical disc. Scale bar = 50 μm.

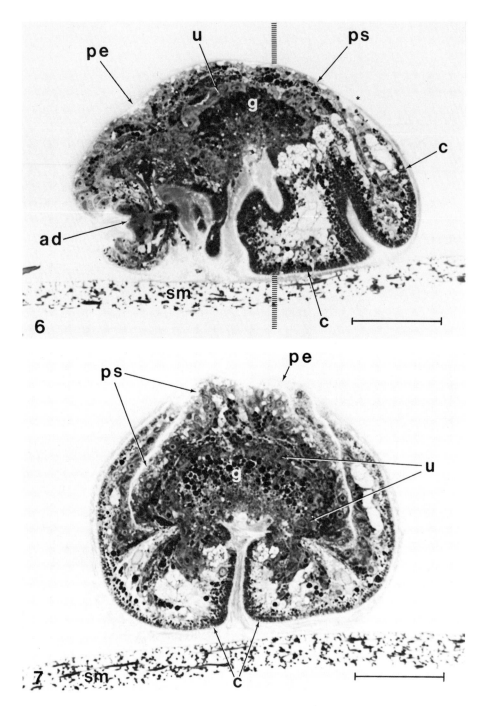

Fig. 6. Median section of a metamorphosing larva with most of the corona involuted. The internal sac (g) and the subjacent layer of undifferentiated cells (u) have moved into the center of the metamorphosing larva above the inrolling corona (c). The other abbreviations are the same as in Fig. 5. The dashed line indicates the level of section in Fig. 7. Scale bar = 50 μm.

Fig. 7. Transverse section of a metamorphosing larva at the level indicated in Fig. 6. The internal sac (g) and the subjacent layer of undifferentiated cells (u) are located in the center of the metamorphosing larva above the oral margin of the involuting corona (c). The pellicle (pe) is visible above the exposed pallial sinus epithelium (ps). Scale bar = 50 μm.

internal sac. According to Hondt,[7] in *Bowerbankia imbricata* these cells represent the neck cells of the internal sacs of other bryozoan larvae, but have an unknown function. In the larva of *Bowerbankia gracilis* these cells constitute the entire internal sac.

A spectrum of structural complexity is apparent in the internal sacs of various marine bryozoans. The most complex internal sac occurs in larvae of the superfamily Cellularioidea, where it consists of three distinct regions (neck, wall, and roof) with specific roles.[1] Only two regions have been described in the internal sac of the cyphonautes larva of *Electra pilosa*[10,15] (superfamily Malacostegoidea). The internal sac is simplest in vesicularioid larvae. In *Bowerbankia gracilis* it is not differentiated into discrete regions, but consists of a glandular epithelium similar in structure and function to the neck region of more complex internal sacs. Its role in ancestrula formation is correspondingly diminished, and it forms only an ephemeral attachment disc during metamorphosis. In this respect the metamorphosis of *Bowerbankia gracilis* resembles that of *Victorella muelleri* and *Bulbella abscondita*[16] (superfamily Paludicelloidea).

The fate of the layer of undifferentiated cells subjacent to the internal sac in *Bowerbankia gracilis* has not been followed. The corresponding cells in closely related species were considered to be part of the internal sac by previous authors,[9,10] but in *B. gracilis* they do not form an epithelium. In the larva of *Bowerbankia imbricata* these cells have been identified by Hondt as invaginated infracoronal cells that form the epidermis of the ancestrular polypide.[7] While this interpretation is supported by the undifferentiated state of these cells in *B. gracilis*, this mode of polypide formation is quite different from the typical pattern of bryozoan metamorphosis[4] and requires confirmation. The morphogenetic events that follow eversion of the internal sac in *Bowerbankia gracilis* are being investigated.

ACKNOWLEDGMENTS

I am most grateful to Professor Emeritus Robert L. Fernald for his active interest in my personal and professional development as a biologist. His tutelage and guidance during this embryonic period are largely responsible for my current research interests and career goals. The contribution of my committee chairman, Professor Richard A. Cloney, also cannot be overstated. I am thankful for his generous provision of time and counsel and for critically reviewing the manuscript. The subject of this paper was conceived during discussions with Professors Russel L. Zimmer and Robert M. Woollacott, and their ideas and suggestions have been greatly appreciated. The research was conducted at the University of Washington's Friday Harbor Laboratories; I gratefuly acknowledge the Director, Professor A.O.D. Willows, for the use of the facilities. This study was supported by NIH Developmental Biology Training Grant No. 5-TO1-HD00266 to the University of Washington.

REFERENCES

1. Woollacott, R. M. and Zimmer, R. L. (1971) J. Morph., 134, 351-382.
2. Loeb, M. J. and Walker, G. (1977) Mar. Biol., 42, 37-46.
3. Zimmer, R. L. and Woollacott, R. M. (1977) in Biology of Bryozoans, Woollacott, R. M. and Zimmer, R. L. eds., Academic Press, New York, pp. 57-89.
4. Zimmer, R. L. and Woollacott, R. M. (1977) in Biology of Bryozoans, Woollacott, R. M. and Zimmer, R. L. eds., Academic Press, New York, pp. 91-142.
5. Barrois, J. (1886) Ann. Sci. Nat. (Zool. Paleon.), ser. 7, vol. 1, 1-94.
6. Hondt, J. L. d'. (1975) in Bryozoa 1974. Docum. Lab. Geol. Fac. Sci. Lyon, H.S. 3, fasc. 1, 125-134.
7. Hondt, J.L. d'. (1977) Arch. Zool. exp. gén., 118, 211-247.
8. Hondt, J.L. d'. (1976) Bull Soc. Zool. France, vol. 101, suppl. 5, 41-47.
9. Ostroumoff, A. A. (1886) Archs, Slaves Biol., 2, 184-190.
10. Calvet, L. (1900) Trav. Inst. Zool. Univ. Montpellier, N.S., 8, 1-488.
11. Zirpolo, G. (1933) Mem. Accad. Nuovi Lincei, 2, ser. 17, 109-442.
12. Eiben, R. (1976) Mar. Biol., 37, 249-254.

13.   Hasper, M. (1910) J. Mar. Biol. Ass. U.K., 435-436.
14.   Reed, C. G. and Cloney, R. A. (1977) Cell Tiss. Res., 185, 17-42.
15.   Kupelweiser, H. (1905) Zoologica, Stuttg., 47, 1-50.
16.   Braem, F. (1951) Zoologica Stuttg., 102, 1-59.

METAMORPHOSIS OF CELLULARIOID BRYOZOANS

Robert M. Woollacott
Museum of Comparative Zoology, Harvard University, Cambridge, Massachusetts 02138
Russel L. Zimmer
Department of Biological Sciences, University of Southern California, Los Angeles, California 90007

Cellularioids (e.g., *Bugula* spp) are one of the most extensively studied groups of marine bryozoans. Their embryos are brooded and, in most species, receive extensive extraembryonic nutrients via a placenta-like system involving the maternal lining of the brood chamber and the tissue of the presumptive metasomal (internal) sac. The larvae are anenteric and have a short free-swimming period. Comparative studies of gymnolaemate larval structure and metamorphosis indicate that correlations exist between the development of the larval aboral epithelium and metasomal sac and their relative contributions to the adult epidermis. In many bryozoan groups, the aboral epithelia and metasomal sac of the larvae contribute about equally to the body wall epidermis. In cellularioids, the aboral epithelium is reduced and the metasomal sac is massive; in the ancestrula the body wall epidermis derives exclusively from the metasomal sac and the aboral epithelium contributes only the tentacle sheath.

INTRODUCTION

Metamorphosis of marine bryozoans consists of two phases. The first phase involves an intricate series of morphogenetic movements that are completed in a few minutes. These changes produce a "preancestrula" by rearrangement of larval tissues and segregation of purely transitory larval components. The second phase is longer in duration (one to several days) and is characterized by the histogenic differentiation of the first zooid of the bryozoan colony (the ancestrula) from the preancestrula.

The foundations of our current understanding of larval biology and metamorphosis in marine bryozoans were established by Barrois,[1] Calvet[2] and Kupelwieser.[3] Subsequent to these pivotal works, numerous studies have been published concerning bryozoan reproduction. Useful reviews of the literature on larval biology and settlement are provided by Ryland[4,5] and on larval structure and metamorphosis by Nielsen[6] and by Zimmer and Woollacott.[7,8]

This paper focuses on the larva and its transformation into a preancestrula in cellularioid bryozoans. The Cellularioidea is a superfamily of anascan cheilostomes which contains genera such as *Beania, Bicellariella, Bugula, Cabera, Kinetoskias, Scrupocellaria, Synnotum* and *Tricellaria*. The adults form erect, multibranched, tuft-like colonies. Species of this group are common components of fouling communities and occur frequently as epibionts on algae. Larval anatomy and, in some species, metamorphosis, are documented for approximately 25 species in 5 genera.* In this review, we will rely heavily on descriptions of larval anatomy and metamorphosis of *Bugula neritina*[9] and on our previously unpublished observations on this and other species of cellularioids. Sufficient information exists to permit a comparative analysis of larval structures and metamorphosis of cellularioids, but little is presently understood about how the various morphogenetic movements which accompany metamorphosis are accomplished.

---

*Nielsen[6] provides an authoritative listing of the literature on larval anatomy and metamorphosis in gymnolaemate and stenolaemate bryozoans. We add to this list previously unpublished observations on the metamorphosis of *Bugula californica, B. mollis, B. pacifica, Scrupocellaria bertholetti*, and *Tricellaria occidentalis*.

We are giving special emphasis to the metasomal sac: its histology, function in nutrient transport during embryogenesis, and transformations at metamorphosis. The metasomal sac is a shared component of both stenolaemate and gymnolaemate bryozoan larvae. In cellularioids, the metasomal sac is highly regionated and participates in metamorphosis to form the extracellular and epidermal components of the ancestrular body wall. The extensive involvement of the metasomal sac in cellularioid reproduction contrasts sharply with its role in certain other bryozoan groups. In the previous chapter, Reed[10] discussed the metasomal sac in vesicularioid ctenostomes, a group in which development of the sac is comparatively reduced and the larval aboral epithelium forms the body wall epidermis of the ancestrula.

LIFE CYCLE OF GYMNOLAEMATE BRYOZOANS

In order to discuss cellularioid metamorphosis, certain general features of gymnolaemate reproduction must be outlined. The gymnolaemate life cycle differs from that of many groups of marine invertebrates in which the common mode of development involves production of a planktotrophic larva that settles and metamorphoses directly into a juvenile. Three factors have been instrumental in altering this basic pattern.[8] First, most gymnolaemates (and other bryozoans) retain their embryos and release short-lived anenteric larvae. Planktotrophic larvae such as the well-known Cyphonautes occur only in some species of the stoloniferan and halcyonelloid ctenostomes and in some representatives of the malacostegoid cheilostomes (e.g., species of *Electra* and *Membranipora*). Second, the polypide of the ancestrula is formed by budding from the cystid wall. Therefore, the digestive tract and eversible lophophore of the ancestrula neither develop from transformation of the larval feeding apparatus and digestive tract (when present) nor do they contain elements of embryonic entoderm. Instead, these structures arise exclusively from ectoderm and ectomesoderm through the process of polypide replacement, a phenomenon usually associated with senescence and rejuvenation in these colonial organisms. Third, there is an appearance early in ontogeny of structures critical to the adult phase of the life cycle. Heterochrony, in this case acceleration of somatic features, is evidenced by the precocious development of the epidermal and mesodermal blastemas and of the metasomal sac. These structures are not known to have any function in the larva. The blastemas constitute the anlage of the ancestrular polypide and the metasomal sac contributes, at least initially, the attachment platform and, in most cases, a major portion of the ancestrular body wall.

CELLULARIOID EMBRYOGENESIS

With rare exception, cellularioid embryos are retained one at a time within a brood chamber. The embryonary is actually an external compartment that is isolated from the surrounding sea water. The chamber is formed by the maternal autozooid and a polymorphic heterozooid which is derived from the autozooid distal to the maternal zooid.[2,11]

Embryogenesis has been examined in five species of *Bugula* (*B. calathus*,[12] *B. simplex*,[2] *B. avicularia*,[13] *B. flabellata*[14] and *B. neritina*[15]). Although there are discrepancies among these accounts, development proceeds in a fundamentally similar manner. The brief description given here is based primarily on the detailed study of *B. flabellata* by Correa.[14] Cleavage is biradial and nearly equal. Gastrulation occurs during transition from the 60- to 64-cell stages by delamination of 4 cells into the interior at the vegetal pole of the embryo. These mesentodermal cells multiply as a solid mass that cavitates to form a small archenteron with no opening to the exterior. Continued cell proliferation eventually obliterates the gut vesicle and blastocoel. Coelomic compartments are not developed. The ring of 16 supra-equatorial blastomeres of the 60- and 64-cell stages divides once to form a 32-cell band. These cells become heavily ciliated and constitute the larval locomotory organ, the corona. During embryogenesis, the coronal cells of cellularioids elongate in an aboral-oral direction to form most of the surface of the larva. The definitive corona contains from 32 cells (*B. flabellata*) to over 300 cells (*B. neritina*[9]). Most of the vegetal ectoderm (the presumptive metasomal sac tissue) becomes specialized for the uptake of extraembryonic nutrients (see below). The next structures to differentiate are the apical disc and neuromuscular system. Subsequently, the pyriform complex is elaborated. This structure, unique to gymnolaemate larvae, contains sensory and glandular components which are involved in substrate selection and temporary attachment at settlement. Near the end of embryogenesis the epidermal and mesodermal blastemas differentiate at the aboral pole and the presumptive metasomal sac tissue invaginates at the opposite pole. Before release of the larva, the latter tissue undergoes a complex histogenic redifferentiation correlated with its role during and after metamorphosis.

In six of the seven cellularioids studied there is, during the course of embryogenesis, a large increase in volume of the embryo (50- to 1000-fold depending on the species).[2,16,17,18*] As the embryos are filled with tissue, this increase reflects an addition of mass. In five of the six species with embryonic growth (*Bicellariella ciliata, Bugula avicularia, B. flabellata, B. neritina,* and *B. simplex*), the nutrients required for this extensive growth are derived through a placenta-like system involving hypertrophy of the maternally derived portion of the ovicell. The embryonic component of the system is the incipient metasomal sac in *B. neritina* and apparently in the other four species as well. The ultrastructural events of extraembryonic nutrition have been documented only in *B. neritina*, in which the ovum measures 35 μm, but the larva about 300 μm.[17] In this species, growth of the embryo occurs only during a period of time in which the presumptive metasomal sac ectoderm and the maternal lining of the brood chamber are in apposition. These two tissues are specialized as the embryonic and maternal components of the placenta-like system. Ultrastructural studies demonstrate that exocytosis from cells of the hypertrophied maternal lining occurs concomitantly with pinocytotic uptake by cells of the future metasomal sac. At the conclusion of nutrient transfer, the metasomal sac tissue of the now full-sized embryo invaginates and the maternal lining of the ovicell undergoes morphological regression. Thus, the first important role of the metasomal sac cells in *B. neritina* is as the recipient tissue of extraembryonic nutrients. Because the larvae are incapable of feeding, these reserves, combined with the small amount of yolk, must provide all the resources necessary for development of an ancestrula with the capability of feeding.

In *Synnotum aegypticum*,[18] the sixth cellularioid with embryonic growth, brooding is within the maternal coelom, not an ovicell. The ovarian follicle is reported to persist around the enlarging embryo and to be the proximate source of extraembryonic nutrients; the site of uptake was not specified. Thus, placenta-like systems involving the lining of the ovicell and the vegetal ectoderm of the embryo are common, but not ubiquitous in the Cellularioidea.

Although most other cheilostomes are lecithotrophic, at least three species of ascophorans also have extensive growth of the embryo while it is in the ovicell.[16] Whether either the ovicell or the incipient metasomal sac is involved in nutrient transfer to the embryos in these unrelated species remains unknown.

CELLULARIOID LARVAE

Structure

The larvae of different species in the Cellularioidea have varying shapes, ranging from barrel- to pear-like (Figs. 1-3). Additionally, the appearance of an individual larva changes frequently as it swims. For example, *B. neritina* (Fig. 4) will transform from barrel- to dumbbell-shaped, depending on the degree of contraction of the equatorial muscle band. In general, larvae measure from 200 to 400 μm along the oral-aboral axis and from 100 to 200 μm in diameter at the

*The one exception is *Kinetoskias smitti*,[16] a morphologically specialized form which lives on soft substrates.

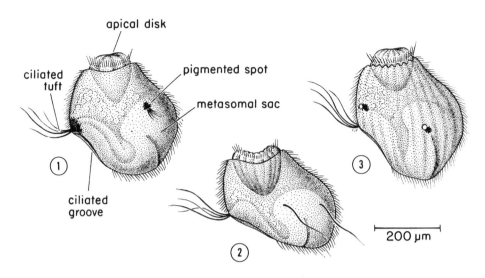

Fig. 1. *Scrupocellaria reptans*. Left-lateral surface view of larva (redrawn from Barrois[1]).

Fig. 2. *Bicellariella ciliata*. Left-lateral surface view of larva (redrawn from Barrois[1]).

Fig. 3. *Bugula plumosa*. Left-lateral surface view of larva (redrawn from Barrois[1]).

equator. Most of the surface is covered by the ciliated corona which limits the oral and aboral epithelia to small polar fields. The most conspicuous surface structures aside from the corona are the apical disc and the pyriform complex. Some species possess pigmented "eyespots."

In a comparative analysis of gymnolaemate larval anatomy, Zimmer and Woollacott[7] proposed a classification of larval types based on relative development of larval structures. We determined that all known cellularioid larvae were contained within a category designated Type AEO/ps (coronate larvae with expanded coronas that are aboral, equatorial, and oral [AEO] in position and with small pallial sinuses [ps]). Larvae of this type lack shells, are anenteric, have an expanded corona which extends both aborally and orally from the equator, and possess a small invaginated pallial epithelium. Although Type AEO/ps larvae occur sporadically within

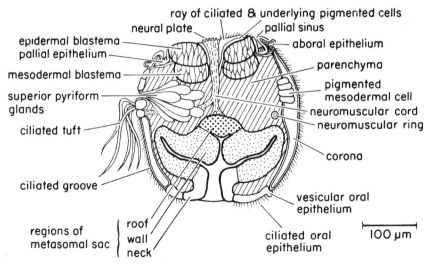

Fig. 4. *Bugula neritina*. Left-lateral view of midsagittal section of larva (redrawn from Woollacott and Zimmer[9]).

other groups of gymnolaemates, their uniform occurrence in cellularioids suggests that major features of larval anatomy and, consequently, metamorphosis follow a consistent pattern in this superfamily. The proposed system for categorizing larvae was not intended to identify particular species, although frequently species can be distinguished on the basis of size, coloration, number of coronal cells, and positioning and number of pigmented spots.

Two aspects of larval structure are especially important in relating larval to adult anatomy. These features are the aboral epithelia and the metasomal sac. On completion of metamorphosis, a combination of these two tissues will form the body wall epidermis of the ancestrula. The degree to which each tissue participates in morphogenesis of the adult epidermis can be predicted on the basis of its relative development in the larva. Analysis of the literature and our own observations indicate that larvae of different species of cellularioids uniformly have small aboral epithelia and massive metasomal sacs. The descriptions given below are based on *B. neritina* because, at present, extensive cytological information exists only for this species.

The aboral epithelia are epidermal tissues situated between the aboral margin of the corona and the apical disc (Fig. 5). In cellularioids, this field is reduced to a narrow ring-shaped zone owing to the expansion of the corona. Its components include: aboral vesicular epithelium, vesicular connecting cells, pallial epithelium, and apical-connecting cells. The pallial tissue, although the most extensive of the aboral epithelia, is invaginated and effectively sealed off from the surface within a narrow furrow. Pallial cells contain prominent Golgi bodies and numerous microtubules.

The metasomal sac is massive in cellularioids and occupies almost half the volume of the larva. The opening of the sac is located at the center of the oral field and is surrounded by the oral ciliated epithelium (Fig. 6). The sac is differentiated into three regions: neck, wall, and roof (Fig. 4). The neck zone (Fig. 6) is situated nearest the surface of the larva and is composed of cells containing large vesicles filled with electron-dense material. These vesicles stain intensely with Periodic acid Schiff acetylation (Fig. 7). The remainder of the cytoplasm contains some mitochondria but few other organelles. The wall region, which lies adjacent to the neck zone, constitutes the major portion of the metasomal sac. The epithelium of this region is columnar and is reflected on itself, effectively obliterating the lumen of the sac. The organelles of wall-region cells are highly segregated into distinct zones (Fig. 8). There is extensive basal development of rough endoplasmic reticulum. The nucleus separates this basal region from a zone containing prominent Golgi complexes and numerous vesicles with electron-dense contents. The apices of these cells are solidly filled with large electron translucent vesicles. Histochemical tests indicate that the vesicles of the intermediate zone are PAS-positive and the apical vesicles are PAS-negative. The roof region is the most interior portion (that part originally at the center of the invagination). Its epithelium is more strongly columnar than that in the wall region. Ultrastructural studies demonstrate that its cells also contain extensive organelles for synthesis and packaging of secretory materials. Two types of secretory cells occur: one with smaller vesicles with electron-dense contents, and the other with larger vesicles which contain a small electron-dense core but are mostly electron translucent. That these two cell types are not distributed uniformly may suggest the roof region is more complex than we previously reported.[9]

Behavior and Settlement

Ryland[4,5,19] has recently reviewed the experimental aspects of behavior and settlement of cellularioid and other marine bryozoan larvae. Consequently, no detailed attention to these topics is necessary here except to provide a general background for events prior to metamorphosis.

Release of larvae from the brood chamber is triggered by light, but nothing is known of the mechanism mediating this response. The larvae are anenteric and, therefore, considered to be "shortterm." Calvet[2] concluded that the total larval period ranged from 15 minutes to 6 to 8

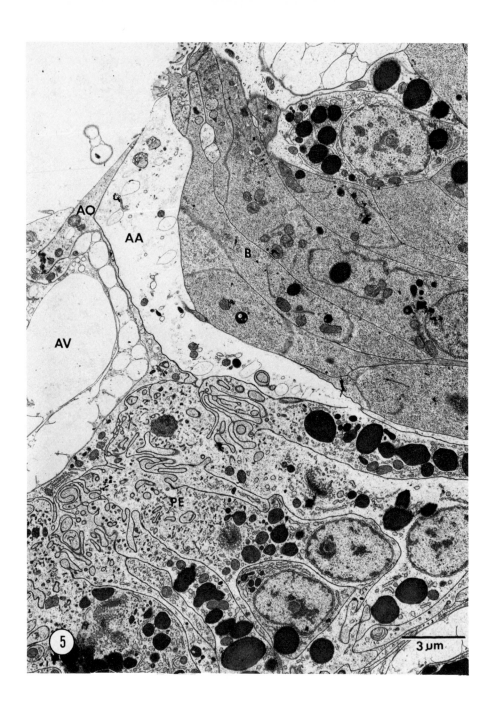

Fig. 5. *B. neritina* larva. Portions of the aboral epithelia and apical disc. AA, apical-connecting cell; AO, aboral-connecting cell; AV, aboral-vesicular cell; B, epidermal blastema with nonciliated surface connection on apical disc; PE, pallial epithelium.

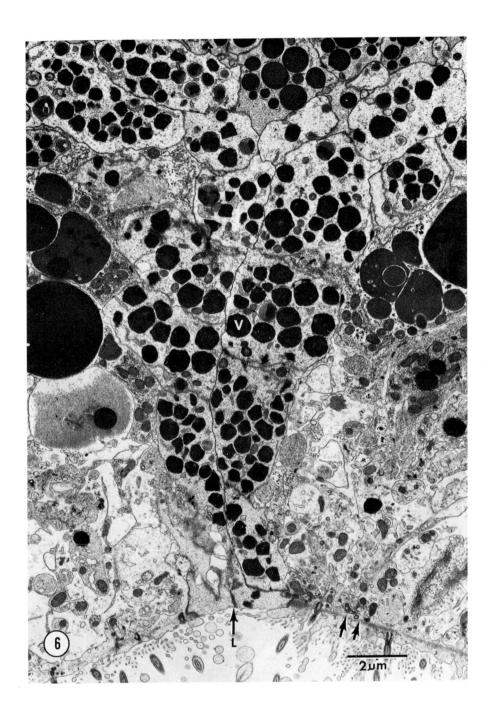

Fig. 6. *B. neritina* larva. Opening (L) of the neck region of the metasomal sac at the larval surface. The neck region cells are filled with numerous electron-dense vesicles (V). The cells adjacent to the opening of the sac are part of the oral ciliated epithelium. The band of microfilaments (double arrow) is visible near the apical margins of these cells.

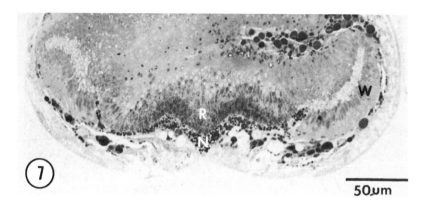

Fig. 7. One-micron section of vegetal hemisphere of *B. neritina* larva stained with Periodic acid Schiff acetylation to demonstrate PAS sensitivity of different regions of the metasomal sac. N, neck; W, wall; R, perimeter of roof.

hours, depending on the species. Although larvae of individual species vary widely in their responses to environmental factors, all larvae have an initial free-swimming phase followed by a thigmotactic period of substrate selection.

Most species are reported to be positively phototactic during the free-swimming period. There is morphological evidence for the existence of potential photoreceptoral organs in larvae of *B. neritina*,[20] but similar studies have not been extended to other species. There exists neither a published action spectrum for phototaxis nor physiological confirmation of photoreceptoral capacity of any "eyespot."

Little is understood about the events in transition from the free-swimming to substrate selection phases of larval existence. Grave[21] reported that the minimum free-swimming period in *B. flabellata* is 4 hours. In our experiences with *B. neritina*, the larvae do not immediately initiate metamorphosis upon release even when presented with an appropriate substrate and, generally, do not begin substrate selection until several hours after liberation from the brood chamber. These observations suggest that "early" larvae may lack competency for metamorphosis. Presumably, even in these nonfeeding larvae, this transition process is associated with the trade-off between dispersal capability, no matter how limited, and substrate selection opportunities. Unfortunately, the maximum possible duration of larval life is undocumented. Preliminary studies on the energetics of *B. neritina* larvae by W. Vernberg and Crisp (reported by Crisp[22]) indicate that the average respiration rate for larvae of mean dry organic weight of 1.5×10^{-6}gm/individual is 15×10^{-6}ml O_2/ hr/individual (oxygen tension and temperature were unspecified). Similar rates of respiration were recorded for individuals during metamorphosis. Naturally, it is impossible to speculate from these data about the partitioning of energy reserves and how they are expended. These studies, however, demonstrate the feasibility of continuing quantitative investigations of metabolism in relationship to larval longevity and to the development of a metamorphosed individual capable of feeding.

Once the substrate selection phase begins, larvae swim close to surfaces of objects and appear to "test" the substrate with cilia of the vibratile plume. During this period, larvae frequently cease swimming and form tentative attachments between the pyriform complex and substrate surface. These attachments are temporary, but sufficiently strong for larvae not to be easily dislodged mechanically from the substrate. The larvae, however, can quickly effect dissolution of the adhesive and continue swimming. Loeb and Walker[23] have demonstrated that this temporary cementing agent is an acid mucopolysaccharide. Such testing reactions and temporary attachments may be repeated numerous times before final settlement occurs. Permanent attachment is accomplished by eversion of the metasomal sac and constitutes the first event of meta-morphosis.

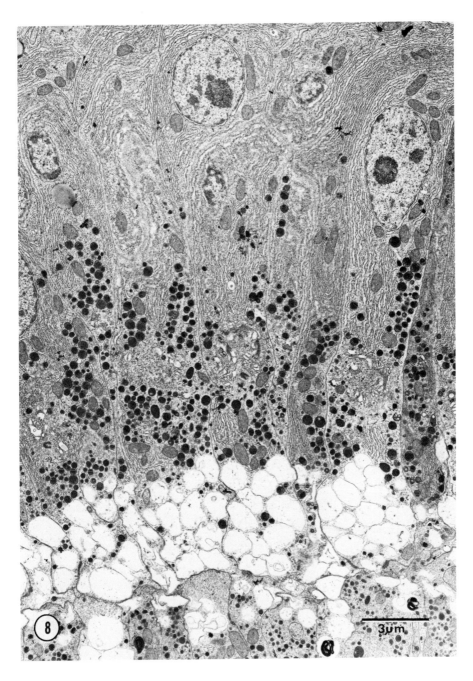

Fig. 8. *B. neritina* larva. Wall region cells of the metasomal sac. Cells are differentiated into a basal (top) zone of rough endoplasmic reticulum, middle "packaging" zone containing Golgi complexes and numerous electron-dense vesicles and apical (bottom) region filled with electron transparent vesicles.

CELLULARIOID METAMORPHOSIS

Metamorphosis involves first a rapid series of morphogenetic rearrangements of tissues followed by a slower phase of histogenic differentiation. Dissection of the events in metamorphosis is difficult for two reasons. First, the larvae are small but morphologically complex, and, second, the initial movements are intricate and occur within a few minutes. It is not surprising, therefore, that the literature contains contradictory accounts of metamorphosis both within and between species of cellularioids (Table 1). Despite these historical problems, the advent of plastic embedding techniques permitting one-micron sections for light microscopy and the use of scanning and transmission electron microscopy have facilitated a better understanding of the basic steps in cellularioid metamorphosis.[8,9]

In this paper we will consider only the sequence of movements which transform the larva of *B. neritina* into its preancestrula (Figs. 9-11). The subsequent histogenesis of the ancestrula involves elongation of the preancestrula cystid and differentiation of the polypide (Figs. 12-13).

Metamorphosis begins with a massive contraction of the equatorial neuromuscular ring which causes eversion of the metasomal sac. The roof of the sac forms the attachment disc anchoring the metamorphosing individual to the substrate (Fig. 14). After eversion of the sac, the wall region temporarily remains in a reflected position, as in the larval stage. Ultrastructural analysis of individuals during the stage illustrated in Figure 11 (approximately one minute into metamorphosis), indicates that the neck-region cells have released the contents of the numerous vesicles. This material is believed to form the extracellular calyx which surrounds the metamorphosing individual and also it may contribute adhesive for permanent attachment. It is not known if the roof-region cells contribute to this adhesive material. However, the ultrastructural studies of cells during the stage illustrated in Figure 12 (approximately 15 minutes into metamorphosis) provide evidence that exocytosis of material from the small electron-dense vesicles is occurring during this stage and that numerous large electron translucent vesicles still remain in the cytoplasm (Fig. 15).

TABLE 1

Reported Origins of the Body Wall Epidermis in Cellularioids

Species	Origins	Reference
Bugula calathus	PE/MS[b]	Vigelius[12]
B. californica	MS[a]	Mawatari,[24] personal observations
B. flabellata	PE/MS	Barrois[1]
B. mollis	MS	Personal observations
B. neritina	PE/MS	Calvet[2]
	PE/MS	Lynch[25]
	MS	Mawatari[15,26]
	MS	Woollacott and Zimmer[9]
B. pacifica	MS	Personal observations
B. simplex	PE/MS	Calvet[2]
	Corona	Grave[22]
B. stolonifera	PE/MS	Calvet[2]
B. sp.	MS	Braem[27]
Scrupocellaria bertholetti	MS	Personal observations
S. sp.	MS	Braem[27]
Tricellaria occidentalis	MS	Mawatari,[28] personal observations

[a]Exclusively from the metasomal sac (MS).
[b]Combination of pallial epithelium (PE) and metasomal sac (MS).

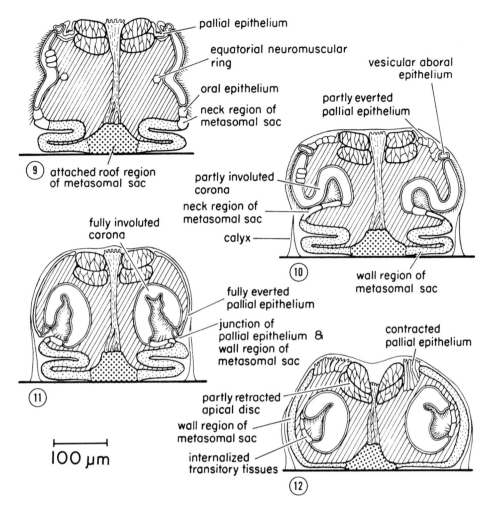

Figs. 9, 10, 11, and 12. *B. neritina* metamorphosis. Figures 9-12 are of median longitudinal sections at four stages in formation of the preancestrula (from Zimmer and Woollacott[8]).

The next events in metamorphosis are evagination of the pallial epithelium and a concomitant movement of certain transitory larval tissues into the interior of the incipient preancestrula (Fig. 11). Nothing is known of the mechanisms underlying these events. Ultrastructural studies of the larva indicate, however, that a series of muscles extends from the junction of the epidermal and mesodermal blastemas to the aboral vesicular epithelium. Conceivably, a contraction of these muscles could force the pallial epithelium to evaginate. The pallial epithelium continues to expand and form the surface of the transforming individual as the inrolling of other epidermal tissues proceeds. These movements continue until the aboral vesicular epithelium, corona, oral vesicular epithelium, oral ciliated epithelium and neck region of the metasomal sac are involuted. The pallial cells now form a delicate squamous epithelium which extends from the apical disc to the wall region of the everted metasomal sac (Fig. 11). At this point in metamorphosis, the margins of the wall region and the pallial epithelium sever their connections with adjacent epithelia and fuse with each other. This event finalizes internalization of the transitory larval tissues.

59

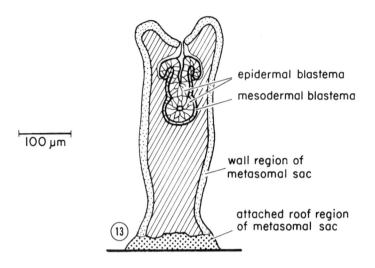

100 μm

epidermal blastema

mesodermal blastema

wall region of metasomal sac

attached roof region of metasomal sac

13

Fig. 13. Diagram of median longitudinal section of *B. neritina* individual in transformation from preancestrula to ancestrula.

The pallial epithelium now begins a progressive contraction until eventually it is reduced to a small ring around the apical disc. As contraction proceeds, the wall region is brought out of its reflected position and ultimately forms the entire free surfaces of the preancestrula (Fig. 12). Uplifting of the wall region by contraction of the pallial epithelium is thought to be mediated by a microfilament-generated movement,[9] but there is no ultrastructural documentation to support this assertion.

Formation of the preancestrula and completion of the initial phase of metamorphosis is accomplished by retraction of the apical disc and invagination of the contracted pallial epithelium. Retraction of the disc occurs by contraction of muscles which extend between the neural plate and roof of the metasomal sac. As the disc is pulled inward, the ciliated and pigmented cells forming the radii of the apical disc dehisce and are shed as transitory tissues (in *B. pacifica*).

The events in transformation of *B. neritina* larvae into their preancestrulae are documented by extensive light and electron microscopy. Many of the previous reports which indicate alternative interpretations of metamorphosis in this and in other species of cellularioids (Table 1) are from the early literature and were based exclusively on observations of living material or on limited histological data. As we have confirmed that the same events documented for *B. neritina* also occur in *B. mollis, B. pacifica* and *S. bertholetti*, we conclude that reports implicating the corona and pallial epithelium in formation of the adult epidermis are most likely in error. On the basis of similarities of larval anatomy and our observations of metamorphosis, we suggest that metamorphosis in the Cellularioidea is more uniform than previously indicated.

ACKNOWLEDGMENTS

We are grateful to Ms. Georgia Hsiau, Ms. Susan Hunt, Mr. David Lawrence and Ms. Joy Mulholland for their generous technical assistance. Mr. Laszlo Meszoly skillfully prepared the illustrations redrawn from the literature. Original research reported in this paper was supported by contract no. N00014-78-C-0064 from the Office of Naval Research with Harvard University and through grants from the National Science Foundation.

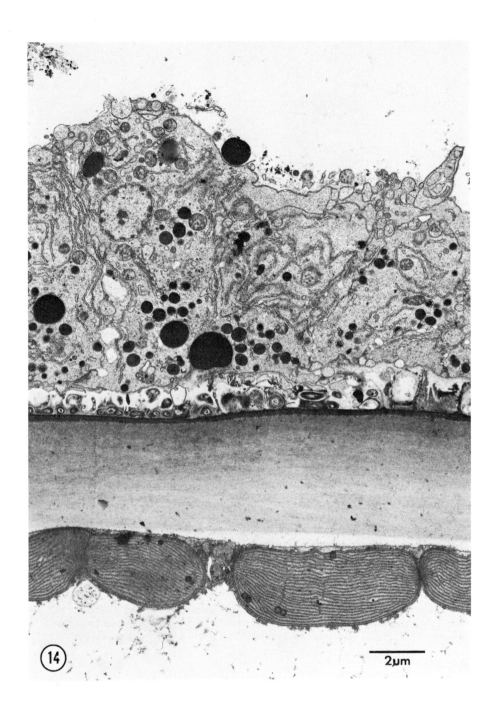

Fig. 14. *B. neritina* preancestrula. Electron micrograph showing attachment between the metasomal sac (top) and substrate (bottom) with numerous bacteria occurring within the interface. The substrate is a frond of *Rhodymenia* sp.

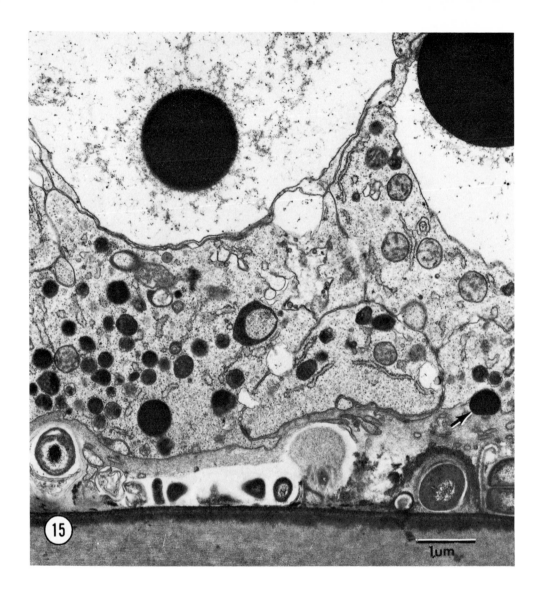

Fig. 15. *B. neritina* preancestrula. Algal cell wall (bottom)–roof region (top) interface showing bacteria within this zone and an extracellular matrix presumably formed in part by secretion (at arrow) of material from cells of the roof. Some large vesicles remain within roof region cells at this stage.

REFERENCES

1. Barrois, J. (1877) Trav. Stn. Zool. Wimereux 1, 1-305 .
2. Calvet, L. (1900) Trav. Inst. Zool. Univ. Montpellier [N.S.] 8, 1-488.
3. Kupelwieser, H. (1905) Zoologica (Stuttgart) 47, 1-50.
4. Ryland, J.S. (1974) Thalassia Jugoslavica 10, 239-262 ·
5. Ryland, J.S. (1976) Adv. Mar. Biol. 14, 285-443.
6. Nielsen, C. (1971) Ophelia 9, 209-341.
7. Zimmer, R.L. and Woollacott, R.M. (1977) in Biology of Bryozoans, Woollacott, R.M. and Zimmer, R.L. eds., Academic Press, New York, pp. 57-89.
8. Zimmer, R. L. and Woollacott, R. M. (1977) in Biology of Bryozoans, Woollacott, R. M. and Zimmer, R.L. eds., Academic Press, New York, pp. 91-142.
9. Woollacott, R.M. and Zimmer, R.L. (1971) J. Morphol. 134, 351-382.

10. Reed, C. (1978) in Metamorphosis and Settlement of Marine Invertebrate Larvae, Chia, F.S. and Rice, M.E. eds., Elsevier North-Holland, New York, pp. 41-28.
11. Woollacott, R.M. and Zimmer, R.L. (1972) Marine Biology 16, 165-170.
12. Vigelius, W.J. (1886) Mitt. Zool. Stn. Neapel 6, 499-541.
13. Marcus, E. (1938) Univ. Sao Paulo, Fac. Filos., Cienc. Let., Bol. Zool. 2, 1-196.
14. Correa, D.D. (1948) Univ. Sao Paulo, Fac. Filos., Cienc. Let., Bol. Zool. 13, 7-71.
15. Mawatari, S. (1951) Misc. Rep. Res. Inst. Nat. Resour. (Tokyo) 19-21, 47-54.
16. Marcus, E. (1938) Univ. Sao Paulo, Fac. Filos., Cienc. Let., Bol. Zool., 2, 1-196.
17. Woollacott, R.M. and Zimmer, R.L. (1975) J. Morphol. 147, 355-378.
18. Marcus, E. (1941) Archos Cirurg. Clin. Exp. 5, 227-234.
19. Ryland, J.S. (1977) in Biology of Bryozoans, Woollacott, R.M. and Zimmer, R.L., eds., Academic Press, New York, pp. 411-436.
20. Woollacott, R.M. and Zimmer, R.L. (1972) Z. Zellforsch Microsk. Anat. 123, 435-469.
21. Crisp, D.J. (1974) Thalassia Jugoslavica 10, 103-120.
22. Grave, B.H. (1930) J. Morphol. 49, 355-379.
23. Loeb, M.J. and Walker, G. (1977) Marine Biology 42, 37-46.
24. Mawatari, S. (1946) Misc. Rep. Res. Inst. Nat. Resour. (Tokyo) 9, 21-28.
25. Lynch, F.L. (1947) Biol. Bull. (Woods Hole) 92, 115-150.
26. Mawatari, S. (1946) Shigen Kagaku Kenkyujo Tanhe 9, 1-15 (in Japanese).
27. Braem, F. (1951) Zoologica (Stuttgart) 102, 1-59.
28. Mawatari, S. (1951) Misc. Rep. Res. Inst. Nat. Resour. (Tokyo) 22, 9-16.

A REVIEW OF METAMORPHOSIS OF TURBELLARIAN LARVAE

Edward E. Ruppert

The Smithsonian Institution, Fort Pierce Bureau, R.R. 1, Box 196, Fort Pierce, Florida, 33450

Trochophore-type larvae are recognized and described for Turbellaria Catenulida and Polycladida. Larval anatomy is described briefly. Settlement involves loss of the photopositive response and exploration of the substratum. Metamorphosis is marked by the resorption of the larval arms, degeneration of the frontal organ, dorsoventral flattening, multiplication of eyes, and development of parenchymal muscles and the definitive plicate pharynx. A transition from ciliary particle feeding to muscular engulfment of food also is correlated with metamorphosis. Patterns of ciliary beat are documented on the larval arms and in and around the larval stomodaeum. The epidermis, protonephridia, epidermal and cerebral eyes, frontal organ and larval arms are characterized ultrastructurally. Ciliary and rhabdomeric eyes are described. The similarity of the larval arms of the Müller's larva to feeding tentacles in tentaculates is noted. The desirability of making a detailed comparison of the frontal organs of Müller's and Götte's larvae with larval organs of attachment known in other phyla is discussed.

INTRODUCTION

Primary, trochophore-type larvae[1,2] have been documented in two orders of Turbellaria, the Poycladida and the Catenulida.[10,23] In addition, Jägersten discovered a turbellarian larva with a statocyst containing two statoliths and he suggested that this animal was the larva of an unidentified species of the order Nemertodermatida,[1] a group in which the adults are known to possess characteristically a statocyst with two statoliths.[3] If these orders are placed into the current phylogenetic framework, then the Polycladida is understood to be at the base of the divergence of the higher Turbellaria.[4,5] The Nemertodermatida is interpreted as an early evolutionary line in Turbellaria, with possibly some relationship to the Acoela[6,7,3,4,5,8]; and, the Catenulida is recognized as an early independent divergence from the main line of turbellarian evolution.[9] The occurrence of trochopore-type larvae in these "primitive" groups—*i.e.*, the polyclads, the catenulids and perhaps the nemertodermatids—suggests that a complex life cycle with a trochophore larva could be fundamental to Turbellaria, as Jägersten previously has argued.[1]

There are few investigations of settlement and metamorphosis in Turbellaria. Serious life history studies of Turbellaria with a complex life cycle have been confined largely to the Polycladida and these emphasize early development.[10,11,12,13,14] Müller first collected larvae from the plankton and determined that they metamorphosed into Turbellaria Polycladida.[15] The larval type that he described is now known as "Müller's larva." Dalyell and also Girard were the first to rear Müller's larvae from eggs deposited in laboratory culture vessels, but both failed to keep them alive through metamorphosis.[16,17] Götte, studying the larva of *Stylochus pilidium*, correctly recognized a second type of larva with only four arms in the Polycladida.[18] The independent existence of this larva (Götte's larva) was questioned by Lang, who wrote the only completed treatise on the Polycladida and described in detail the Müller's larvae of *Yungia* and *Thysanozoon*.[10] Although rejected by Lang, Götte's larva was established as a second larval type in the Polycladida by Hofker,[19] Pearse,[20] and notably by Kato.[13,21] Until recently, Götte's larva was believed to be limited to the acotylean genus *Stylochus*, a genus also known to contain species having a Müller's larva,[15] with a pelagic larva lacking arms[22] and with direct development.[10] Anderson, however, discovered Götte's larvae in the acotylean genus *Notoplana* and was able to rear them through metamorphosis.[14]

Müller's larvae are known to occur in both suborders of Polycladida. Götte's larva is known only from the Acotylea. The wide distribution of Müller's larva as well as the morphological series in the genus *Stylochus* of Müller's larva—Götte's larva—direct development, suggest that the eight-armed Müller's larva is the basic larval type of the Polycladida. This conception is also endorsed by Lang who developed it formally into an idealized archetype.[10]

A freshwater, pelagic larva was discovered by Reisinger in 1924 who determined that it developed into a turbellarian of the order Catenulida.[23] Luther's larva, named in honor of Alexander Luther, is reported only from the genus *Rhynchoscolex*. Reisinger described its anatomy and metamorphosis. No subsequent investigations of this interesting larva have been made.

This review attempts to assemble the salient features of turbellarian larval anatomy, settlement and metamorphosis. Some of my own observations of the larvae of Polycladida are also integrated into the general framework of this paper.

MATERIAL AND METHODS

The new information in this paper about Müller's and Götte's larvae is derived from individuals collected in the plankton. Three specific larvae were considered in some detail: a Götte's larva resembling closely the larva of *Stylochus uniporus* depicted by Kato[13]; a Müller's larva resembling closely the larva of *Planocera multitentaculata* also shown by Kato[13]; and a Müller's larva resembling larvae of *Yungia* and *Thysanozoon* depicted and described by Lang.[10] The first two types of larvae were collected routinely in the Indian River near Vero Beach, Florida in September and October 1977. The third larva and one other, a brilliant magenta Müller's larva, were obtained from plankton tows made offshore near Fort Pierce, Florida during September and October of 1977. These three larval types will be referred to as the estuarine Götte's larva (EGL), the estuarine Müller's larva (EML) and the oceanic Müller's larva (OML).

Living larvae were maintained in culture dishes with daily sea water changes. Preparation of material for light and electron microscopy is given in Ruppert and Shaw.[24]

Determinations of directions of ciliary beat were made on living animals by direct observation. Structural observations were made from glutaraldehyde-osmium fixed specimens examined as 1 μm thick sections, as thin sections examined with TEM and as whole animals examined with SEM.

LUTHER'S LARVA OF TURBELLARIA, CATENULIDA

A discussion of adult anatomy of species of Catenulida can be found in papers by the following authors: Reisinger,[23] Karling,[3] Sterrer and Rieger,[9] Doe and Rieger[25] and Rieger.[30]

The Luther's larva of *Rhynchoscolex* was collected by Reisinger in the spring of the year, from the end of March through the beginning of April, near Graz, Austria (Fig. 1A-C). This vermiform, pelagic larva occurred infrequently in limnic habitats.

The larva is 800 μm long, 30-35 μm wide and circular in transverse section. The anterior end of the body has a weakly delimited precerebral rostrum which is encircled by several concentric, ciliary bands (Fig. 1B-C). Although cilia occurred on the remainder of the body, Reisinger observed that the ciliary bands on the rostrum functioned as the principal locomotory organs of the larva. He noted that the larva swam while suspended vertically in the water column, rotating slowly about its long axis. The cilia in these trochal bands measured up to 4.5 μm in length. There is a 6 μm wide statocyst located frontal to the brain between the two anterior nerve cords. The statocyst encloses a single statolith having cilia arising from the inner wall of the statocyst. The structure of the mouth, pharynx, brain and sensory plates is similar to that of the adult of *R. simplex*. The gut consists only of a compact "syncytial" cylinder. A large number of cells, designated "stem-cells" by Reisinger, is located between the body wall and the "syncytial" cylinder. These cells are assumed to be neoblasts.

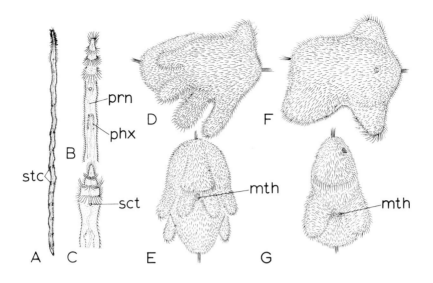

Fig. 1. Turbellarian larvae. A, Luther's larva of *Rhynchoscolex*. B, anterior end of Luther's larva. C, contracted, anterior end of Luther's larva (A-C after Reisinger[23]). D, E, lateral and ventral views of an early Müller's larva (OML). F, G, lateral and ventral views of a Götte's larva (EGL).

The metamorphosis of Luther's larva involves the loss of the statocyst and the reduction of the ciliary bands on the rostrum. The larva reaches a length of 1.0 to 1.8 mm during this period, the rostrum becomes more well developed and, according to Reisinger, the larva quickly assumes the definitive adult body form. The adult ciliation is complete at this time and, in spite of the reduction of the trochal bands, the larva remains capable of swimming, although with difficulty, in a spiral path. Subsequently a period of asexual reproduction by paratomy begins and produces characteristically two zooids. At most, two additional divisions are considered possible; further asexual divisions are assumed to be obviated by the limited and fixed number of "stem-cells." Reisinger states that asexual reproduction ends several weeks after the April-May larval period and that now benthic worms begin to develop their reproductive organs.

MÜLLER'S AND GÖTTE'S LARVAE OF TURBELLARIA, POLYCLADIDA

The two characteristic larvae of the Polycladida are the eight-armed Müller's larva (Fig. 1D-E) described by Müller[15] and known from the Cotylea and Acotylea, and the four-armed Götte's larva (Fig. 1F-G) described by Götte,[18] known only from the Acotylea. Lang, in his treatise on the polyclads,[10] considered that Götte's larva of *Stylochus pilidium* would have been observed to develop additional arms, to become a Müller's larva, if its development had been folowed to metamorphosis. Hofker,[19] Pearse,[20] and particularly Kato,[13] established clearly, by following development up to metamorphosis, that the Götte's larva of the Acotylea, Craspedommata did not pass through an eight-armed stage and that it was a distinct larval type in the polyclads.

Recently, Anderson has described a four-armed larva in the schematommatan family Leptoplanidae (Fig. 6C) that closely resembles Götte's larva.[14] He has brought the relationship of Götte's and Müller's larvae into persepctive with this discovery by suggesting, as one possibility, that the schematommatan Götte's larva and the Götte's larva of *Stylochus* may be independently derived from Müller's larva and therefore convergently similar to one another. If this hypothesis is correct, then the Götte's larva, although a distinct larval type, could also be interpreted as an intermediate form between indirect development with a Müller's larva, *e.g.*, *Stylochus luteus*, and direct development, *e.g.*, *Stylochus neapolitanus.*

The fully developed Müller's larva is a 0.5 to 1.8 mm long, bilaterally symmetrical larva with 8 arms, although as few as 7 and as many as 10 arms have been noted.[13,1] The larvae are often pigmented in various hues from brown to orange to magenta. Anteriorly, a tuft of sensory cilia and associated gland cells called the frontal organ, are present. Another tuft of sensory cilia extends from the posterior end of the body. The epidermis is completely ciliated. Prototrochal cells with long cilia are located on the arms and along ciliary bands connecting the arms. The distribution of prototrochal cells can be understood as a circular preoral band interrupted by the eight arm-like outgrowths of the body wall (Fig. 1D-E).

The fully developed Götte's larva is a small (*ca.* 100 μm) transparent to whitish brown larva with 4 arms. It too has anterior and posterior sensory cilia, a densely ciliated epidermis and an interrupted ring of prototrochal cells (Fig. 1F-G).

The arms of the Müller's larva form an anterior, middorsal projection; an anterior, midventral projection; a pair of ventrolateral projections, and two pairs of lateral projections. The midventral arm is usually spatulate and caudally directed, forming a structure similar to a preoral hood. One ventrolateral arm is located on each side of the circular mouth opening. The dorsal and especially the lateral arms are directed posteriorly.

The four arms of Götte's larva are noticeably shorter and blunter than those of the Müller's larva. The midventral arm is spatulate and directed slightly toward the posterior end of the larva. The ventrolateral arms are located on each side of the circular mouth. The larva is laterally flattened anteriorly and nearly circular in transverse section posteriorly.

The epidermis of all larvae investigated consists of ciliated epidermal cells (Fig. 5A), ciliated prototrochal cells (Fig. 5B), rhabdite cells (Fig. 5A), mucus cells (Fig. 5A), pigment cells (according to Lang[10] and Fig. 7A), monociliated sensory cells (Fig. 5B, E), ciliated photoreceptor cells (EGL, EML, Fig. 9B), chemoreceptor cells, some gland cells with secretion products resembling those of viscid gland cells,[4] and intraepithelial neurons. The epidermis of the arms contains only prototrochal cells, ciliated epidermal cells, monociliated sensory cells and neurons.

The brain anlage forms from an invagination of two masses of ectodermal cells according to Lang,[10] or as a single group of cells according to Kato[13] (Fig. 2).

The frontal organ of these larvae is intimately associated with the brain (Fig. 2). The frontal organ of the EML consists of a cluster of monociliated sensory cells surrounded by a circle of gland cell necks, supported peripherally by microtubules (Fig. 3). The necks of these gland cells open near the bases of the ciliated sensory cells. The sensory cells are located in the brain. The gland cell necks extend posteriorly, dorsal and ventral to the neuropile, to cell bodies above, below and behind the brain. The latter abut the basal portions of the entodermal cells.

The frontal organ of Götte's larva (EGL) is similar to that of the EML except that the ciliated cells are multiciliated and the gland cell necks lack peripheral microtubules (Fig. 4).

The terminal sensory organ has not been studied in detail. It is clear, however, that it consists of ciliated sensory cells and that gland cells are absent (Fig. 2).

The mouth is circular and is formed during the stomodaeal invagination (Fig. 7B-D). The larval foregut is heavily ciliated (Fig. 8A, C). The effective strokes of the cilia of the anterior, lateral and posterior cells are toward the interior of the body (Fig. 8C, D). A single circlet of peculiar ciliated cells is found at the junction of the stomodaeum and larval gut of all the larvae investigated (Fig. 8A-C). This ring of cells produces cilia that are inflated distally into club-shaped structures. A typical 9+2 array of microtubules extends from the basal body to the tip of the cilium but between the axoneme and expanded ciliary plasmalemma are numerous supernumerary microtubules (Fig. 8B).

The gut occupies most of the trunk of the larva in all types investigated (Figs. 2, 8C). It extends anteriorly up to and somewhat dorsal to the brain and posteriorly into the caudal lobe of the body. The lumen is large and the cells are densely ciliated. The cells contain basal nuclei in all forms and large lipid spheres in Götte's larva. The anterior and dorsalmost entodermal cells characteristically are packed with granular endoplasmic reticulum.

A thin layer of circular and longitudinal muscles is developed immediately below the basal lamina of the epidermis (Fig. 5C).

The "parenchyma" contains the protonephridia (flame bulbs, Fig. 5D), nerve cells, secretory cells, muscle cells, pigment-containing cells and undifferentiated cells. Koopowitz[26] reports that there is no subepidermal nerve plexus in the adult polyclads that he observed. Instead, a well-developed submuscular plexus is present. In EGL and EML, sections posterior to the brain reveal neuropile-like clusters of intraepithelial neurons associated with the bases of the sensory cells in the prototroch (Fig. 5B, E). Two dorsal and two ventral bundles are present in the EGL.

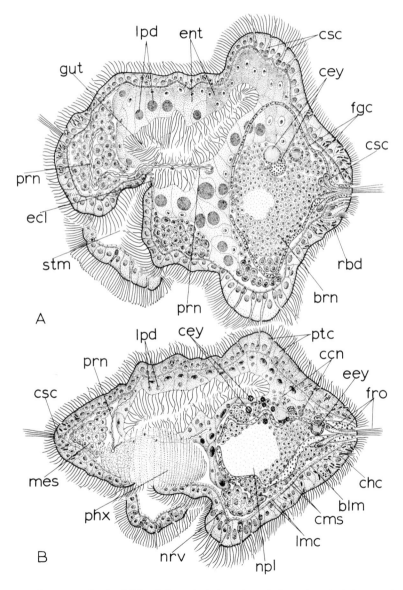

Fig. 2. Anatomy and metamorphosis of Götte's larva (EGL). A, lateral view of a pelagic larva. B, lateral view of a demersal larva during metamorphosis. Reconstruction based on serial transverse sections 1 μm thick.

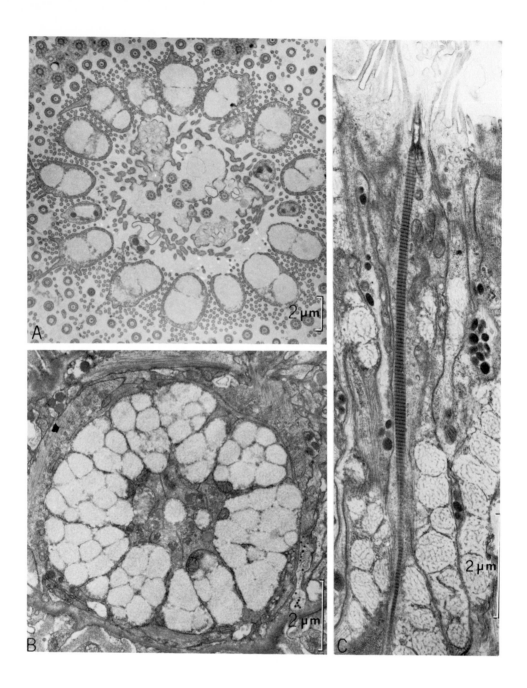

Fig. 3. TEM sections of the frontal organ of Müller's larva (EML). A, transverse section of the extreme anterior tip of the frontal organ, showing circle of gland necks, central sensory cells and microvilli. Note peripheral microtubules in gland necks. B, transverse section of the frontal organ approximately midway between the anterior tip of the larva and the bases of the anteriormost epidermal cells. C, sagittal section through the frontal organ and larva from the bases of the epidermal cells to the anterior end of the larva. Note the long primary rootlet of the cilium of one sensory cell and one of the several secondary rootlet fibers.

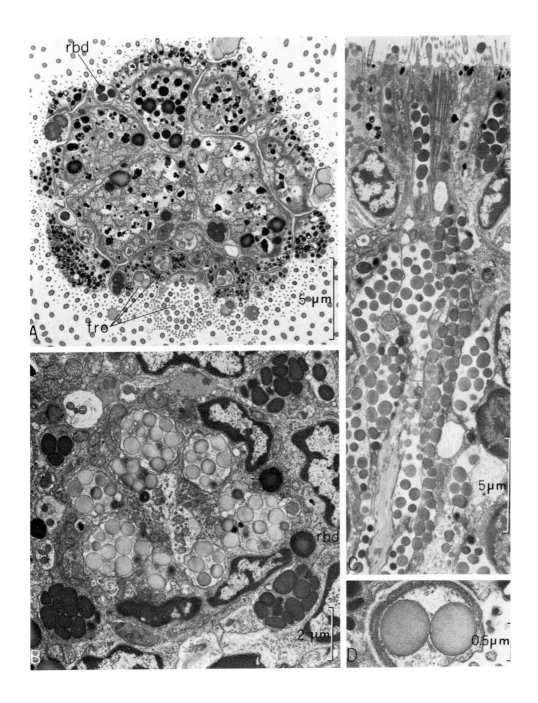

Fig. 4. TEM sections of the frontal organ of Götte's larva (EGL). A, slightly oblique transverse section through the anterior tip of the larva and frontal organ. B, transverse section through the frontal organ approximately midway between the anterior tip of the larva and the bases of the anteriormost epidermal cells. Note the ciliary rootlets in the centrally located sensory cells. The nuclei of the sensory cells are located peripheral to the circle of gland necks. C, slightly parasagittal section through the anterior tip of the larva and frontal organ. The two nuclei at the top of the figure are epidermal cell nuclei located just above the basal lamina. Note the involvement of longitudinal muscle with the frontal organ. D, enlargement of one gland neck showing membrane-bound granules.

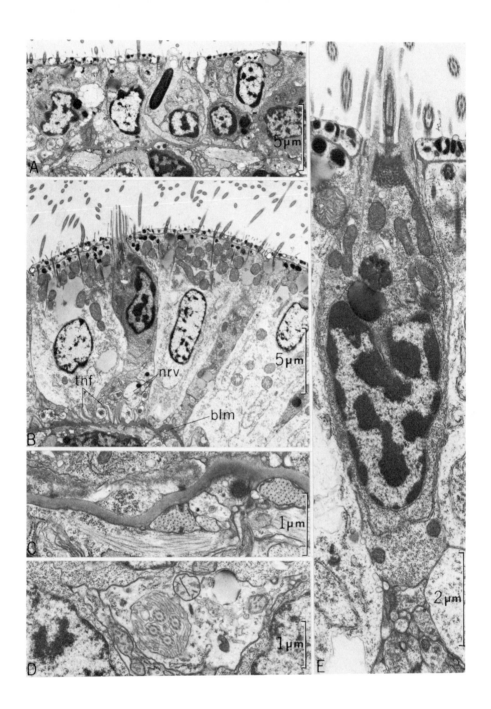

Fig. 5. TEM sections of Götte's larva (EGL). A, transverse section through a portion of the regular epidermis showing a rhabdite forming cell. B, transverse section of prototrochal cells showing one of the characteristic sensory cells. Note the large number of mitochondria and lipid droplets just below the terminal web among the ciliary rootlets. C, enlargement of the epidermal basal lamina. Plumes of tonofilaments, above the basal lamina, anchor the epidermal cells; longitudinal and circular muscles and one neuron are seen immediately below the basal lamina. D, transverse section through a flame bulb located in the ventrolateral parenchyma, near the mouth. E, enlargement of a prototrochal sensory cell. Note the cone-like arrangement of rootlet fibers and the neurons basal to the cell.

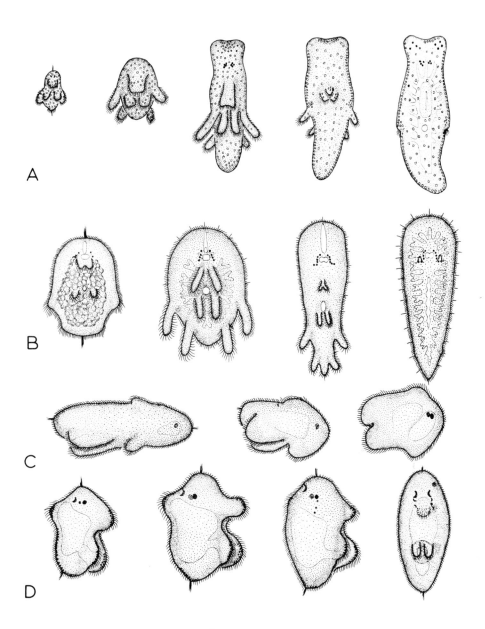

Fig. 6. Metamorphosis of polyclad larvae. A, Müller's larva of *Yungia aurantiaca* (first 4 drawings) and *Thysanozoon brocchi* (last drawing) in ventral views (all after Lang[10]). Left to right: larva soon after hatching, young pelagic larva, definitive larval form, beginning of arm resorption, juvenile worm, B, intracapsular Müller's larva of *Planocera reticulata* in ventral view (after Kato[13]). Left to right: early embryo beginning to develop larval arms, definite Müller's larval stage, larva during metamorphosis, hatched larva. C, Götte's larva of *Notoplana australis* in lateral view (after Anderson[14]). Left to right: pelagic larva after hatching, planktonic larva, demersal larva in late metamorphosis. D, Götte's larva from the Indian River (EGL) in lateral views. Left to right: young larval stage, definite Götte's larva, larva beginning metamorphosis, late metamorphosis.

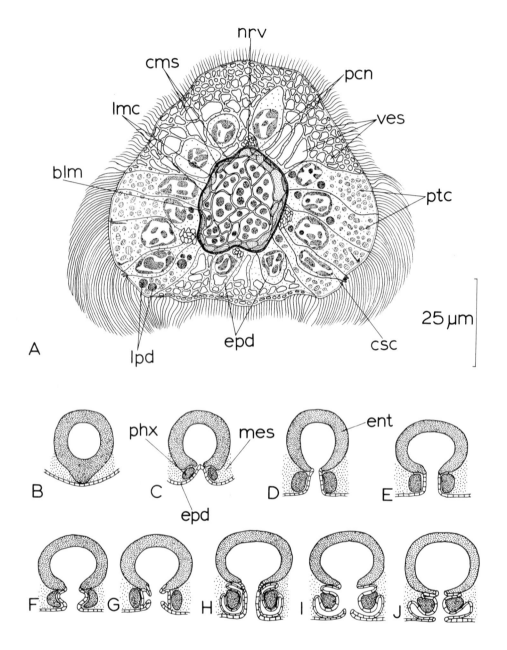

Fig. 7. A, transverse section through a ventrolateral arm of a Müller's larva (OML) drawn from a montage of electronmicrographs and 1 μm thick sections. B-J, development of the plicate pharynx of *Yungia* and *Thysanozoon* (transverse sections after Lang[10]). B-C, formation of the stomodaeum. D, mouth and gut communicate just after hatching. E, simple or larval pharynx stage. The blocks of mesoderm on each side of the stomodaeum, the developing pharynx, form a ring of outer circular muscle, inner longitudinal muscle and numerous gland cells. F-H, folding of the stomodaeal walls to form the pharyngeal sac during the period when the larva has reached its definitive form (Fig. 6A). I-J, development of the definitive pharynx completed by the end of metamorphosis (Fig. 6A).

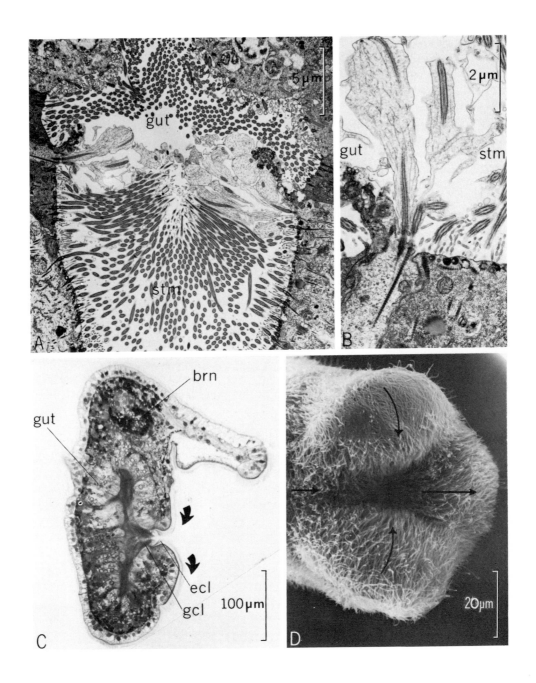

Fig. 8. A, transverse TEM section through the stomodaeum and gut of Götte's larva (EGL). Note ring of club-shaped cilia at the junction of the stomodaeum and gut. B, enlargement of club-shaped cilia showing supernumerary microtubules peripheral to the axoneme. C, median sagittal section of an oceanic Müller's larva (OML). The direction of the effective strokes of cilia around the mouth is indicated by arrows. The swelling at the base of the midventral arm is an artifact. D, SEM showing a ventral view of the mouth (center) and ventrolateral arms of Götte's larva (EGL). The directions of the effective strokes of cilia about the mouth are indicated with arrows.

An anastomosis was observed between the two ventral clusters of neurons just anterior to the mouth in the EGL. In EGL, a pair of flame bulbs, located laterally in the parenchyma, extends posteriorly from the anterior end of the gut. Each of these collecting tubes is joined by another protonephridium, located near the dorsal body wall in the region of the mouth. The two flame bulbs on each side of the body open in common on the ventrolateral body wall just behind the mouth (Fig. 2).

SETTLEMENT OF MÜLLER'S AND GÖTTE'S LARVAE

Virtually nothing is known about the normal settling behavior of turbellarian larvae. Although it has been reported often that larvae "in culture" lose their photopositive response, become demersal, glide slowly over the substratum, whirl in circles, come to a halt and contract their bodies into various shapes, Lang[10] as well as Selenka[11] argued convincingly that this behavior or at least much of it was pathological.

The EGL were kept in culture dishes for just over two weeks and sea water was changed daily. At the end of that period, the larvae had settled to the bottom of the culture vessel. They had become smaller, somewhat dorsoventrally flattened, and the larval arms were partially resorbed. Their behavior, however, was considered pathological, similar to that described by Lang.[10] The specimens fixed and sectioned from those cultures were observed to have continued development toward metamorphosis but many necrotic cells were also noted.

Anderson, however, was able to rear the Götte's larvae of *Notoplana* through metamorphosis.[14] He indicates that settling behavior begins about one week after hatching. "The larva becomes demersal and loses its positive response to light. Swimming becomes slower and is interspersed with periods of temporary settlement and exploration of the substratum. . . . Finally, the metamorphosing larva settles."

METAMORPHOSIS OF MÜLLER'S AND GÖTTE'S LARVAE

The metamorphosis of polyclad larvae involves a gradual alteration of external body form and a continuation of development of internal features correlated with settlement and a change in feeding habit. In general, the frontal organ and the larval arms are gradually lost, the body becomes dorsoventrally flattened, adult tentacles develop, the simple, larval pharynx becomes a plicate pharynx, the gut develops diverticula, eyes increase in number and parenchymal muscles develop. The caudal adhesive organ develops during this period in the Cotylea. Cilia may be lost from the gut of some Acotylea.

Lang describes changes in the external form of the Müller's larva of *Thysanozoon* and *Yungia*.[10] These larvae were observed to settle to the bottom of the culture dishes where they remained rather motionless. The body flattened and became somewhat pointed posteriorly and rounded anteriorly. Tentacular swellings became more noticeable and the bases of the tentacles were observed to be slightly hollowed out. The arms quickly became smaller, degenerating to nipple-shaped projections, but still supporting long cilia. They finally disappeared altogether but the position they occupied remained visible for some time owing to an absence of pigment and a slight interruption of the body contours at these points. The dorsal ciliary bands that had previously joined the middorsal arm with the two dorsolateral arms persisted as white bands well after the complete resorption of the arms. On the ventral side of the body, the mouth, free of larval arms, was surrounded by a whitish halo representing the pharynx. The adhesive organ was already developed at some distance behind the mouth (Fig. 6A).

Kato described the metamorphosis of the intra-capsular Müller's larva of *Planocera reticulata*.[13] Special features of this particular larva are:

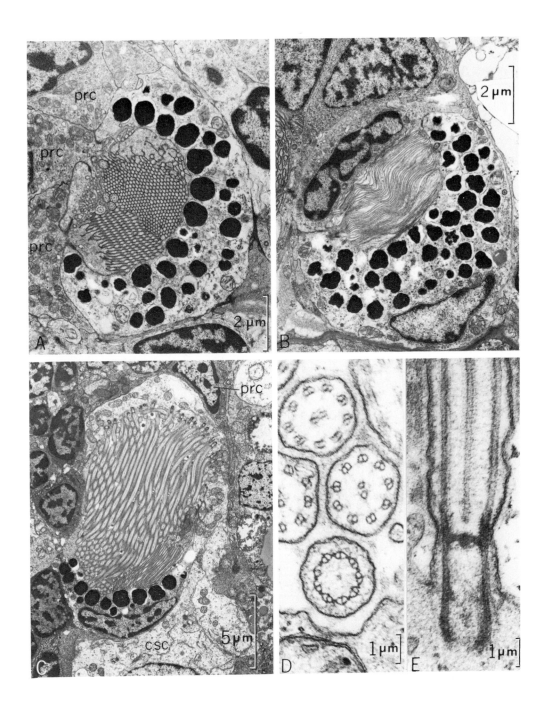

Fig. 9.   TEM sections of eyes of polyclad larvae. A, longitudinal section of cerebral eye of Müller's larva (EML). B, longitudinal section through the epidermal eye of Müller's larva (EML; identical in structure to the epidermal eye of EGL). Note the stacked membranes arising from cilia produced by the pigment cell and the lens-like structure of the nucleus of the cell over the opening of the optic cup. The basal lamina of the epidermis is at the bottom of the figure. C, cerebral eye of Götte's larva (EGL) in slightly parasagittal section. See text for description. D-E, cilia of the cerebral eyes of Götte's larva (EGL),. Note absence of ATP-ase containing arms on peripheral doublets.

The body is flat and oval, lacking the dorsal process and with [an] elongately [sic] bifurcated antero-ventral process; the intestinal cavity is narrow, but with well developed intestinal branches. The intra-capsular Müller's larva (Pl. LX) undergoes a complete metamorphosis in the egg shell in about ten days [beginning 15 days after oviposition]. . . . The body becomes very thin and elongate; the anterior half of the body especially grows well. Six pairs of eyespots are formed by repeated divisions of the first two pairs. . . . The anterior and posterior tufts of sensory flagella vanish. . . . The lobed processes [arms] gradually dwindle and are absorbed into the inside of the body.

He mentions that the plicate pharynx is well developed and that the intestinal branches are almost completely formed. Also, "The frontal organ is at its maximum growth." The larva then hatches from the egg case, settles to the bottom of the culture dish and, "One or two days after hatching the organ entirely disappears. . . ."(Fig. 6B).

The metamorphosis of Götte's larva was observed by Kato for *Stylochus*[13] and by Anderson for *Notoplana*.[14] Kato reports that the arms simultaneously degenerate and are absorbed by the body. "The body becomes oval or spherical in shape, and then gradually flattens." First one eye, then the other is reported to divide forming a total of four eyes. Finally, "Sensory flagella on both ends of the body vanish." Anderson describes the metamorphosis of the larva of *Notoplana* as beginning after the first week of larval life with a reduction in the external ciliation. In addition, "The swimming cilia become shorter, the anterior tuft of stationary cilia is lost and the cilia of the posterior tuft become fewer in number. . . ." Then, "Gradually, during the second week of larval life, the ciliated lobes become reduced . . . the amount of yolk in the larva is also reduced. . . ." During the third week, "the cilitated lobes have become quite small . . . . The metamorphosing larva settles, absorbs its remaining ciliated cells, transforms from bilaterally flattened to dorsoventrally flattened and completes the transition to a juvenile flatworm" (Fig. 6C).

Jennings[27] has noted that the intestine of the adult acotylean *Leptoplana* lacks cilia. If this is also true of *Stylochus* and *Notoplana,* then the loss of cilia from the gut is probably correlated with metamorphosis in these groups.

The development of the muscular, plicate pharynx in the Müller's larva, which reaches its definitive form at metamorphosis, is described in detail by Lang[10] and is summarized here in Figure 7. My own observations confirm Lang's description. Furthermore, I believe that the metamorphosis of these larvae involves not only cumulative morphological changes in the structure of the foregut, but also a transition from ciliary feeding to the adult habit of muscle mediated sucking and/or engulfment of food. The following evidence bears on this possibility.

Historically, there is the observation of Lang that the yolk spheres in the entodermal cells were expended in young Müller's larvae.[10] My observations of oceanic Müller's larvae (2 spp.) confirm Lang's description. These observations suggest that a mode of feeding other than lecithotrophy is occurring in these larvae. Jägersten, observing some exceptionally large larvae in the plankton,[1] offered the opinion that Müller's larvae must be planktotrophic. Kato made the following observation of the intra-capsular Müller's larva of *P. reticulata*, "If some of 6 or 7 embryos in an eggshell die and collapse owing to some cause or other, the broken small yolk granules and tissues are transported by the inhalent current through the mouth into the intestinal cavity of the surviving embryo in the same shell; such an embryo becomes very large in size. Lang observed a similar phenomenon in *S. neapolitanus.* . . ."[13] Lang noted that the mouth and stomodaeum break through to the gut shortly after hatching of the Müller's larva. Kato mentioned that the mouth, stomodaeum and gut are well developed and intercommunicating at the hatching of the Götte's larva. There is, therefore the structural possibility of planktotrophy in these larvae.

New indirect evidence of ciliary feeding is derived from the determination of ciliary beat in and around the mouth, the direction of ciliary beat on the arms, the anatomy of the larval arms and their relationship to the mouth, and the occurrence of modified cilia at the junction of the stomodaeum and gut.

The cilia on the ventral body wall anterior to the mouth and those lateral to the mouth on the arms have effective strokes toward the mouth. The cilia posterior to the mouth, on the ventral body wall, beat away from the mouth. The cilia on the anterior, lateral and posterior walls of the stomodaeum beat with effective strokes toward the interior of the body. The pattern of ciliary beat in and about the mouth is sufficient to result in an incurrent flow from three sides and an excurrent flow from the fourth side (Fig. 8A, C-D). The junction of the stomodaeum and gut is delimited by a single ring of ciliated cells bearing expanded, club-shaped cilia (Fig. 8A-B). Both longitudinal and transverse sections through the mouths of Müller's (OML, EML) and Götte's (EGL) larvae have revealed these specialized cilia. All of the club-shaped cilia appear to beat, and the direction of beat appears to be toward the gut. The structure of these cilia suggests that they may be functioning as flippers or valves at the stomodaeum-gut junction perhaps assisting or regulating the input of food items.

The arms of the Müller's larva have the same general structure as the feeding tentacles in entoprocts or ectoprocts (Fig. 7A). Each arm is roughly triangular in shape, longitudinal rows or prototrochal cells are located on the lateral surfaces of each arm, and ciliated sensory cells are located laterofrontally. The long lateral cilia of the prototrochal cells beat toward the frontal surface of each arm and the frontal cilia beat in the direction of the body of the larva. This structural arrangement creates a water flow through the arms toward the body where the ciliary tracts mentioned above could carry particles to and into the mouth. Figure 8D depicts directions of ciliary beat around the mouth of the Götte's larva.

Götte's larva, in addition to being smaller than Müller's larva and having shorter larval arms, contains yolk spheres in the gut cells longer into its larval life than does Müller's larva. This difference indicates that Götte's larva may be primarily lecithotrophic although there remain the structural and functional possibilities for facultative particle feeding. The acquisition of lecithotrophy in Götte's larva is consistent with observations that Götte's larvae occur in genera known also to contain direct developing, hence lecithotrophic species.

The accumulated indirect evidence points strongly to the possibility that the larvae of polyclads feed on material carried into their guts by a ciliary current augmented by the activity of specialized appendages during their larval lives. If this hypothesis can be demonstrated experimentally, then the metamorphosis of these larvae must involve the important transition from one mechanism of feeding (ciliary) to another (muscular) mode of ingestion.

The development of eyes in the Müller's larva is described by Lang.[10] He states that three eyes are present in the early larvae of *Yungia* and *Thysanozoon*, 2 anterior cerebral eyes and one posterior epidermal eye. The posterior eye is reported to sink into the mesoderm and all subsequent cerebral eyes are considered to be derived from divisions of that eye. The additional tentacular and marginal eyes are described as originating from the paired anterior eyes. The Müller's larva of *Planocera multitentaculata* has three eyes. Kato describes the formation of two, initial eyes in the epidermis that sink subsequently into the mesoderm.[13] The third eye forms as the result of the division of one of the first two eyes. In the EML and the EGL that I have examined, the youngest individuals had three eyes each (Fig. 2A): two symmetrically located cerebral eyes and one asymmetrically located epidermal eye. Additional cerebral eyes were present in older larvae of both the EML and EGL but the asymmetrical eye remained singular and located in the epidermis (Fig. 2B). The fine structure of these epidermal eyes is identical in both types of larvae.

Each epidermal eye consists of two cells. A pigmented cell in the shape of a cup rests on the basal lamina of the epidermis and opens toward the exterior of the animal. The concave surface of the pigment cell is ciliated. These cilia are modified to form flattened membranes oriented perpendicular to the direction of incident light. This cell, therefore, is interpreted as the photoreceptoral cell. The second cell encloses the open end of the optic cup and is joined peripherally to the pigmented cell by a zonula adhaerens and a septate desmosome.

The nucleus of this covering cell is a planoconvex organelle that fits neatly over the opening of the optic cup with its convex surface directed toward the exterior of the larva. The nucleus appears to function as a positive lens (Fig. 9B). These eyes were found only in animals collected in the Indian River estuary. The did not occur in oceanic larvae.

The cerebral eye of the Götte's larva (EGL) consists of a pigment cell forming a cup, several multiciliated cells forming an expanded bulb of cilia about the opening of the cup, and several cells producing a rhabdomeric layer of parallel microvilli within the cup. These cilia are oriented parallel to the direction of incident light and may function as waveguides. The perikarya of the ciliated cells are located behind the optic cup and the perikarya of the rhabdomeric cells are found in front of the cup (Fig. 9C). The cerebral eye of the EML is similar in construction except fewer cilia are present in the optic cup (Fig. 9A), and a ciliary rootlet was noted in one of the rhabdomeric cells.

The detailed structure of eyes in these larvae (EGL, EML) indicates that both epidermal and cerebral eyes are definitive organs with different structural bases. The cerebral eyes are not derived from epidermal eyes by a simple sinking of the latter into the mesoderm. Such an event may occur, however, when a developing ectodermal eye, a presumptive cerebral eye, invaginates along with other cells to form the anlage of the brain and frontal organ.

The significance of the frontal organ during the larval life and/or metamorphosis of these larvae has not been determined. Kato's interpretation that the organ is used by the larva like a battering ram to break the egg shell so that the larva might escape[13] is not supported well by its ulstrastructure which reveals it as a complex glandulo-sensory organ (Figs. 3, 4). It is also clear, contrary to other reports,[13,21] that it occurs well developed in both the Müller's and Götte's larvae. Since the frontal organ is so intimately associated with the cerebral ganglia and eyes of the larvae, two questions immediately arise: (1) What if any similarity does the frontal organ of polyclad larvae have with frontal glands of other turbellarians? and (2) What if any similarity does the frontal organ have with organs of larval attachment to the substratum known in other phyla of Spiralia, *e.g.,* the loxosomella larva of Entoprocta? In the latter, the frontal organ or larval brain, contains a pair of ciliated cerebral eyes and is surrounded by a ring of gland cells.[28,29] This description compares favorably with that given for the brain, cerebral eyes and frontal organ of polyclad larvae in this paper.

The transition from a circular or slightly laterally flattened larva to the dorsoventrally flattened adult form is correlated, at least, with the progressive development of dorsoventral and diagonal "parenchymal" muscles. No information was obtained on the mechanism of resorption of the larval arms during metamorphosis, but my limited observations indicate that neither the prototrochal cells nor their cilia are lost. An hypothesis is that these cells are gradually altered as they are incorporated into the juvenile epidermis.

ABBREVIATIONS

blm, basal lamina; brn, brain; ccn, calcareous concretions; cey, cerebral eyes; chc, chemosensory cell; cms, circular muscle cells; csc, ciliated sensory cell; ecl, expanded, club-shaped cilia; eey, epidermal eyes; ent, entodermal cells; epd, epidermis; fgc, frontal organ gland cells; fro, frontal organ; gcl, gut cilia; gut, gut; lmc, longitudinal muscle cells; lpd, lipid droplets; mes, mesoderm; mth, mouth; npl, neuropile; nrv, nerve, neuron; pcn, pigment cell necks; phx, pharynx; prc, photoreceptoral cell; prn, protonephridium; ptc, prototrochal cells; rhd, rhabdites; sct, statocyst; stc, stem cells, neoblasts; stm, stomodaeum; tnf, tonofilaments; tnt, tentacles; ves, electron lucent vesicles.

ACKNOWLEDGMENTS

This project was supported by a Smithsonian Postdoctoral Fellowship. The enthusiasm and conversation of Mary Rice and Kevin Eckelbarger were appreciated during this investigation.

Some plankton samples were kindly provided by Pamela Blades and by Joseph Murdoch. The cooperation of Harbor Branch Foundation's plankton program, directed by Marsh Youngbluth, is appreciated. Julianne Piraino assisted with the operation of the SEM. June Jones skilfully and patiently typed the manuscript.

REFERENCES

1. Jägersten, G. (1972) Evolution of the Metazoan Life Cycle, a Comprehensive Theory, Academic Press, London and New York, 282 pp.
2. Hyman, L. H. (1951) The Invertebrates. II. Platyhelminthes and Rhynchocoela. The Acoelomate Bilateria, McGraw Hill Book Co., New York, 550 pp.
3. Karling, T. G. (1974) in Biology of the Turbellaria, Riser, N. W. and Morse, M. P. eds., McGraw Hill Book Co., New York, pp. 1-16.
4. Tyler, S. (1976) Zoomorphologie, 84, 1-76.
5. Tyler, S. (1977) Mikrofauna Meeresboden, 61, 271-286.
6. Ax, P. (1963) in The Lower Metazoa, Dougherty, E. C. ed., Univ. California Press, Berkeley and Los Angeles, pp. 191-224.
7. Sterrer, W. (1966) Nature, 210, 436.
8. Tyler, S. and Rieger, R. (1975) Science, 188, 730-732.
9. Sterrer, W. and Rieger, R. (1974) in Biology of the Turbellaria, Riser, N.W. and Morse, N. P. eds., McGraw Hill Book Co., New York, pp. 63-92.
10. Lang, A. (1884) Fauna u. Flora Golfes Neapel., XI, 688 p.
11. Selenka, E. (1881) Zoologische Studien. II. Zur Entwicklungsgeschichte der Seeplanarien. Ein Beitrag zur Keimblätterlehre und Dezendenz Theorie, Leipzig, 36 pp.
12. Surface, F. M. (1907) Proc. Acad. Nat. Sci. Philadelphia, 58, 514-559.
13. Kato, K. (1940) Jap. J. Zool., 8, 537-574.
14. Anderson, D. T. (1977) Austr. J. Mar. Freshwater Res., 28, 303-310.
15. Müller, J. (1850) Arch. Anat. Physiol. von J. Müller, pp. 485-500.
16. Dalyell, J. G. (1853) in The Powers of the Creator Displayed in the Creation, or, Observations on Life, Amidst the Various Forms of the Humbler Tribes of Animated Nature, II, London, pp. 95-106.
17. Girard, A. (1946) Bull. Soc. Sci. Neuchâtel 2, 300-308.
18. Götte, A. (1882) in Abhandlungen zur Entwicklungsgeschichte der Thiere. 1 Heft. Untersuchungen zur Entwicklungsgeschichte der Würmer. Beschreibender Theil, Leipzig, pp. 1-58.
19. Hofker, J. (1930) Z. Morph. Ökol. Tiere 18, 189-216.
20. Pearse, A. S. (1938) Proc. U.S. Natl. Mus. 86, 67-98.
21. Kato, K. (1957) in Invertebrate Embryology, Kume, M. and Dan, K. eds., Nolit, Belgrade, pp. 125-143.
22. Lytwyn, M. W. and McDermott, J. J. (1976) Mar. Biol., 38, 365-372.
23. Reisinger, E. (1924) Z. Morph. Ökol. Tiere 1, 1-37.
24. Ruppert, E. E. and Shaw, K. (1977) Zool. Scr., 6, 185-195.
25. Doe, D. and Rieger, R. (1978) Zoomorphologie, (in press).
26. Koopowitz, H. (1974) in Biology of the Turbellaria, Riser, N. W. and Morse, M. P. eds., McGraw Hill Book Co., New York, pp. 198-212.
27. Jennings, J. B. (1974) in Biology of the Turbellaria, Riser, N. W. and Morse, M. P. eds., McGraw Hill Book Co., New York, pp. 173-197.
28. Nielsen, C. (1971) Ophelia 9, 209-341.
29. Woollacott, R. M. and Eakin, R. M. (1973) J. Ultrastr. Res., 43, 412-425.
30. Rieger, R. M. (1978) Zoomorphologie (in press).

MORPHOLOGICAL AND BEHAVIORAL CHANGES
AT METAMORPHOSIS IN THE SIPUNCULA

Mary E. Rice

Department of Invertebrate Zoology, National Museum of Natural History, Smithsonian Institution, Washington, D.C. 20560, and Smithsonian Institution, Fort Pierce Bureau, R.R. 1, Box 194-C, Fort Pierce, Florida 33450

Pelagosphera larvae, collected from open-ocean plankton, were reared in the laboratory to sexually mature adults and identified as *Golfingia misakiana* (Ikeda, 1904). From spawnings of these adults, embryos were reared through the trochophore stage to the young pelagosphera. Metamorphosis is described from the trochophore to the pelagosphera and from the pelagosphera to the juvenile. Behavioral changes at metamorphosis are noted and a procedure is reported for inducing metamorphosis of the pelagosphera in the laboratory. Literature on metamorphosis in the Sipuncula is reviewed.

INTRODUCTION

Developmental patterns of sipunculans, now known for 20 species, fall into four basic patterns (Table 1). One is direct development in which the egg develops into the juvenile without passing through a swimming larval stage. Usually the embryo hatches from one or more enveloping egg coats into a vermiform stage—a crawling, nonfeeding form with plastic body shape—which undergoes a gradual transition into the juvenile. More frequently development of sipunculans is indirect, including one or two swimming larval stages. Some species have only one pelagic larva, a lecithotrophic trochophore which swims for a short time before settling and transforming into the juvenile. The majority of species have two larval stages, a trochophore and a pelagosphera. The trochophore stage, which is always lecithotrophic, is followed by the pelagosphera, which may be either lecithotrophic or planktotrophic. The lecithotrophic pelagosphera swims for a short time, from 2 to 9 days in different species, before transforming to the crawling, vermiform stage.[1] The planktotrophic larva, on the other hand, may remain in the plankton for as long as 6 to 7 months,[1,2] during which time it increases in size considerably before a final metamorphosis into the juvenile form.

Two metamorphoses are recognized in sipunculan development. The first is metamorphosis of the trochophore. In those species in which the trochophore is the only larval stage, this metamorphosis ends pelagic existence. In other species metamorphosis of the trochophore results in another larval stage, the pelagosphera. The second metamorphosis is from the pelagosphera larva to the bottom-dwelling juvenile.

Of the two larval types in the Sipuncula, the trochophore, has the same essential features as that of other Protostomia.[1,3,4,5,6,7,8,9] An equatorial band of ciliated prototroch cells divides the larva into anterior and posterior hemispheres, the former bearing a pair of dorsal eyespots and an apical rosette with long apical cilia. In the posterior hemisphere a ventral stomodaeum is located just beneath the prototroch and a lateral band of mesoderm lies on either side of the rudiment of the gut. There is no protonephridium.

The second larval form, the pelagosphera, is unique to the Sipuncula.[1,10] Characteristic features of the pelagosphera are a prominent band of metatrochal cilia and a reduced prototroch. The body is regionated into head, "thorax" or metatrochal collar, trunk, and, frequently, a terminal attachment organ.[1,7,10] In the planktotrophic pelagosphera, there is a completed gut, consisting of esophagus, bulbous stomach, and looped intestine opening into a dorsal anus

TABLE 1

Patterns of Development in the Sipuncula[a]

DIRECT DEVELOPMENT

   I. EGG → WORM

     Golfingia minuta[5]
     Phascolion cryptus[11]
     Themiste pyroides[1]

INDIRECT DEVELOPMENT

   II. EGG → TROCHOPHORE → WORM

     Phascolion strombi[5]
     Phascolopsis gouldi[4]

   III. EGG → TROCHOPHORE → LECITHOTROPHIC PELAGOSPHERA → WORM

     Golfingia elongata[6]
     Golfingia pugettensis[1]
     Golfingia vulgaris[4]
     Themiste alutacea[11]
     Themiste lageniformis[b][12]
     Themiste petricola[13]

   IV. EGG → TROCHOPHORE → PLANKTOTROPHIC PELAGOSPHERA → WORM

     Aspidosiphon parvulus[9]
     Golfingia misakiana
     Golfingia pellucida[9]
     Paraspidosiphon fischeri[11]
     Phascolosoma agassizii[1]
     Phascolosoma antillarum[11]
     Phascolosoma perlucens[11]
     Phascolosoma varians[11]
     Sipunculus nudus[14]

---

[a] Modified from Rice 1976.[9]
[b] A pelagic trochophore stage is lacking.

in the middle of the trunk. Two organs associated with the mouth and presumably used in feeding are the protrusible buccal organ and the lip glands, which open through a pore on the lower lip at the base of the mouth. A prominent ventral nerve cord extends from the posterior trunk to the lip, and a dorsal, bilobed brain is present in the head. A pair of nephridia opens ventro-laterally in the anterior or middle trunk. Two pairs of retractor muscles, which function to withdraw the head into the trunk, traverse a spacious coelom containing numerous coelomocytes.

Metamorphosis of the trochophore to the pelagosphera usually results in a loss or reduction of the prototroch; a rupture in the egg envelope overlying the stomodaeum which opens to form the ventral ciliated surface of the head; an elongation of the post-prototrochal body; expansion of the coelom; formation of the metatroch as the functional locomotory organ; and frequently, the formation of a terminal attachment organ. When the resulting pelagosphera is planktotrophic, the gut is completed at metamorphosis with the opening of the mouth and anus.

The events of larval development and metamorphosis in the Sipuncula have been enumerated and reviewed in a series of previous articles.[1,7,8,9] Much of the previous information on metamorphosis has been concerned with metamorphosis of the trochophore. Only a few studies have considered metamorphosis of the planktotrophic pelagosphera to the juvenile.[7,10,14,15,16]

In this paper recent studies will be reported on the metamorphosis of an open-ocean pelagosphera, reared in the laboratory and identified as *Golfingia misakiana* (Ikeda 1904).* This larva, previously designated as *Baccaria oliva* by Häcker in 1898,[17] and later as Type C by Hall and Scheltema in 1975,[15] has been reared through metamorphosis to sexually mature adults. From spawnings of these adults, embryos have been reared through the trochophore stage to the young pelagosphera. This is the first time that breeding adults of Sipuncula have been reared from planktotrophic larvae and, with the exception of Hatschek's observations on *Sipunculus nudus*,[14] that metamorphosis has been observed in both trochophore and pelagosphera larva in a single species with planktotrophic larval development. Efforts to rear the young pelagosphera to the larger and older larval form found in the plankton have been unsuccessful. Observations have been made on behavioral changes of this species at metamorphosis and a technique devised for inducing metamorphosis in the laboratory. The final discussion in the paper will review the literature on metamorphosis in the Sipuncula, including a consideration of recent observations on *Golfingia misakiana*.

MATERIALS AND METHODS

Pelagosphera larvae were collected in surface plankton tows in the Florida Current 20 to 25 miles offshore from Fort Pierce, Florida over bottom depths of 200 to 270 m, with a net 3/4 m in diameter and mesh of 125 μm. The tows, each lasting from 15 to 20 minutes were made from the RV *Gosnold* of the Harbor Branch Foundation, Inc. or, weather permitting, from a smaller 22-ft boat. Sipunculan larvae were sorted immediately on return to the laboratory.

Substratum, when provided, consisted of a silty-mud collected from offshore stations in areas known to be inhabited by *Golfingia misakiana*. It was sieved through a mesh of 100 to 200 μm and placed in plastic dishes of 500 to 1000 ml capacity to a depth of 1 cm. Being larger than substratum particles, specimens could be easily removed for observation or fixation by sieving the substratum through a screen of appropriate mesh size. An inlet tube inserted into the dish above the level of the substratum and an outlet tube at a higher level allowed passage of fresh sea water through the dish without disturbing the substratum. Water was run through the dishes for 10 to 20 minutes each day, usually 6 days a week.

Specimens to be used for light microscopy and scanning electron microscopy were fixed in 2.5% glutaraldehyde, buffered with Millonig's phosphate buffer adjusted to an osmolality of 1000 milliosmols by the addition of sodium chloride. Prior to fixation, larvae or juveniles were anesthetized in 10% ethanol in sea water for approximately 5 minutes or until their heads remained extended. If not extended, heads could sometimes be forced to protrude by gentle pressure on the trunk with an applicator stick, broken to form a fine point. Specimens to be used for light microscopy were embedded in polyester, sectioned at 4 to 6 μm and stained in Mallory's stain, or post-fixed in 2% osmium tetroxide with Millonig's phosphate buffer, embedded in Epon for sectioning at 1 μm and stained with Richardson's stain.[18] For scanning electron microscopy, specimens were dried in a critical point dryer with liquid CO_2 and coated in a sputtering unit with gold-palladium.

Adults, reared from larvae, were removed periodically from the substratum and examined for coelomic gametes which, when present, are visible through the thin body wall of the

---

*Identification as *Golfingia misakiana* is based on the following characters of the adult: average ratio of introvert length to trunk, 4 to 1; usually 6 to 8 tentacles dorsal to mouth; introvert hooks usually with 5 basal spinelets; 4 retractor muscles, dorsals usually attached near level of nephridiopores, ventrals slightly posterior to anus; bilobed nephridia attached anterior to anus; spindle muscle attached posteriorly.

introvert. Specimens with gametes ranged in age from 9 months to two and one-half years. They were maintained in fingerbowls without substratum until spawning occurred. Fertilized eggs were transferred to tall covered petri dishes for culturing of embryos and larvae.

LARVAL DEVELOPMENT AND METAMORPHOSIS OF THE TROCHOPHORE

Development from the fertilized egg through metamorphosis of the trochophore to the young pelagosphera larva of *Golfingia misakiana* is illustrated in photographs of living embryos and larvae in Figures 1 to 5. At fertilization the egg is ovoid in shape with a central apical depression and measures 70 x 88 μm. There is a thick egg envelope composed of 3 distinct layers which are perforated by pores. Within 8 hours, at 23°C, prototrochal cilia and the apical tuft make their appearance. The embryo begins to swim, at first remaining near the bottom but later swimming throughout the water column. By 24 hours the features of the trochophore are fully developed. Enclosed by the egg envelope, the trochophore has about the same size and shape as the egg. A pair of red eyespots is located dorsolaterally above the prototroch, and an apical groove, separating cytoplasm from egg envelope surrounds the cells bearing the apical tuft. The prominent equatorial prototroch with its numerous long cilia is expanded ventrally to form a lobe, beneath which lies the ciliated stomodaeal opening leading to the esophagus. Rudiments of stomach and intestine are distinguished in living larvae by a white pigmentation.

Certain changes preliminary to metamorphosis are apparent at 3 to 3½ days of age. The trochophore begins to elongate posterior to the prototroch. A small bulge at the posterior extremity marks the position of the future terminal organ and depressions in the egg envelope at the level of stomodaeum and rectum indicate the sites where the envelope will rupture to form the mouth and anus. The three areas of the gut are well defined and the lumen has formed. The buccal organ is visible and actively protrusible beneath the egg envelope. The coelom, which will undergo later expansion at metamorphosis, is visible as a narrow slit traversed by long retractor muscles.

Metamorphosis of the trochophore to the pelagosphera larva occurs on the fourth day after fertilization and takes place over a period of 6 to 8 hours (Fig. 6). At metamorphosis the trochophore elongates, attaining a length of 150 μm when fully extended. The mouth and anus are opened to the exterior by rupture of the overlying egg envelope, thus completing the gut, and the terminal attachment organ is formed. The eversion of the entire stomodaeal area results in the formation of the ventral ciliated surface of the head and the disruption of the ventral portion of the prototroch. The head is rotated backward, the former apex assuming a more dorsal position and the apical tuft is soon lost. At the same time the retractor muscles become functional, so that the head and terminal organ can be withdrawn into the trunk. The postprototrochal egg envelope is transformed into the larval cuticle, losing its porosity and lamellation. The pretrochal envelope of the head is sloughed off gradually over a period of several days and replaced by a thin underlying cuticle. The entire body becomes quite extensible, with the capacity for considerable elongation and contraction. A metatrochal lobe is formed, delimited posteriorly by the sphincter muscle, and metatrochal cilia appear on the inner side of the lobe nearest the head. Metatrochal cilia are apparent in the sectioned material, although they are difficult to distinguish from prototrochal cilia in living larvae. Unlike young pelagosphera of other species, the dorsal and lateral prototroch is not reduced and metatrochal cilia combine with prototrochal cilia to perform a locomotory function.

At the time of metamorphosis the larva descends to the bottom of the dish, at first swimming along the bottom, and making temporary attachments with the terminal organ. Although most frequently attached to the substratum, the larva may swim or crawl along the bottom or it may move along the bottom on its head, with ventral surface of the head applied to the substratum and posterior extremity directed upward.

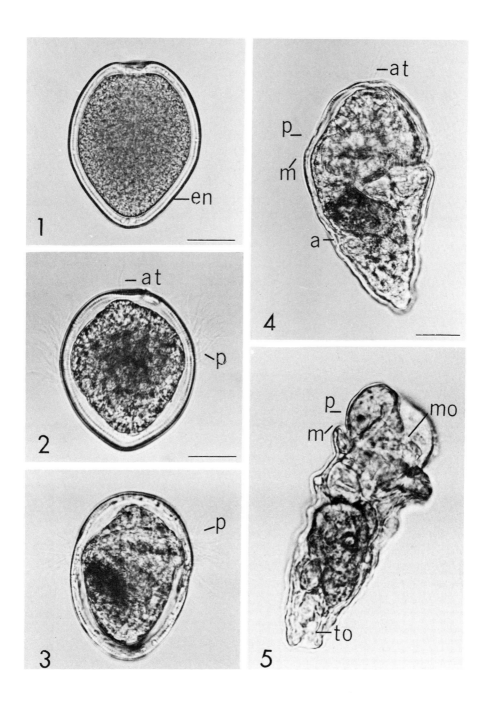

Figs. 1-5. Early development and metamorphosis of the trochophore of *Golfingia misakiana*. 1. Recently spawned, unfertilized egg. 2. Early trochophore, approximately 24 hours. 3. Trochophore, 30 hours. 4. Premetamorphosis, 3½ days. Regions of future mouth are indicated by depressions on larval cuticle. 5. Young pelagosphera, 5½ days. Terminal organ retracted. Scale, 25 μm.

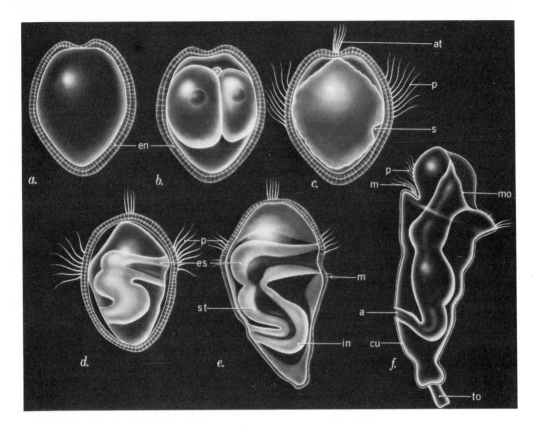

Fig. 6. Diagrammatic representation of metamorphosis of the trochophore of *Golfingia misakiana*, showing opening of gut; formation of coelom, terminal organ, metatroch, and ventral ciliated surface of head; and transformation of egg envelope to larval cuticle. a. Unfertilized egg. b. Two-cell stage. c. Early trochophore, one day. d. Trochophore, 3 days. e. Premetamorphosis, 4 days. f. Young pelagosphera, 5 days.

Cultures of early pelagosphera larvae, reared from the fertilized egg, have survived in the laboratory as long as 30 days. The oldest larvae reached a maximum length of 170 μm and were either attached or swimming close to the bottom of the dish.

MORPHOLOGY OF THE PELAGOSPHERA LARVA

Although they have the same basic morphological features, older pelagosphera larvae, collected from the oceanic plankton, are strikingly different from the younger larvae reared in the laboratory (Figs. 7a, 7b, 14, 20). Oceanic larvae range in size from 0.5 to 0.8 mm and the body is generally white or pinkish in color. The green nephridia and the black esophagus, stomach, and intestine are readily visible through the body wall. The shape and proportions of the older larva differ in that the head is relatively smaller and the thorax or metatrochal collar is distinctly delimited by the postmetatrochal sphincter and is more extensible. The trunk is less attenuated posteriorly, being either cylindrical or ovoid, depending on the state of contraction, and the terminal organ is proportionally reduced. Details of the external and internal morphology of the oceanic pelagosphera larva are outlined below, with notations regarding other differences from the younger larva.

The ciliated ventral head of the oceanic pelagosphera is bisected by a black-pigmented median groove. This groove broadens basally to include the mouth opening and the transverse groove through which the buccal organ is protruded (Fig. 10). The pattern of ciliation of the ventral

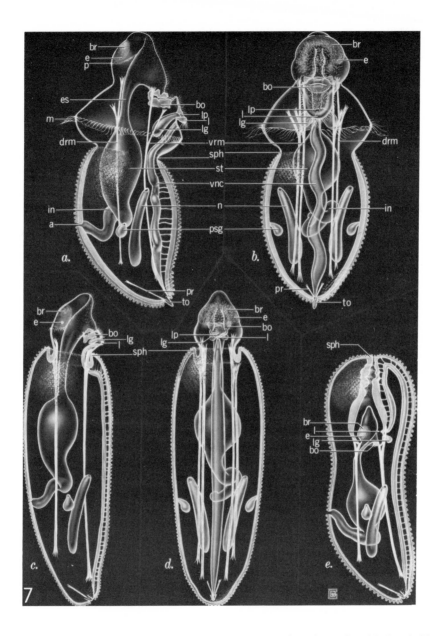

Fig. 7. Illustrations showing initial metamorphosis of the pelagosphera larva of *Golfingia misakiana* from open-ocean plankton. a. Lateral view of larva. b. Frontal view of larva. c. Two days in substratum, lateral view. d. Two days in substratum, frontal view. e. Three days in substratum, lateral view. Head retracted.

head continues around the buccal groove to surround the lip pore of the lower lip, covering about one-half of the surface of the lip. The lower lip is more clearly delimited from the thorax than in the younger larva. The ciliated portion of the lip is raised and separated from the remainder of the lip by a semicircular groove. The remainder of the crescent-shaped lip is devoid of cilia, except at the margin where the cilia are longer and less dense. A horseshoe-shaped band of short cilia on the dorsal head, presumably a remnant of the prototroch of the larva, merges anterolaterally with the ventral cilia of the head. The two small eyespots of the oceanic larva, dark red to black in color, are located in the same position on the head as the larger eyespots of the young pelagosphera.

The thorax, defined as the region of the body between the head and postmetatrochal sphincter, includes the metatrochal band[10] (Figs. 7a, 7b, 8, 14). Unlike the weakly ciliated band of the early pelagosphera of *Golfingia misakiana,* the metatroch of the older larva is formed of dense cilia and is the primary locomotory organ of the larva. The entire thorax is capable of great distension and may be fully distended when the larva is swimming. When the larva is contracted, both head and thorax may be withdrawn within the trunk.

The anterior margin of the trunk is formed by the postmetatrochal sphincter. Unlike the cuticle of the remainder of the body and that of the early pelagosphera, the cuticle of the trunk is covered with small papillae (Fig. 19). They are approximately 10 µm in height and 8 µm wide at the base, tapering to a rounded apex of 5 µm in diameter. Beneath the rounded cap of the papilla, the column characteristically bears two or three projecting ridges. Sensory-secretory organs, observed with the scanning electron microscope, are scattered among the papillae. On the anterior trunk, immediately posterior to the sphincter, the papillae are smaller and spaced farther apart.

In contrast to its prominence in the early pelagosphera, the terminal organ in the oceanic larva is relatively small and usually retracted. Even when extended, it is used for attachment only infrequently, and even then the attachment is weak and the animal easily dislodged.

Major features of the internal morphology of the oceanic larva of *Golfingia misakiana* observed in sectioned material are illustrated in Figures 7a, 7b, and 8. There is a long esophagus, a large, bulbous stomach, and an intestine. The intestine forms two or more loose loops before ascending anteriorly to the short rectum and anus in the mid-dorsal trunk. Two organs associated with the mouth and common to all planktotrophic pelagosphera larvae are the buccal organ, a protrusible muscular organ at the base of the lower lip, and a pair of lip glands opening through a common pore on the lip. Two tubular nephridia, approximately 1/3 the length of the trunk, are suspended in the coelom from ventrolateral attachments anterior to the anus. A posterior sacciform gland is present in a lateral position on either side of the anus. There are four retractor muscles, two dorsal and two ventral. Posteriorly the dorsal retractors attach to the body wall somewhat below the anus while the ventral retractors attach to the body wall farther posteriorly. Anteriorly both dorsal and ventral retractors split, each muscle with one branch attaching to the mid-premetatrochal body wall. The dorsal retractors continue into the head, attaching just posterior to the brain. The anteriormost attachments of the ventral retractors are on either side of the mouth region, at the base of the lobes of the ventral head. Two short retractors of the terminal organ attach posteriorly to the dorsal body wall. A median, unsegmented ventral nerve cord is connected to the body wall along its length by numerous nerves. The cord extends from the posterior extremity of the body anteriorly and attaches to the

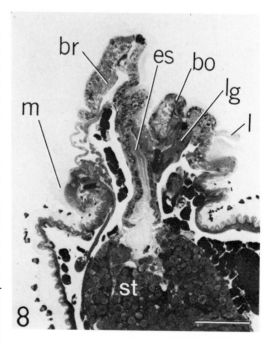

Fig. 8. Sagittal section through anterior body of oceanic pelagosphera of *Golfingia misakiana.* 1 µm, Epon-embedded, Richardson's stain. Scale, 50 µm.

body just beneath the lower lip. Circumesophageal connectives are difficult to follow, but presumably are basiepidermal, extending along the ventral lobes of the head to the dorsal brain. The brain is subepidermal and bilobed with a prominent central neuropile. The lateral eyespots occur within the outermost posterior portion of the brain.

MORPHOLOGICAL CHANGES AT METAMORPHOSIS OF THE PELAGOSPHERA

Complete metamorphosis of the pelagosphera of *Golfingia misakiana* from larva to juvenile requires approximately 16 days at 23°C under laboratory conditions. It begins with retraction of the larval head and ends with complete extension of the newly formed introvert and juvenile head, bearing terminal tentacles (Figs. 10-18, 20-25). Whereas changes in external morphology have been studied for the entire period of metamorphosis, internal changes were noted only during the first three days.

Metamorphosis begins with construction of the postmetatrochal sphincter, preventing the extension of the retracted head. This may occur after the larva burrows into substratum, or, in the absence of substratum, after a prolonged period in laboratory containers (see **Behavioral Changes**). Approximately two days after construction of the sphincter, several major morphological changes are apparent. The body becomes more elongate and the head, which at this time may still occasionally be extruded past the tightened sphincter of the anterior trunk, is narrower and more pointed (Figs. 7c, 7d, 11, 12, 15). The lobes of the ventral head are reduced, the head is more flattened dorsoventrally and the lower lip has regressed (Figs. 11, 12). Sectioned material shows the head coelom to be diminished (Fig. 9). Although the prototroch is still present, the metatroch has undergone considerable change. The entire metatrochal area or thorax is narrower and the metatrochal cilia are reduced in size and patchy in distribution. In some individuals the cilia are entirely lost and in others they occur only in a limited mid-dorsal region. The epidermis on either side of the metatrochal band has thickened and the premetatrochal body appears shortened.

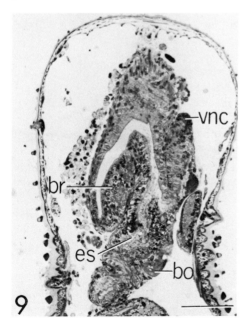

Fig. 9. Sagittal section through anterior body of metamorphosing pelagosphera of *Golfingia misakiana* after 3 days in substratum. Head is retracted. 1 μm, Epon-embedded, Richardson's stain. Scale, 50 μm.

At three days the larval head can no longer be extended from its retracted position (Figs. 7e, 9, 16). The sphincter of the anterior trunk is tightly contracted and the two pairs of thickened retractor muscles extend to the posterior trunk from their attachments at the base of the head. Anteriorly the dorsal retractors attach to the body wall at the base of the brain and the ventral retractors attach on either side of the esophageal opening. The branch of each retractor which formerly attached above the metatroch in the larva now appears fused with the inverted premetatrochal body wall and, in part, with the main muscles at the base of the head.

The head, as seen in sectioned material, is further reduced in size and more pointed in shape. The head coelom is completely obliterated, effecting the apposition of epidermis of the ventral head and the brain. The brain is in a more posterior position in the head and its two lobes, now closer together, enclose the central neuropile. The ventral nerve cord, attached to the body wall by numerous nerves, extends anteriorly to the closed sphincter, then posteriorly to follow the inverted body wall, ending, as in the larva, in a subepidermal attachment beneath the lip pore. At this 3-day stage the lip pore is still present but the lip glands and buccal organ are reduced in size. The epidermis of the inverted thorax has inreased in thickness and, along with its overlying cuticle, has become greatly elongated with numerous infoldings.

At 5 days the newly forming tubular introvert can be extruded for a short distance beyond the constricted sphincter. By relatively rapid growth, the inverted larval thorax gives rise to the introvert of the juvenile and, as the new introvert is pushed outward, the epidermis is unfolded and stretched. (Fig. 17).

At 7 days the body shape is similar to that of juvenile and adult (Fig. 18). The anterior trunk is slender, tapering into the narrow introvert. The anterior swelling of the trunk has disappeared, although the position of the sphincter can still be detected by a slight constriction. The newly formed introvert that is extended beyond the sphincter has a thin and transparent body wall and lacks the cuticular papillae characteristic of the trunk. The terminal organ is lost within the first few days, but internal organs of the trunk remain relatively unchanged during the initial phases of metamorphosis.

Although the introvert continues to grow it is not fully extended until two to three weeks after metamorphosis begins (Figs. 22, 23, 24). At this time the transformed head, bearing 4 short terminal tentacles, is finally extruded and the extended introvert is two to three times the length of the trunk. With extension of the head, transformation to the juvenile is complete.

As in the adult, the tentacles are dorsal to the mouth and the dorsal tentacles are always shorter than the ventral tentacles (Figs. 13, 26). Outer surfaces of the tentacles are ciliated and on the inner surfaces there are scattered clumps of cilia. A ciliary band surrounds the base of the tentacular crown and the mouth opening (Fig. 26). Posterior to the ciliary band the cuticle is smooth and thin, and forms a distinctive neck region which is set apart from the remainder of the introvert by a cuticular fold. The most anterior introvert bears 6 rows of hooks, each hook usually with 5 basal spinelets (Fig. 27). The rows of hooks alternate with rows of small introvert papillae, quite different from the cuticular papillae which are still present on the trunk. Scattered hooks, simple in form and lacking the basal spinelets, occur more posteriorly on the introvert. Minute cuticular folds or ridges give the surface of the cuticle a finely striated appearance (Fig. 24). The brain, visible through the wall of the introvert, bears 4 eyespots. Two eyespots are lateral, posterior and black, and two are red, smaller, more median and anterior; the former are the larval eyespots and the latter the newly formed eyes of the adult. The long esophagus, stretching the length of the introvert, has lost its pigmentation. Many juveniles of this age, but not all, show a partial loss of pigmentation in the gut. The expanded stomach is no longer obvious and the intestine forms 2 to 3 loose coils, recurving to the mid-dorsal anus. With an expansion of the coelom of the trunk, the body wall appears thinner than that of the larva and the cuticular papillae are farther apart. The nephridia are lighter green in color, but otherwise not noticeably changed. Because of the great extensibility and changes in shape, absolute size measurements are difficult. The extended body is estimated to be 1.5 mm in length, about 2/3 of which is introvert. In the adult the ratio of introvert to trunk is 4 to 1, the number of tentacles has increased to 6 or 8, and there are many more rows of hooks.

BEHAVIORAL CHANGES AT METAMORPHOSIS OF THE PELAGOSPHERA

Changes in behavior precede the morphological changes of metamorphosis in the pelagosphera larva of *Golfingia misakiana*. When metamorphosis is about to begin, the larva elongates and contracts, stretching to its full length in a long cylindrical form. Usually the metatroch and frequently the head are retracted during elongation. Such elongation can be a normal behavioral

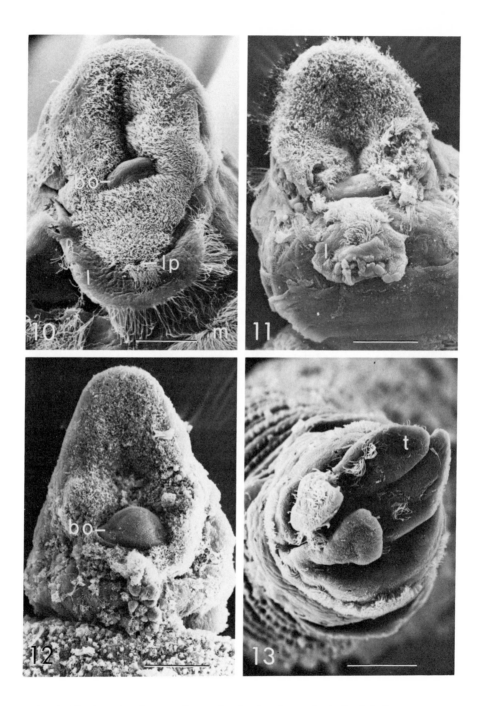

Figs. 10-13. Scanning electron micrographs showing changes in the head during metamorphosis of the pelagosphera of *Golfingia misakiana*. 10. Head of larva. Ventral view. 11. Head of metamorphosing larva after two days in substratum. Ventral view. Note beginning regressions of lip. 12. Head of metamorphosing larva after two days in substratum. Ventral view. Metamorphosis has progressed beyond that specimen in 11; head is narrower and pointed, ventral lobes are reduced, lip has completely regressed. 13. Head of juvenile, 8 weeks. Apical view showing four tentacles. Scale, 50 μm.

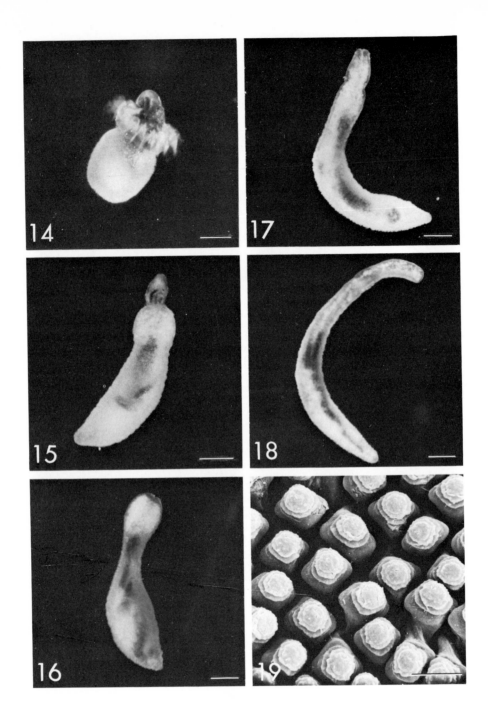

Figs. 14-19. Metamorphosis of the oceanic pelagosphera of *Golfingia misakiana*. Photographs of live specimens. Scale, 200 μm. 14. Larva swimming with metatroch extended. 15. Beginning metamorphosis after two days in substratum. Head is narrower, metatroch retracted. 16. "Bulbous" stage after three days in substratum. Postmetatrochal sphincter is constricted and head permanently retracted. 17. Five days in substratum. Note eversion of newly formed introvert beyond constricted sphincter. 18. Seven days in substratum. Shape resembles juvenile. Head is retracted. 19. Scanning electron micrograph of cuticular papillae of the oceanic pelagosphera larva of *Golfingia misakiana*. Scale, 10 μm.

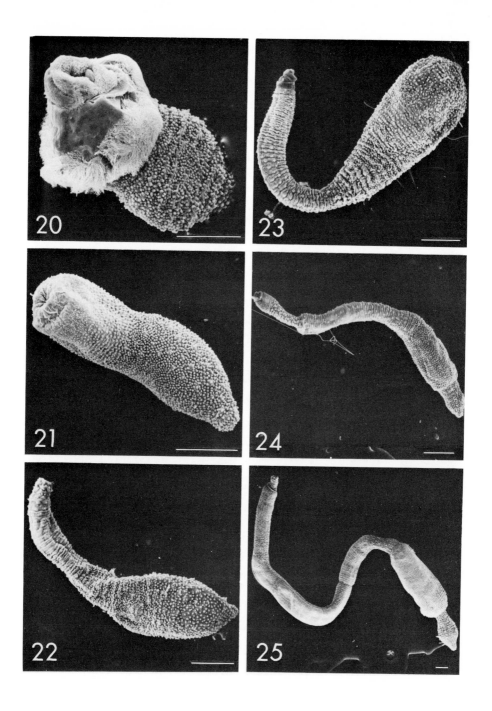

Figs. 20-25. Scanning electron micrographs showing metamorphosis of the pelagosphera of *Golfingia misakiana* and juvenile development to the adult. Scale, 200 μm. 20. Larva. 21. Five days in substratum. 22. Ten days in substratum. 23. Sixteen-day juveniles. Head extended. 24. Three-week juvenile. 25. Adult, 9 months after metamorphosis.

pattern of all larvae; however, it occurs with greater frequency when metamorphosis is imminent. Once this behavior has begun, the larva can still revert to the more typical larval activities of extending the head and metatroch and of swimming. The actual onset of metamorphosis is marked by constriction of the postmetatrochal sphincter which prevents extrusion of the retracted head. For convenience this can be termed bulbous behavior because of the anterior swelling which results as the head is pressed forward against the closed sphincter. Once this behavioral change has been established, the process of metamorphosis is initiated and cannot be reversed. A series of morphological changes over a period of two to three weeks, described in the previous section, then give rise to the definitive juvenile form.

When placed on a substratum of fine sand and silt (see **Materials and Methods**) larvae begin what appear to be exploratory movements over the surface. They elongate and stretch with the head extended and the metatroch usually but not always retracted. They may also progress over the substratum in the manner of an inchworm, or with the ventral head applied to the substratum and the posterior end pointed upward. When the ventral surface of the head is on the substratum, sand grains may be passed by a ciliary current through the ventral groove to the lower lip. During this passage they are adhered together by secretions, presumably from the ventral head and lip glands. At the region of the lip pore the agglutinated sand grains are directed away from the larva by ciliary action and the occasional protrusion of the buccal organ. Tracks of curled strands of adhering sand grains are left wherever the larva has moved over the surface. There is no evidence that the larva ingests this sand; thus the behavior could be interpreted as a testing of the substratum. Another characteristic behavioral pattern is the approach of anterior and posterior ends, the larva assuming the shape of a C or doughnut, usually with the head retracted. The body then rotates around a central axis, remaining in place on the substratum. This movement allows the many sensory-secretory organs scattered over the surface of the trunk to contact the substratum, and could serve as another means for testing. If conditions are favorable, the larvae will burrow, either soon after contact or after a period of several hours of activity on the surface. If conditions are not appropriate, larvae may continue exploratory activities or, more commonly, remain quiescent on the substratum, periodically leaving the bottom to swim through the water. Larvae may show similar exploratory movements on the bottom of a glass dish, except that the metatroch is more frequently extended and rotating behavior is not usually observed.

Metamorphosis of a small percentage of larvae will occur in glass dishes over a one- to two-month period after collection, with no treatment other than frequent changes of sea water and the addition of phytoplankton for food. In an attempt to increase the number of metamorphosed individuals available for study, larvae were provided with substratum collected from the same station as adults of this species. When the substratum was meticulously selected from that immediately surrounding burrows of adults of *Golfingia misakiana*, larvae responded to it by burrowing and subsequent metamorphosis. This observation suggested that adults might produce a metamorphosis-inducing factor and led to the series of 6 tests described below and summarized in Table 2. The tests were designed to establish a procedure for inducing metamorphosis of large numbers of larvae for rearing and studies of metamorphosis.

The first 4 tests were intended to ascertain the effect of the presence of adults in the substratum on larval metamorphosis (Table 2). The substratum used in these and other tests was obtained as described in **Materials and Methods**, sieved through a 100 μm screen, and added to plastic dishes of 1000 ml capacity to a depth of 1 cm. Fifty larvae were introduced to each dish and at the end of 3 or 4 days they were removed by sieving and the number of metamorphosed specimens counted. The number of dishes in any one test was limited by the availability of larvae. The relatively low percentage recovered in some preliminary tests indicated the desirability of using no less than 50 larvae for each dish. The results of the first 4 tests showed that when

96

TABLE 2

Influence of Adults on Metamorphosis of Pelagosphera Larva of *Golfingia misakiana*

With Substratum		Percent Metamorphosis (3-4 days)[a]							
Treatment of Substratum	Treatment of Sea Water	Test 1	Test 2	Test 3	Test 4	Test 5	Test 6	X̄	σ
Without adults	None	24.0	46.8	69.7	30.4	17.8	25.0	35.6	±19.4
Adults present (added with larvae)	None	94.0	93.8	87.2	54.8			82.5	±18.7
Adults present (added 3 days prior to larvae)	None	91.8	87.0	88.9	57.9			81.4	±15.8
Adults removed (added 3 days prior to larvae, then removed)	None			84.4	78.3			81.4	±4.3
Without Adults	Exposed to Adults[b]			93.5	86.6	63.6	93.3	84.3	±14.1
Without Substratum									
—	None							3.2	
—	Exposed to Adults[b]							0.0	

[a] Figures listed for the 6 separate tests represent the percent of metamorphosed larvae of the total live specimens recovered 3 to 4 days after placement in test conditions.

[b] Adults were placed in sea water without substratum for a minimum of 3 days and removed prior to use of the water in the experiments.

adults were present in the substratum, whether introduced simultaneously with the larvae or 3 days prior, the percentage of metamorphosed larvae was considerably increased over the controls.

The effect of the past presence of adults in the larval medium, either substratum or sea water, was investigated in tests 3 through 6. When adults were placed in substratum for 3 days, then removed before introduction of larvae, the percentage of metamorphosed larvae again increased (Tests 3, 4, Table 2). Similarly, when sea water, previously inhabited by adults, was added to untreated substratum, the number of metamorphosed larvae far exceeded that in the controls (Tests 3-6, Table 2). The water was prepared by placing adults (20 animals per 500 ml sea water) in a dish without substratum at least 3 days prior to the test. The high percentage of metamorphosed larvae in media previously inhabited by adults suggests that adults may release some factor into the sea water which is effective in inducing larval metamorphosis. That substratum makes some significant contribution to metamorphosis is indicated by the absence or low rate of metamorphosis in the absence of substratum.

DISCUSSION

Review of Metamorphosis in the Phylum

Metamorphosis of the Trochophore. Features common to metamorphosis of the trochophore, regardless of the form that results, are formation or expansion of the coelom, elongation of the body, shedding or transformation of the egg envelope, and regression of the prototroch. When, as in the majority of species, the resultant form is a pelagosphera, metamorphic changes also include elaboration of the metatroch and usually the formation of a terminal attachment organ. The retractor muscles begin to function either just before or soon after metamorphosis. Metamorphosis of *Golfingia misakiana* differs from other known species in that the prototroch

97

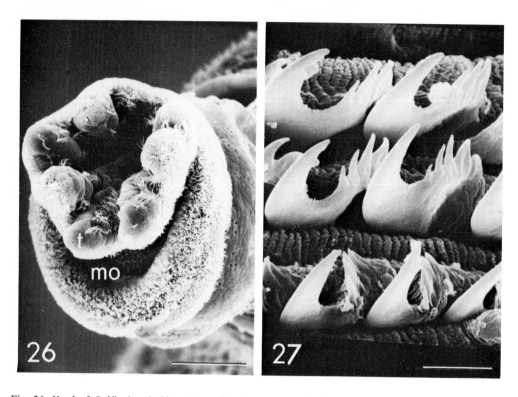

Fig. 26. Head of *Golfingia misakiana* 7 months after metamorphosis. Apical view showing tentacles. Scanning electron micrograph. Scale, 20 μm.

Fig. 27. Introvert hooks of juvenile of *Golfingia misakiana*. Unknown age. Scanning electron micrograph. Scale, 5 μm.

is retained at metamorphosis and continues to function in locomotion along with the newly formed metatroch.

The cells of the prototroch generally undergo a marked regression at trochophoral metamorphosis. In *Golfingia vulgaris* and *Phascolopsis gouldi* the entire cytoplasm and nuclei of the prototroch cells are passed into the coelom.[4] In *Golfingia pugettensis* the cells release lipid and yolk granules into the coelom, then degenerate.[1] The prototroch thus serves an apparent nutritive function, although its significance for the total nutrition of the developing larva may vary in different species. In species with planktotrophic larvae the prototroch cells contain relatively less yolk and release it prior to trochophoral metamorphosis. In *Phascolosoma agassizii* it is released into a prototrochal cavity beneath the prototroch, and by the time the trochophore metamorphoses into a planktotrophic larva the yolk has been entirely utilized.[1] The trochophore of *Sipunculus nudus* lacks an equatorial band of prototrochal cilia, but a ciliated "serosa" surrounds the entire embryo and is considered homologous to the prototroch of other sipunculans.[3] Granules are discharged from the cells of the serosa into an inner "amniotic cavity."[14]

In all species with planktotrophic pelagosphera larvae, with the exception of *Sipunculus nudus,* the egg envelope of the posttrochal trochophore is retained as the cuticle of the pelagosphera. At metamorphosis of the trochophore the egg envelope is stretched as the larva elongates, losing its porosity and lamellation. The pretrochal egg envelope is sloughed off and replaced by a thin underlying layer which forms the cuticle of the head. The ventral, pretrochal egg envelope ruptures and the ciliated stomodaeal area opens outward to form the ventral ciliated surface of the head. In *Sipunculus nudus* the egg envelope is cast off at metamorphosis along with the underlying layer of ciliated cells.[14] The egg envelope of *Golfingia vulgaris* and *Phascolopsis gouldi* is also shed at metamorphosis.[4] In all other species studied having both lecithotrophic larvae and direct development, the posttrochal egg envelope is transformed into the cuticle of the larva or vermiform stage.

At metamorphosis of the trochophore into the pelagosphera larva, the metatroch generally becomes the functional locomotory organ. Although present prior to metamorphosis in *Golfingia elongata,*[6] *G. pugettensis,*[1] and *G. vulgaris,*[4] it becomes more fully developed as the prototroch cells regress, and assumes a primary role as the locomotory apparatus. The metatroch of *Sipunculus nudus* is present shortly before metamorphosis, but not functional until the egg envelope is shed.[14] In other species with planktotrophic development, it is usual for the metatroch to become ciliated at metamorphosis.

Metamorphosis of the Pelagosphera. The end of the pelagosphera stage in both lecithotrophic and planktotrophic larvae is marked by the loss of metatrochal cilia. The lecithotrophic pelagosphera then passes through what has been termed a "vermiform stage,"[1] a transition period during which the yolk is absorbed, the gut completed, the body and introvert elongated to assume the adult habitus, and tentacular lobes formed.[1,8,9] The duration of the stage varies among species, lasting only 10 to 11 days in *Themiste lageniformis,*[16] and as long as 5 to 7 weeks in *Golfingia pugettensis.*[1] Initially the shape of the small worms is generally plastic; they crawl in the manner of an inchworm, but later become less mobile, their major activity being extension and retraction of the anterior end.

In contrast to the lecithotrophic pelagosphera which is pelagic for only a few days, the planktotrophic pelagosphera may remain in the plankton from one month, in the case of *Sipunculus nudus,*[14] up to as long as 6 to 7 months in other species.[1,2] In studies of *Sipunculus nudus,* Hatschek[14] reported that larvae at the age of one month sank to the bottom and, over a period of one to two days lost the metatroch, buccal organ and lip glands. At the same time the head became proportionately smaller, the mouth moved anteriorly and, on either side of the mouth, a tentacular lobe was formed. In an unidentified pelagosphera, observed in the laboratory, Jagersten[10] reported a similar metamorphosis, noting the regression of the lip and the formation of three pairs of tentacles from the rim of the mouth. Hall and Scheltema[15] reared 4 pelagosphera larvae from open-ocean plankton samples through metamorphosis to juveniles, assigning one to the genus *Aspidosiphon,* but were unable to identify the others. They described the larvae and juveniles, but gave no details of metamorphic events.

The most detailed observations of metamorphosis of pelagosphera larvae are reported in a recent study by Williams[16] comparing the lecithotrophic pelagosphera of *Themiste lageniformis* with three oceanic planktotrophic larvae of unknown species. In all 4 larvae it was noted that the head is permanently retracted during metamorphosis and head epidermis fused to the wall of the introvert. Initial retraction in *Themiste lageniformis,* the species described in greatest detail, is attributed to a constriction of the postmetatrochal sphincter, which prevents extension of the head. Permanent retraction is then assured by dorsal and lateral fusion of the epidermis of the head and introvert with the dissolution of intervening cuticular layers. Lack of ventral fusion results in a tube which becomes the anterior portion of the esophagus. Four lobes grow out around the anterior opening of the tube, or the mouth, and later give rise to the tentacles. The process of head fusion was found to be similar in oceanic pelagosphera larvae except that in two of the three larval types the anterior esophagus and anteriorly directed mouth are formed by a fusion of the lobes of the ventral head. Dorsally a slit resulting from incomplete fusion was proposed to be an ocular tube. Less information was available on the third planktotrophic pelagosphera, labeled Type P, which, in contrast with the other two larvae with smooth cuticles, had a rough or papillated cuticle, similar to that of *Golfingia misakiana.* Fusion of the ventral lobes of the head of the Type P larva was not observed.

Metamorphosis of Golfingia misakiana. With the information reported here on *Golfingia misakiana,* it has been possible for the first time to compare the morphology and behavior of early planktotrophic larvae with that of older oceanic larvae of a single species. Previously, young planktotrophic larvae of various species reared in the laboratory have been described as benthi-pelagic or demersal, usually attached to the substratum by a prominent terminal organ.[1,7,11] However, such larvae had never been reared to the fully grown larval form and metamorphosis had not been observed. For *Golfingia misakiana* we now know that the young pelagosphera has a relatively prominent terminal organ and behaves as other young pelagosphera larvae by remaining close and usually attached to the bottom of dishes in the laboratory. In the later pelagic larva occurring in the surface plankton of the open ocean, benthic features such as the terminal organ have been reduced, and the metatroch, a highly specialized swimming structure, has been greatly elaborated. Thus the older pelagosphera is well adapted for a long planktonic existence in the ocean. Scheltema[2,19] has referred to such larvae as teleplanic and has hypothesized trans-Atlantic transport of pelagosphera species. He emphasized their importance in the widespread dispersal of species and their significance as genetic carriers between widely separated populations. Because young demersal pelagosphera larvae of *Golfingia misakiana* have not been reared in the laboratory to the older form, it is not known whether the young larvae, if exposed to an appropriate substratum, would metamorphose without undergoing the more prolonged planktonic phase.

Metamorphosis of the trochophore of *Golfingia misakiana* is similar to that of other species with planktotrophic larvae except that the prototroch is not reduced. However, in older pelagospheras it is diminished and must therefore regress at some point during the growth of the larva.

Morphological changes during metamoprhosis of the pelagosphera larva of *Golfingia misakiana* are similar in many respects to those described by Williams[16] for *Themiste lageniformis* and three unidentified planktotrophic pelagospheras, in that the postmetatrochal sphincter contracts, preventing extension of the head. However, not reported by Williams are the reduced size of the head, obliteration of the coelomic cavity of the head, and regression of the lobes of the ventral head. Also, unlike the other species, no fusion of head epidermis with introvert, or of the ventral lobes of the head with one another has occurred at the end of three days, the oldest stage critically examined in sectioned material of *Golfingia misakiana.* Both the pre- and postmetatrochal epidermis of the thorax thickens, and the thorax grows rapidly to form the long introvert characteristic of this species. Observations of living specimens show that at 5 days the newly forming introvert presses outward beyond the sphincter. It increases daily in its extensible length, as the constriction of the sphincter becomes less obvious. The retention of the head and the most anterior introvert within the body for a period of at least two weeks may also be related to a delayed development in length and extensibility of the retractor muscles. Finally, when the head is fully extended the brain is attached to the wall of the esophagus and maintains contact with the head epidermis within the circle of tentacles, presumably the location of the cerebral pit of the adult. The various processes by which these changes are achieved have not been defined at this time. More definitive explanations will be forthcoming with ultrastructural studies. Differences in metamorphic changes in larvae studied by Williams and those of *Golfingia misakiana* might be attributed to differing morphologies of the adults; such correlations must await the determination of adult affinities for all of the larvae under consideration.

Tests designed to induce metamorphosis of the pelagosphera larva of *Golfingia misakiana* for purposes of rearing revealed that metamorphosis was increased by the presence of adults, or by media previously occupied by adults. While successful in establishing a procedure for rearing larvae through metamorphosis, these tests at the same time raised a number of questions regarding a possible metamorphosis-inducing factor. Is a chemical factor released by the adults to which the larvae respond, or does the response depend on some associated microfloral assemblage, as has been demonstrated for certain other infaunal organisms?[20] Is there a species-specific inducing factor, or will the larvae respond equally well to other species, particularly those occupying the same habitat? What role does the substratum play in enhancing metamorphosis? Does it provide only an inert medium which facilitates burrowing, or are organic components and the presence of microorganisms important? What is the ecological significance, if any, of these laboratory responses?

Considerable information has been accumulated on factors influencing larval settlement and metamorphosis for several groups of marine invertebrates. This literature has been the subject of recent, intensive reviews by Crisp[21,22] and Scheltema.[20] No previous studies, however, have been made on sipunculans. Whether the response of larvae of *Golfingia misakiana* to the presence of adults, as suggested in the present study, can be considered a true gregarious response, as defined by Crisp[22] and Scheltema[20] for other invertebrates, must await definitive proof of species specificity. Invertebrates for which gregarious responses have been demonstrated in the past are sessile organisms, such as oysters,[23,24] barnacles[25,26] and serpulid worms,[27] whose larvae settle on solid substrata, or tube-dwelling sabellariid polychates whose larvae settle on previously constructed tubes.[28,29] Studies of settlement-inducing factors of infaunal organisms, particularly polychaetes, have focused on physical attributes of the substratum and on bacterial-algal films on substratum particles.[20] In burrowing organisms the effect of adult factors on the substratum or on the surrounding water column generally has not been considered. The nature of the adult-related factor demonstrated in this study to promote metamorphosis of *Golfingia misakiana,* and other interacting factors influencing settlement and metamorphosis, will be pursued in future experimental investigations.

ABBREVIATONS

a, anus; at, apical tuft; bo, buccal organ; br, brain; cu, cuticle; drm, dorsal retractor muscle; e, eye; en, egg envelope; es, esophagus; in, intestine; l, lip; lg, lip gland; lp, lip pore; m, metatroch; mo, mouth; n, nephridium; p, prototroch; pr, posterior retractor muscle; psg, posterior sacciform gland; sph, postmetatrochal sphincter; s, stomodaeum; st, stomach; t, tentacle; to, terminal organ; vnc, ventral nerve cord; vrm, ventral retractor muscle.

ACKNOWLEDGMENTS

The author acknowledges with appreciation the able assistance of Julianne Piraino, Douglas Putnam and Joseph Murdoch in the collection and sorting of material, microtechnology, photography, and other aspects of this research. Illustrations 6, 7, 8, 9, and 10 were made by Carolyn Gast, scientific illustrator at the National Museum of Natural History, Washington, D. C.

The research was made possible by support from the Fort Pierce Bureau of the Smithsonian Institution and the facilities of the Harbor Branch Foundation, Inc., Fort Pierce, Florida.

REFERENCES

1. Rice, M. E. (1967) Ophelia, 4, 143-171.
2. Scheltema, R. S. and Hall, J. R. (1975) in Proceedings of the International Symposium on the Biology of the Sipuncula and Echiura, Vol. I, Rice, M. E. and Todorovic, M. eds., Naučno Delo Press, Belgrade, pp. 103-115.
3. Gerould, J. H. (1903) Mark Anniversary Volume, 439-452.
4. Gerould, J. H. (1907) Zool. Jahrb., Anat., 23, 77-162.
5. Åkesson, B. (1958) C.W.K. Gleerup, Lund, Unders. över Öresund 38, 1-249.
6. Åkesson, B. (1961) Ark. Zool., 13, 511-531.
7. Rice, M. E. (1973) Smithson. Contrib. Zool., No. 132, 1-51.
8. Rice, M. E. (1975) in Reproduction of Marine Invertebrates, Vol. II, Giese, A. and Pearse, J. eds., Academic Press, New York, pp. 67-127.
9. Rice, M. E. (1976) Amer. Zool., 16, 563-571.
10. Jägersten, G. (1963) Zool. Bidr. Uppsala, 36, 27-35.

11.   Rice, M. E. (1975) in Proceedings of the International Symposium on the Biology of the Sipuncula and Echiura, Rice, M. E. and Todorovic, M. eds., Naučno Delo Press, Belgrade, pp. 141-160.
12.   Williams, J. (1972) Amer. Zool., 12, 723.
13.   Amor, A. (1975) Physis, Sec. A, 34, 359-370.
14.   Hatschek, B. (1883) Arb. Zool. Inst. Univ. Wien Zool. Sta. Triest, 5, 61-140.
15.   Hall, J. R. and Scheltema, R. S. (1975) in Proceedings of the International Symposium on the Biology of the Sipuncula and Echiura, Vol. I, Rice, M. E. and Todorovic, M. eds., Naučno Delo Press, Belgrade, pp. 183-197.
16.   Williams, J. A. (1977) Functional Development in Four Species of the Sipuncula. Doctoral Dissertation, Univ. of Hawaii, pp. 1-218.
17.   Hācker, V. (1898) Ergeb. Plankton-Exped. Humboldt-Stiftung, 2, 1-50.
18.   Richardson, K. C., Jarett, L., and Finke, E. H. (1960) Stain. Technol. 35, 313-323.
19.   Scheltema, R. S. (1975) in Proceedings of the International Symposium on the Biology of the Sipuncula and Echiura, Vol. I, Rice, M. E. and Todorovic, M. eds., Naučno Delo Press, Belgrade, pp. 199-210.
20.   Scheltema, R. S. (1974) Thalassia Jugoslavica, 10, 263-296.
21.   Crisp, D. J. (1974) in Chemoreception in Marine Organisms, Grant, P. T. and Mackie, A. M. eds., Academic Press, New York, pp. 177-265.
22.   Crisp, D. J. (1976) in Adaptation to Environment: essays on the physiology of marine animals, Newell, R. C. ed., Butterworths, London, pp. 83-124.
23.   Cole, H. A. and Knight-Jones, E. W. (1949) Minis. Agric. and Fish., Fish. Invest., Ser. 2, 17, 1-39.
24.   Bayne, B. L. (1969) J. mar. biol. Assoc. U.K., 49, 327-356.
25.   Daniel, A. (1955) J. Madras Univ., Sect. B, 25, 97-107.
26.   Knight-Jones, E. W. (1955) Nature, 175, 266.
27.   Knight-Jones, E. W. (1951) J. mar. biol. Assoc. U.K. 30, 201-222.
28.   Wilson, D. P. (1968) J. mar. biol. Assoc. U.K., 48, 387-435.
29.   Wilson, D. P. (1970) J. mar. biol. Assoc. U.K., 50, 33-52.

SETTLEMENT AND METAMORPHOSIS IN THE ECHIURA: A REVIEW

John Pilger

Smithsonian Institution, Fort Pierce Bureau, Route 1, Box 194-C, Fort Pierce, Florida 33450

Two types of settlement and metamorphosis are distinguished. Males of the family Bonellidae have specialized attachment structures used during settlement; they undergo an abbreviated, neotenic metamorphosis. Information on settlement of echiurans other than bonellids is limited. Metamorphosis of the trochophore proceeds through the loss of trochal bands and protonephridia, as well as the transformation of the gastrointestinal valve and the pre- and posttrochal lobes into adult structures. The phenomenon of sex determination in the Bonellidae is reviewed.

INTRODUCTION

Echiurans occur in benthic habitats in shallow subtidal to hadal ocean depths[1] and in some areas, they represent a significant component of the benthic community.[2,3] Filter feeding species such as *Urechis caupo* are important in their ability to direct planktonic energy to the marine benthos.[4] Most echiurans feed on the nutrients in deposited sediments and thus play a major role in recycling energy in benthic communities.[5]

The Echiura are suitable subjects for the study of reproduction and development.[6] Early workers described the development of species in the genera *Echiurus,*[7,8] *Bonellia*[9,10,11] and *Lissomyema* (as *Thalassema*).[12] The detailed work of Newby[13,14] on *Urechis caupo* added to the knowledge of echiuran development. Recent investigations have dealt primarily with the cytological, ultrastructural, and biochemical aspects of gametogenesis (see Gould-Somero[15] for a review). Information on the settlement and metamorphosis of echiuran larvae is limited almost exclusively to the early studies. Neither electron microscopy nor advanced methods of light microtechnique have been employed in the investigation of echiuran larval development and metamorphosis.

The metamorphosis of an echiuran trochophore is a gradual transformation during which the structural and functional integrities of the individual are maintained. Metamorphoses of species in the genera *Echiurus, Lissomyema,* and *Urechis* are remarkably similar. *Bonellia,* on the other hand, represents a notable exception to the normal pattern of echiuran development.

The family Bonellidae is characterized by a striking sexual dimorphism in which the males are small neotenic forms living in the uterus of a female.[9] The larvae are sexually indifferent until settlement when their metamorphosis into a male or female form is determined either environmentally[16] or genetically.[17] Because settlement and metamorphosis of bonellid larvae are distinct among the Echiura, they will be considered separately.

SETTLEMENT OF ECHIURANS (EXCLUDING BONELLIDAE)

Within 24 hours after fertilization the gastrulae of *Lissomyema,*[12] *Urechis caupo,*[14] and *Ikedosoma gogoshimense*[18] begin to swim. Shortly afterward the gastrulae develop into planktotrophic trochophores. These larvae are equipped with two major ciliary bands. The largest and most powerful of these is the prototroch. This may be a single preoral band as in *Urechis,*[14] or it may be subdivided ventrally into two or three smaller bands as in *Echiurus*[8,19] and *Lissomyema.*[12] The second preoral band, the telotroch, is located just anterior to the anus. In *Urechis* this is subdivided into two distinct bands. Short cilia commonly cover the preoral lobe, and in some species a neurotroch is present extending from the mouth to the anus along the ventral midline.

The reaction of echiuran larvae to environmental cues is not fully understood. *Echiurus abyssalis* larvae are known to rise to the surface from depths as great as 1900 m.[19] This response probably is a negative geotropism since light levels at that depth are negligible and the larvae do not have eyes.

The larvae live for 2 to 3 months in the plankton before settling.[5,12,14,18] Impending settlement can often be recognized by contraction or flexing of the larval body.[8,12,21] This indicates that the body musculature is developed and that the crawling ability is imminent.

It has been shown that the larvae of *Listriolobus pelodes* settle in mid- to late spring, but no other information on seasonal settlement of echiuran larvae is available for comparison.[5]

It is generally believed that reliance upon chance encounters at the time of larval settlement is highly uneconomical.[22] An increasing number of investigations illustrate that invertebrate larvae are well equipped with sensory organs which might be used in habitat selection.[23] Two lines of evidence suggest that echiuran larvae conform to this generalization.

Indirect evidence that echiuran larvae may select appropriate substrates is found in the distribution of adults of *Listriolobus pelodes* near Santa Barbara, California.[2] Occurring in a large aggregation, they are distributed in a pattern which corresponds to a geographical area characterized by organically rich sediments of the finest particle size found anywhere on the Southern California mainland shelf.[24] Because the adults of this species cannot move over significant distances[5] and therefore cannot be considered effective in the formation of the aggregation, it is assumed that the larvae select the appropriate substrate.

Until recently, direct evidence for substrate selection by echiuran larvae was lacking. However, unpublished observations by Suer,[25] working with *Urechis caupo* larvae, support the widely held view of Wilson[26,27] that the nutrient properties of sediments influence the settlement of benthic marine invertebrate larvae. Suer[25] has found that *Urechis* larvae settle preferentially on sediment having a high organic content.

After settlement, the larvae of *Lissomyema*[12] and *Urechis*[14] are able to swim for 4 or 5 days. Undoubtedly, this is the result of a gradual reduction of the prototroch and telotroch, as swimming becomes progressively more feeble. Ultimately, when most, if not all trochal cilia are lost, the larvae become confined to a benthic existence and crawl using the body musculature and ventral cilia.[8,12,14,21] The burrowing behavior of the juvenile has not been described, but since the body muscles are functional by this time, burrowing is assumed to be similar to the peristaltic method used by the adults.[5,28]

METAMORPHOSIS OF ECHIURANS (EXCLUDING BONELLIDAE)

Metamorphosis of an echiuran trochophore into the juvenile form is not a radical event in the animal's life history but rather it is a gradual process during which the structural and functional integrities of the animal are maintained. As a result, this stage of echiuran development has been given different interpretations in the literature. Newby[14] broadly interpreted metamorphosis as predating settlement and including the development of all of the adult organs. Hatschek[8] and Conn,[12] on the other hand, considered metamorphosis to be the loss and transformation of larval structures. Although the details vary between and within animal groups, I believe that there is an underlying definition which may be applied to all animals. Metamorphosis should first be defined as a time-frame that brackets the end of larval life and the beginning of a juvenile existence. Within this frame, metamorphosis involves the loss or transformation of one or more larval characters or structures. Thus, metamorphosis is a component of the developmental process, but is distinguishable from organogenesis. In this paper, metamorphosis of the Echiura is reviewed on the basis of the preceding definition.

Settlement of the trochophore larva marks the beginning of echiuran metamorphosis. At this time, the larva undergoes structural and functional changes which result in the juvenile form and adult habits. These changes include the loss of the trochal bands and protonephridia as well as the transformation of the gastrointestinal valve, and the pre- and posttrochal lobes into adult structures.

The reduction of the prototroch and telotroch is the salient event in the change from a pelagic to a benthic existence. Apparently, the loss of the cilia is not rapid since *Urechis*[14] and *Lissomyema*[12] are able to swim for 4 to 5 days after settling. However, the body muscles already have differentiated, and the newly settled larva is well equipped for crawling and burrowing.

The larvae of *Echiurus*[8] and *Listriolobus pelodes*[29] possess a pair of protonephridia near the ventral midline posterior to the mouth. During metamorphosis, these structures retrogress and are replaced in the juvenile by metanephridia. The larvae of *Urechis* and *Lissomyema* do not have protonephridia at any time during their development.

Located in the gut of the *Urechis* trochophore is a large gastrointestinal valve, separating the stomach from the intestine[14] (Fig. 1A). This structure forms a nearly complete barrier between the two areas, the only opening being a ciliated hole at the ventral midline. A ciliated furrow in the intestine runs from the gastrointestinal valve opening to the anus. During premetamorphosis, the dorsal and lateral margins of the gastrointestinal valve move posteriorly and ventrally. This enlarges the larval stomach and forms a caecum (Fig. 1B). The posterior wall of the caecum ruptures during metamorphosis so that the stomach opens directly into the intestine (Fig. 1C). The former gastrointetstinal valve lies on the ventral floor of the midgut and forms the dorsal wall of the ciliated groove. The groove is thus converted into a ciliated tube which opens to the midgut at both ends. Although the tube eventually closes, Newby[14] argued that it later reopens, and therefore is the primordium of the midgut siphon (Fig. 1D). Conn[12] also considered the ciliated intestinal groove to be the primordium of the siphon in *Lissomyema*. On the other hand, Baltzer, studying *Echiurus*, stated that the ciliated intestinal groove and the siphon are not related. Newby[14] dismissed Baltzer's observation as incorrect because he had worked only with whole mounts.

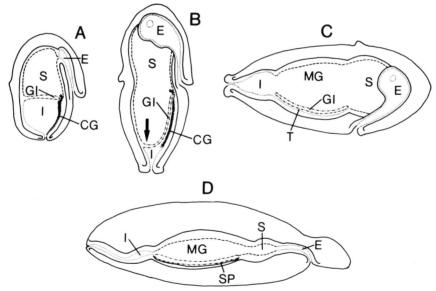

Fig. 1. The transformation of the larval gastrointestinal valve into the primordium of the adult midgut siphon in *Urechis caupo*. A. 4-day old larva, the gastrointestinal valve separates the larval stomach from the intestine. B. 35-day old larva, the dorsal and lateral margins of the gastrointestinal valve are shifted posteriorly. C. 50-day old larva, the gastrointestinal valve has ruptured dorsally making a linear pathway from the midgut to the intestine. The valve lies on the floor of the midgut transforming the ciliated groove into a ciliated tube. D. 60-day old larva, the siphon primordium lies along the ventral side of the midgut. I, intestine; MG, midgut; S, larval stomach; SP, siphon primordium; T, ciliated tube. Arrow indicates the region of the caecum where the gastrointestinal valve will rupture (based on Newby[14]).

One of the most noticeable changes during metamorphosis is the formation of the adult body shape. This is accomplished, anteriorly, by the lengthening of the preoral lobe to form the proboscis. During elongation, the muscle mass of the preoral lobe increases dramatically and reduces the blastocoel to a network of interconnecting cavities beneath the ciliated ventral epithelium (Fig. 2).

In the larva the blood vessels, circumesophageal nerve ring, and part of the esophagus extend into the preoral lobe. As the lobe elongates the nerve ring and blood vessels are drawn out with it and can be identified in the adult proboscis (Fig. 2). The esophagus does not lengthen; instead it is excluded from the pretrochal region.[12] Although proboscides of filter feeding echiurans such as *Urechis* are very small, they are derived from the larval preoral lobe in essentially the same manner as those of deposit feeders.[14]

The short cilia covering the preoral lobe of the larva become restricted to the ventral surface of the proboscis and are used in proboscis extension and food gathering.[5,30,31]

Concurrent with the formation of the proboscis, the postoral region expands to form the body of the adult. Conn[12] believed that in *Lissomyema,* the expansion was due to a sudden influx of water through the anal vesicles. Newby[14] reported that an increase of the coelomic fluid is responsible for the postoral expansion of *Urechis,* but he did not speculate on the source of the fluid.

The metamorphosed juvenile looks essentially like a small adult. Its body is transparent and characteristically undergoes peristaltic contractions. The proboscis actively explores the surrounding sediment for food.

SETTLEMENT OF BONELLIDAE

In contrast to other echiurans, bonellids produce large yolky eggs. These are deposited in masses of gelatinous strings after fertilization by dwarf males living in the female's uterus.[9]

The larvae hatch after two days and begin crawling about using the short cilia which cover the body. By the end of the third or fourth day the prototroch and telotroch have developed and a small green-pigmented swimming trochophore has formed. Bonellid trochophores are lecithotrophic and have a midgut without a mouth or anus.[9] Eyes are present in *Bonellia viridis* but are lacking in *B. fuliginosa.* Further, bonellid larvae differ from those of other echiurans by having a specialized attachment structure which is used at the time of settlement.

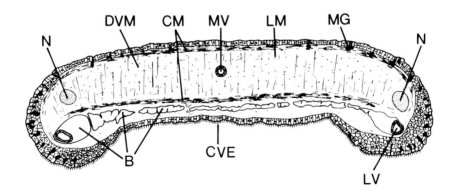

Fig. 2. Transverse section through the proboscis of an adult *Listriolobus pelodes.* The blastocoel from the larval preoral lobe is retained as a string of small cavities beneath the ventral epithelium. B, blastocoel remnants; CM, circular muscles; CVE, ciliated ventral epithelium; DVM, dorsoventral muscles; LM, longitudinal muscles; LV, lateral blood vessel; MG, mucus glands; MV, median blood vessel; N, nerve loop.

Baltzer[16] observed that, prior to settlement, the larvae of *B. viridis* exhibit a strong but short positive phototaxis. The egg masses of this species usually are deposited in rock crevices; hence the photoresponse may be important for the emergence of the larvae and their subsequent dispersal.

Time is an important factor in bonellid settlement relative to sexual development. The larvae may settle at any time from the third day to 3 or 4 weeks after fertilization.[11] However, as shown by Baltzer[32] and Leutert,[33] larvae settling on an adult female before their sixteenth day will tend to become males while those settling later or on a substrate free of female influence tend to become females.

As mentioned previously, bonellid larvae are equipped with specialized attachment structures which are used at the time of settlement. However, the role played by these structures in settlement is known only in those larvae which settle on an adult female and are destined to become males.

The attachment structure of *Bonellia viridis* consists of a large bilobed gland complex located on the ventral surface just behind the prototroch.[34] At the time of settlement, the larva makes contact with the proboscis of an adult femlae and becomes attached by secretions from the anterior region of the complex. After the initial bond is formed, the region of secretory attachment enlarges posteriorly to the midbody area. The pretrochal lobe and the posterior section of the body do not adhere (Fig. 3A). Metamorphosis begins shortly after attachment.

A different type of attachment structure has been reported in the trochophore of *B. fuliginosa*.[35] In this larva a muscular sucker is present midventrally immediately behind the prototroch. Michel[35] found the sucker to be very similar in construction to the acetabulum of the trematodes. When the body of a female *B. fuliginosa* is encountered, strong muscular contractions of the sucker securely fasten the larva to the adult (Fig. 3B).

METAMORPHOSIS OF BONELLIDAE

The extreme sexual dimorphism of bonellid echiurans was first reported in 1868 by Kowalevski[36] and by 1879 the development of the male and female was described.[9] Baltzer[37] noted that a small sexually indifferent trochophore is produced which, depending on the substrate it settles on, will metamorphose into a male or female (Fig. 4). Because male and female metamorphoses are entirely different processes, the larvae remain in a generalized condition until settlement. The adult organs and characters are present in a rudimentary stage in the larvae, that is, at the point where male and female developmental patterns diverge. Thus, most of organogenesis is delayed until the adult sex is determined at settlement.

Female bonellid trochophores metamorphose into juveniles in essentially the same manner as that described for other echiurans.[9,11] Shortly after settlement, the prototroch and telotroch are lost and the short cilia covering the body become restricted to the ventral surface of the preoral lobe (Fig. 4B). This lobe elongates and flattens to form the proboscis. As the proboscis lengthens, the anterior nerve loop is drawn out with it (Fig. 4C). The two larval eyes in *B. viridis* remain associated with the anterior part of the nerve loop, and are displaced to the proboscis tip.[11] Adult *Bonellia viridis* will respond to strong illumination on the proboscis.[31]

The post-trochal region becomes the body of the adult by expansion and increased muscularization. A gastrointestinal valve has not been reported in the larval gut,[9] although a midgut siphon is present in the adult.[38] Female *B. viridis* retain their green pigment over the entire body through metamorphosis and into adulthood.[39]

Male metamorphosis and development begin a few hours after settlement on a mature female.[34] The first change is the loss of the ciliary bands. Loosli[34] reported that in *B. viridis* the prototroch is lost first, followed shortly afterward by the telotroch. Within 13½ hours after settlement both bands are gone but the body remains covered by short cilia (Fig. 4D). Similarly, Michel[35] reported that the trochal cilia of *B. fuliginosa* are lost shortly after settlement.

107

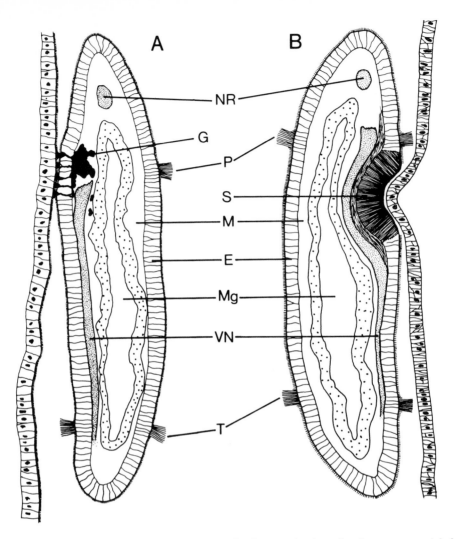

Fig. 3. Sagittal section representations of *Bonellia* trochophores at the time of settlement on an adult female. A. *B. viridis*, secretions from the gland complex firmly attaches the larva to the adult proboscis. B. *B. fuliginosa*, the larva attaches to the body wall of an adult female with a muscular sucker. E, epidermis; G, gland complex and secretions; M, mesoderm; Mg, midgut; NR, anterior nerve ring; P, prototroch; S, sucker; VN, ventral nerve. (A. based on Loosli,[34] B. based on Michel[35]).

The muscular sucker used by *B. fuliginosa* larvae to attach to the body of the female degenerates shortly after settlement.[35] In spite of the loss, the larva remains in firm contact with the adult throughout metamorphosis. Loosli[34] reported that in *B. viridis* the region of glandular attachment expands posteriorly, but the anterior and posterior ends do not adhere (Fig. 3A). The gland complex is retained during metamorphosis, and presumably into adulthood.

In direct contrast to other echiurans, the pretrochal lobe of the metamorphosing bonellid male undergoes a striking reduction. In the larva, the lobe represents approximately one third of the overall body length, but during metamorphosis it becomes virtually nonexistent[34,35] (Fig. 4E). The nerve ring becomes deflected posteriorly as a consequence of this shrinkage. Concurrent with the reduction of the anterior lobe, the posttrochal region elongates, but does not expand.[11] It has been suggested that this elongation may be due to a displacement of tissue from the pretrochal lobe during its reduction.[34] Michel[35] and Loosli[34] reported that the green pigment which covered the anterior lobe of the larva is lost during metamorphosis. In *B. viridis*, some of the pigment becomes incorporated into the subepidermal tissue and migrates to the posttrochal body region, leaving the head as a small white cap.[34]

The pair of eyes of *B. viridis* larvae is lost at some time during male metamorphosis.[11] No eyes are present in the larva or adult of *B. fuliginosa*.[35]

Organogenesis occurs concurrently with metamorphosis. The result is a dwarf male designed solely for reproduction. The small ciliated body consists of a seminal vesicle, a gut without a mouth or anus, and a pair of metanephridia (Fig. 4E). The suppression of adult structures and retention of larval features led Baltzer[11] to consider the bonellid male to be a neotenic form.

FACTORS INFLUENCING THE METAMORPHOSIS OF BONELLIDAE

A discussion of echiuran metamorphosis would be incomplete without considering the problem of sex determination in *Bonellia*. Differentiation into a male or female, and hence male or female metamorphosis, is determined, according to Baltzer,[11,16] by the substrate on which a sexually indifferent larva settles. His experiments showed that up to 92% of the larvae which settled on a clean substrate became females, while the remainder became intersexes and sterile males. In addition, when provided with an opportunity to settle on the proboscis of a female, approximately 70% did so and metamorphosed into males. Those not attaching to the proboscis became females.[37,39,40,41,42] On the basis of these data, Baltzer[11] inferred that male differentiation may be the result of inhibited female development.

Baltzer reasoned that if complete male development requires association with the proboscis, then development possibly is affected by a determining substance which diffuses from the proboscis to the larva. This theory was supported by experiments which showed that aqueous extracts of female proboscides produced males in the majority of larvae tested.[43] Further, heat-stable aqueous extracts of the intestine also were found to be effective in inducing male development if used at a concentration of one part (by weight) dried tissue to 6000 to 9000 parts

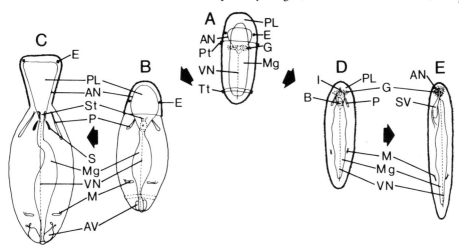

Fig. 4. Metamorphosis of male and female *Bonellia viridis* from a sexually indifferent trochophore. A. a sexually indifferent trochophore. B. an intermediate stage in female metamorphosis; note the loss of the trochal bands and the posttrochal body cilia. C. a juvenile female; the pre-trochal lobe has elongated to form the proboscis; the trunk has enlarged greatly. D. an intermediate stage in male metamorphosis; the trochal bands are lost and the pretrochal lobe has shortened considerably. E. a juvenile male; the pretrochal lobe is reduced to a mere cap; the short body cilia are retained. AN anterior nerve loop; AV, anal vesicle; B, blastopore; E, eye; G, gland complex; I, invagination; M, metanephridium; Mg, midgut; P, protonephridium; PL, pretrochal lobe; Pt, prototroch; S, setae; St, stomodaeum; SV, seminal vesicle; T, telotroch; VN, ventral nerve (based on Baltzer[11]).

sea water. Higher concentrations (1:1000 to 4000) proved inhibitory or fatal to the larvae.[43,44,45] Acetone extracts of the gut were ineffective in inducing male differentiation.[46]

Baltzer's theory proposing a diffusable sex determining substance also was supported by the observation that the completeness of male development is proportional to the amount of time the larva remains attached to the female.[10,42,47,48,49,50] Differentiation of each of the male organs requires a different threshhold of exposure to the proboscis for complete development. For example, reduction of the preoral lobe requires only a weak stimulus, while development of the seminal vesicle and viable sperm necessitates progressively longer exposures.[39,48]

Premature shortening of the length of time on the proboscis slows development and results in the formation of incomplete males, "intersexes."[37,51,52,53] Because the intersexes are most differentiated anteriorly, Baltzer[39] suggested that the secretions from the gland complex may act either as a bridge for the transport of the differentiating substance from the proboscis to the larva, or at least as a means of increasing the larval permeability to the substance.

Another line of investigation into the sex determination problem was pursued by Herbst.[54] He showed that by varying the ionic concentration of certain chemical elements in sea water, complete masculine development could be induced without exposure to a proboscis. For instance, increasing the concentration of copper or potassium resulted in the development of 70 to 80% males.[55,56] Decreasing the concentration of magnesium or sulphate had a similar effect in 90% of the larvae.[57,58] Increasing the acidity of the water also favored male development.[54] On the basis of his experiments he formulated the theory that hydration of the larva results in female development whereas dehydration favors male development.[59,60,61] Unfortunately, the relevance of these factors in nature is not known.

In summary, the studies of Baltzer and Herbst suggest that male development is a result of an interference or inhibition of female development by some unknown factor.

Wilczynski[17,62] took an opposing viewpoint and stated that sex determination is genetically determined during oogenesis. He was unable to discern the chromosomes in the oocytes, but he did claim to find two distinct egg types based on their differential reactivity to a stain which distinguishes male from female tissues. According to his studies the ultimate origin of the egg dimorphism lies with the incorporation by the oocyte of one of its two types of nurse cell nuclei early in oogenesis. Gould-Somero[15] questioned the presence of egg dimorphism as described by Wilczynski since the oocytes he worked with were not homogeneous and contained some obviously degenerate cells.

A recent publication by Leutert[33] re-examined the controversy between Baltzer and Wilczynski. Baltzer's original experiments were repeated using a sample size large enough to be treated statistically. Leutert's results support Baltzer's hypothesis by showing that sex is phenotypically determined in a significant number (43 to 83%) of the larvae. However, the remaining larvae are composed of some whose sex is genetically determined and others whose sex is unalterably indifferent. The fate and ecological significance of sexually indifferent juveniles in natural conditions is not known.

Leutert[33] also investigated oogenesis in *Bonellia viridis* using electron microscopy, and did not find any ultrastructural evidence of dimorphism among the nurse cells. Only a small amount of nurse cell material is passed to the oocyte during oogenesis, and this at an early stage. Further, no evidence of egg dimorphism was found at any stage of oogenesis.

Clearly, the problem of sex determination in *Bonellia* remains unsolved. The available evidence supports the earlier hypothesis that, for the majority of larvae, sex is determined phenotypically. However, a small percentage are unaffected by the determining factors and develop contrary to the phenotypic individuals. It is difficult to speculate whether this difference is due to genetic determination, or to an immunity of the larva to the influence of the substance. Finally, the identity of the determining substance and the nature of its action as a control of metamorphosis and development are unknown.

ACKNOWLEDGMENTS

The assistance of Patricia Pilger at many stages of the manuscript is gratefully acknowledged. This research was supported by a Smithsonian Institution Postdoctoral Fellowship to the author.

REFERENCES

1. Belyaev, G. M. (1966) Akad. Nauk. SSSR, Trudy Inst., Okeaonol., 591, 1-248.
2. Barnard, J. and Hartman, O. (1959) Pac. Nat., 1,1-16.
3. Jumars, P. and Hessler, R. (1976) J. Mar. Res.,34(4), 547-560.
4. Akesson, T. R. (1977) Estuarine and Coastal Mar. Sci., 5, 445-453.
5. Pilger, J. F. (1977) Ph.D. dissertation,Univ. Southern California.
6. Gould, M. C. (1967) in Methods in Developmental Biology, Wilt, F. H. and Wessels, N. K. eds., Crowell, New York, pp. 163-171.
7. Selensky, W. (1876) Morph. Jb., 2, 319-327.
8. Hatschek, B. (1880) Arb. aus der zool. Stat. in Triest., 3, 45-79.
9. Spengel, J. W. (1879) Mitt. Zool. Sta. Neapel, 1, 357-419.
10. Baltzer, F. (1912) Verh. Dtsch. Zool. Ges., 22, 252-261.
11. Baltzer, F. (1925) Publ. Staz. Zool. Napoli, 6, 223-286.
12. Conn, H. W. (1886) Stud. Biol. Lab.Johns Hopkins Univ., 3, 351-401.
13. Newby, W. W. (1932) Biol. Bull., 63, 387-399.
14. Newby, W. W. (1940) Mem. Am. Phil. Soc., 16, 1-213.
15. Gould-Somero, M. D. (1975) in Reproduction of Marine Invertebrates, Vol. III, Giese, A. C. and Pearse, J. eds., Academic Press, New York, pp. 277-311.
16. Baltzer, F. (1931) in Handbuch der Zoologie, Kükenthal, W. and Krumbach, T. eds., Berlin, pp. 62-168.
17. Wilczynski, J. (1960) J. Exp. Zool., 143,61-75.
18. Sawada, N. and Ochi, O. (1962) Mem. Ehime Univ., 4, 437-444.
19. Baltzer, F. (1917) Fauna u Flora Golf Neapel, Berlin Monogr., 34, 1-234.
20. Dawydoff, C. M. (1959) in Traite de Zoologie, Grassé, P. ed., Masson, Paris, pp. 674-717.
21. Baltzer, F. (1914) Verh. schweiz. naturf Ges., 2, 208-212.
22. Crisp, D. J. (1974) in Chemoreception in Marine Organisms, Grant, P. T. and Mackie, A. M. eds., Academic Press, New York, pp. 177-265.
23. Crisp, D. J. (1976) in Adaptation to Environment, Newell, R. D. ed., Butterworths, Boston, pp. 83-124.
24. Allan Hancock Foundation, Univ. Southern California (1965) Calif. State Water Quality Control Board, Publ., 27, 1-323.
25. Suer, L. (in progress) Ph.D. dissertation, Univ. California, Bodega Marine Laboratory, Bodega Bay, California 94923.
26. Wilson, D. P. (1954) J. mar. biol. Ass. U.K., 33, 361-380.
27. Wilson, D. P. (1955) J. mar. biol. Ass. U.K., 34, 531-543.
28. Gislen, T. (1940) Lunds Univ. Arsskr., 36, 1-36.
29. Pilger, J.F. personal observation.
30. Jaccarini, V. and Schembri, P. J. (1977) J. exp. mar. Biol. Ecol., 28, 163-181.
31. Jaccarini, V. and Schembri, P. J. (1977) J. Zool. Lond., 182, 467-476.
32. Baltzer, F. (1937) Wilhelm Roux Arch. Entw. Mech., 136, 1-43.
33. Leutert, R. (1974) J. Embryol. exp. Morph., 32, 169-193.
34. Loosli, M. (1935) Publ. Staz. Zool. Napoli, 15, 16-59.
35. Michel, F. (1930) Publ. Staz. Zool. Napoli, 10, 1-46.
36. Kowalevski, A. (1868) Zap. Kiev.Obshch. Estest., 1, 101-108.
37. Baltzer, F. (1914) Mitt. Zool. Stn. Neapel, 22, 1-44.
38. Stephen, A. C. and Edmonds, S. J. (1972) The Phyla Sipuncula and Echiura, British Museum (N. H.), London, pp. 1-528.
39. Baltzer, F. (1935) The Collecting Net (Woods Hole), 10(3), 1-8.
40. Baltzer, F. (1914) Sber. phys-med. Ges. Wurzburg., 43, 14-19.
41. Baltzer, F. (1928) Verh. Dtsch. Zool. Ges.,32, 273-325.
42. Baltzer, F. (1932) Rev. suisse Zool., 39, 281-305.
43. Baltzer, F. (1926) Rev. suisse Zool., 33, 359-374.
44. Baltzer, F. (1924) Mitt. naturf. Ges. Bern., 1924, 98-117.
45. Baltzer, F. (1925) Rev. suisse Zool., 32, 87-93.

46. Nowinski, W. (1934) Publ. Staz. Zool. Napoli, 14, 110-145.

47. Baltzer, F. (1931) Rev. suisse Zool., 38, 361-371.

48. Baltzer, F. (1937) Rev. suisse Zool., 44, 331-352.

49. Zurbuchen, K. and Baltzer, F. (1936) Rev. suisse Zool., 43, 489-494.

50. Zurbuchen, K. (1937) Publ. Staz. Zool. Napoli, 16, 28-80.

51. Baltzer, F. (1928) Rev. suisse Zool., 35, 225-231.

52. Glaus, A. (1933) Publ. Staz. Zool. Napoli, 13, 39-114.

53. Herbst, C. (1929) Sber. Heidelberger Akad. Will. Math. Naturw. Reihe., 20(16), 1-43.

54. Herbst, C. (1928) Sber. Heidelberger Akad. Will. Math. Naturw. Reihe., 19(2), 1-19.

55. Herbst, C. (1932) Naturwissenschaften, 20, 375-379.

56. Herbst, C. (1935) Wilh. Roux Arch. Entw. Mech. Organ, 132, 567-599.

57. Herbst, C. (1936) Wilh. Roux Arch. Entw. Mech. Organ, 134, 313-330.

58. Herbst, C. (1937) Wilh. Roux Arch. Entw. Mech. Organ, 136, 147-168.

59. Herbst, C. (1937) Wilh. Roux Arch. Entw. Mech. Organ, 135, 178-201.

60. Herbst, C. (1938) Wilh. Roux Arch. Entw. Mech. Organ, 138, 451-464.

61. Herbst, C. (1940) Wilh. Roux Arch. Entw. Mech. Organ, 140, 252-255.

62. Wilczynski, J. (1968) Acta biother., 18, 338-360.

METAMORPHOSIS IN THE OPHELIID POLYCHAETE *ARMANDIA BREVIS*

Colin O. Hermans

Department of Biology, Sonoma State College, Rohnert Park, California 94928

Armandia brevis undergoes more extensive larval development than is known in any other opheliid. Tiny eggs, spawned freely by epitokous females, develop into planktotrophic trochophores. Growth is accompanied by the formation of twenty segments before and nine segments after settlement. The structural differences between pre- and post-metamorphic stages reflect the differences in locomotion and feeding before and after settlement in this species. Metamorphosis changes the slowly swimming, ciliated larvae that feed on plankton into streamlined burrowers that feed on benthic particulate matter and move rapidly, like large nematodes or small amphioxuses, through soft surface sediments. In contrast to the well-known *Ophelia bicornis*, the adults of this species of *Armandia* are widely distributed in shallow marine sediments; their larvae undergo extensive, rapid growth and development in the plankton and do not exhibit highly selective settling behavior.

INTRODUCTION

During settlement, ciliary swimming—the primary mechanism for polychaete larval locomotion—is replaced by attachment, creeping, crawling, or burrowing in post-larval forms. These locomotory changes accompany the extensive structural changes referred to as metamorphosis. The essential differences between larvae and juveniles are locomotory. The digestive system may also be involved, because ciliary feeding is closely integrated with swimming in planktotrophic larvae. In lecithotrophic development ingestion of food usually begins soon after settlement. In either case, extensive changes in digestive structures usually accompany settlement. Development of respiratory and circulatory structures may also be involved.

In many polychaete species development may be direct, without a larval stage. The changes these forms undergo at hatching are usually very similar to metamorphic changes in other species. Thus metamorphosis and hatching tend to be confused in the polychaetes while they are clearly distinguished in many other forms.

Early studies of polychaete larval development focused on the morphological changes accompanying settlement; these established the principal relationships between various larval and adult forms and documented the often remarkable transformations associated with settlement.[1,2] More recently, behavioral and environmental factors in substratum selection became the focus for a number of important studies. The Opheliidae, Sabellariidae, and Serpulidae are the three polychaetous families that are best known for their settling behavior. The important experimental work on larval settling by Wilson, Knight-Jones, and others set the stage for this symposium's papers on metamorphosis in these families of polychaetes.

Study of development and metamorphosis in opheliids began with *Ophelia bicornis*.[3] Wilson determined that the restricted distribution of this species could be accounted for in terms of the settling behavior of the larvae relative to proximate environmental factors. Experiments showed that settling *Ophelia* larvae are not attracted by adults, and that the grade of the sand is an important prerequisite for settlement, but that sand of the correct grade is neutral with respect to the larvae when absolutely clean, repellent when containing dead microorganisms or organic matter, and attractive when containing living microorganisms, bacteria, and yeasts from the normal adult habitat.[4] This species is ideal for studying settling behavior because of its localized distribution and the fact that the larvae are easy to raise. They are ready to settle after only five days and do not feed until after metamorphosis.[3]

Subsequent work on other opheliids, while not experimental, provides information about larvae and juveniles in all three subfamilies: Opheliinae (*Ophelia, Euzonus*), Ophelininae (*Polyophthalmus, Armandia*), and Travisinae (*Travisia*). Dales[5] described development of *Euzonus mucronata* (as *Thoracophelia*), estimating that the larvae are ready to settle with two segments at an age of about ten days. Parke[6] found that cultured larvae of *Euzonus williamsi*, probably a sibling species, are ready to settle when they have three segments, and are about 15 days old and 200 μm long. As in Dales' study, the larvae were not cultured through metamorphosis. Hartmann-Schröder[7] described a number of postmetamorphic juvenile stages in *Ophelia rathkei, Armandia salvadoriana, A. cirrosa*, and *A. bossfeldi*. Retière[8] described some aspects of the direct developmental pattern in *Travisia forbesii*. The young hatch to become freely crawling juveniles with two chaetigerous segments and a length of about 300 μm. At this stage they are very similar in size, general appearance, and number of segments to newly metamorphosed *O. bicornis* and *E. mucronata*. Guerin[9,10] showed that *Polyophthalmus pictus* and *Armandia cirrosa* both settle when they have five segments and lengths of about 300 μm. With the exception of *Armandia brevis*, all opheliids studied thus far become benthic with fewer than six segments and lengths of less than 400 μm. They have either direct development or short, indirect, holoplanktonic development prior to metamorphosis.

MATERIALS AND METHODS

Armandia brevis (Moore) 1906 (=*A. bioculata* Hartman, 1938; Berkeley and Berkeley, 1941) is a common benthic inhabitant of marine coastal waters of the Northeastern Pacific from the Kurile Islands[11] to Mexico.[12] Although I have collected this species in a wide variety of habitats in Washington and Central California, most of the specimens used in this study were from the vicinity of the Friday Harbor Laboratories in the San Juan Islands of Washington. Immature adults were usually collected from muddy intertidal and subtidal substrata. Spawning, epitokous adults were collected at night as they swam toward a light hung in the water from the laboratory dock. Such epitokes could be collected after dark on nearly any evening from June to September. Larvae were collected with a plankton net. Settling and metamorphosing larvae and juveniles were collected by hanging small racks of Petri dishes enclosed in plastic window screening from the laboratory dock below the lowest tide level but several meters above the bottom. The accumulated sediments were examined for specimens.

Gametes were collected from the spawning epitokes. Following fertilization, larvae were raised on the water tables at the laboratory in pyrex baking dishes to which cultured algae, *Phaeodactylum tricornutum* and *Platymonas* sp. were added as food. Larvae that were competent to metamorphose did so after the addition of small amounts of mud from the beach at the laboratory.

Drawings and measurements of larvae were made with a camera lucida calibrated against a stage micrometer. Photomicrographs used under-developed, Kodak, High Contrast Copy Film or Kodak, Contrast Process Ortho cut film in a Wild photomicroscope equipped with Zeiss planachromatic objectives and a Kodak Wratten filter no. 58 (green).

Specimens were narcotized with chloretone prior to fixation. Whole mounts were fixed with either Bouin's fluid or 10% formalin in sea water and stained with chlorazol black E in 70% alcohol. For microtomy, larval specimens were either embedded in Epon or in Tissuemat, melting point 57°C containing 5% piccolyte, after fixation in isotonic, bicarbonate buffered, 1% osmium.

RESULTS

Life History

Sexually mature *Armandia brevis* swim to the surface of the water and spawn at night during the summer months, May to September, at Friday Harbor. These epitokous adults swim for several hours after they have shed their gametes, but eventually sink to the bottom and die.[13,14] Spheroidal eggs are released and fertilized at the metaphase stage of the first meiotic division. They drift passively and individually, sinking slowly in the sea water. Upon fertilization they become spherical, 50 μm in diameter, and complete meiosis. They undergo equal, holoblastic, spiral cleavage. At normal temperatures of 11° to 13°C, they form swimming blastulae within 12 hours and feeding trochophores in less than two days. Stomach contents reveal that larvae normally feed on a wide variety of planktonic material, irregular in both size and shape, although the 50 μm diameter trochophores feed only on nanoplankton.

Segments are added sequentially; they form one at a time from the pygidial growth zone. During the larval phase there is a linear relationship between larval length and segment number (Fig. 1). Larvae are competent to metamorphose with 20 segments. The normal time required to form 20 segments is not known. Under optimal culture conditions 20-segmented larvae were obtained in three to seven weeks. There was considerable variation within cultures and between cultures. The fastest development was obtained in the least crowded cultures. Under suboptimal conditions larvae may develop no more than three segments in two months. Comparisons of segment number and length in planktonic and cultured larvae show that segmentation, growth, and differentiation are closely linked. The rate at which these occur appears to depend upon food supply.

When presented with sediments, the competent larvae undergo rapid morphological change. During metamorphosis growth rate increases relative to segmentation. By the time juveniles reach a length of 5 mm they have 29 segments, the full complement of an adult. Subsequent growth is without segmentation. Postlarval growth is rapid. Empty Petri dishes, placed in the harbor to collect sediments and settling larvae, contained juveniles ranging from 0.8 mm with 20 segments to 2.0 mm with 24 segments after only six days. Other dishes, set out for four days, contained specimens ranging in size from 0.9 mm with 20 segments to 1.1 mm with 22 segments. A water table that had been cleaned of all detritus in mid-July contained reproductively mature worms over 10 mm long at the end of August. These could only have reached the water table as

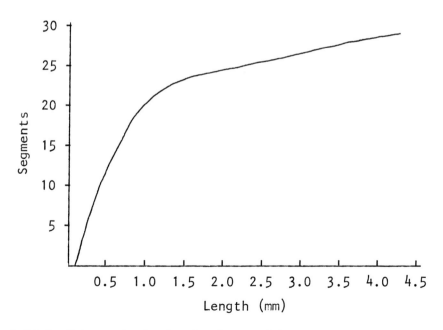

Fig. 1. Relationship between length and number of segments in larval and juvenile *Armandia brevis*.

larvae pumped into the sea water system. If the spawning season starts in early May and ends in early September at Friday Harbor, and it takes as little as three weeks for larvae to be ready to metamorphose and six weeks for juveniles to become reproductive, then the *Armandia brevis* in the Friday Harbor region may go through between two and three generations per year. The sex ratio for this species is usually about 1:1 although one sample of 54 individuals from Westcott Bay, near Friday Harbor, contained exactly two females for every male.

Anatomy of the Adult

The adults of both sexes are spindle-shaped, about 7 to 25 mm long and 1 to 1.5 mm in diameter. The sexes are separate and superficially indistinguishable. The prostomium is conical, terminating in a small palpode (Fig. 2). Three prostomial eyes are visible through the body wall: a bilateral, ventral pair and a single, dorsal ocellus embedded in the supraoesophageal ganglion.[15] Slightly anterior to the mouth a pair of nuchal organs form obvious, heavily ciliated infoldings on the dorsolateral surfaces of the smoothly tapering head. The mouth opens in a ventral depression at the level of the nuchal organs and the first segment. A pair of small, simple or dichotomously branched, ciliated papillae protrude slightly from the mouth. The proboscis, which is protruded to envelope and ingest food particles, is heavily ciliated, fleshy, slipper-shaped and tongue-like. When protruded it is shaped like a slipper with the toe protruding beyond the tip of the head and the heel beneath the first segment. The opening into the digestive tract is in the middle of what would be the arch on the bottom of the slipper.

The trunk of the worm has 29, occasionally 30, chaetigerous segments, each merging smoothly with the next without intersegmental furrows, and giving the body a nematode-like appearance. The cuticle is smooth, thick, and iridescent. Circular muscles are lacking except in the anal funnel; longitudinal muscles are strongly developed into a pair of ventrolateral and a pair of dorsolateral bundles. Several bundles of oblique muscles traverse the coelom in each segment. They originate from the connective tissue sheath surrounding the ventral nerve cord and attach to the lateral body wall between the dorsolateral and ventrolateral muscle bundles. Contraction of these muscles forms grooves down the sides and the ventral surface of the worm. The parapodia are small tubercles arising from the ventral margins of the lateral grooves. Each bears neuropodial and notopodial bundles of simple, slender, tapering, colorless chaetae. The chaetae are undifferentiated except that neuropodial chaetae are about half as long as those in the notopodia. In the last five or six segments the neuropodial are as long as the notopodial chaetae. All parapodia except the first pair bear simple, slender, respiratory cirri, or branchiae about 1 mm long that curve dorsally around the sides of the worm. They contain blood vessels and bear strongly developed bands of cilia that drive currents of water down both sides of the worm. Eleven pairs of ocelli are arranged segmentally, slightly anterior to the parapodia of the seventh to the seventeenth segments.[16] Four pairs of genital openings are situated anteriorly to the parapodia and ventrally to the ocelli of the tenth to the thirteenth segments. The genital openings are slits that open only during spawning. They are not associated with internal tubular structures. Eleven pairs of nephridia contact the body wall in segments fourteen through twenty-four in the same location as the genital openings in the preceding segments. Nephridiopores are lacking and the nephridia contain brownish refractile granules. Each nephridium contacts a segmental blood vessel. The coelom appears to be aseptate. In the middle of the worm the gut is surrounded by a haemal sinus in which blood passes anteriorly. In the ninth segment blood in the sinus enters a dorsal, pulsatory heart-like structure that drives blood forward in a dorsal vessel to the head, and laterally around the gut in two vessels that join midventrally to form the ventral blood vessel. The latter carries blood posteriorly and gives off a pair of vessels in each segment. These contact the nephridia as they pass to the branchiae. At the posterior end of the pharynx, in about the eighth segment, a pair of short tubular diverticula branch from the gut. At the level of the ninth segment the cells lining the gut are distinctly glandular and strongly basophilic. In the remaining segments of the body the gut is uniform in diameter. A typhlosole forms a longitudinal partition extending the

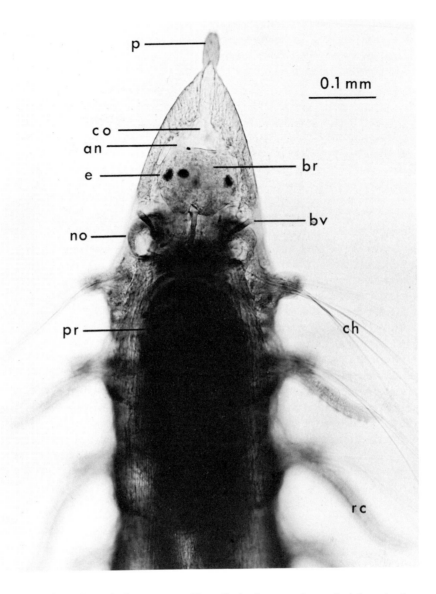

Fig. 2. Dorsal view of anterior end of 29-segmented juvenile that has recently acquired the major features of an adult, including: palpode (p), head coelom (co), anterior nerves (an), brain (br), eyes (e), blood vessels (bv), nuchal organs (no), proboscis (pr), chaetae (ch), and respiratory cirri (rc). Whole mount stained with Chlorazol black which has an especially strong affinity for cilia.

length of the gut. Posteriorly the gut empties into an anal funnel that broadens posteriorly and bears a highly variable number (0-8) of small finger-like papillae on its margin and a long, slender, contractile, anal cirrus originating from the floor of the anal funnel. The latter may be drawn up inside the anal funnel or may protrude from it. The funnel is notched midventrally and usually bears a pair of papillae on either side of the notch. The wall of the funnel contains about six bundles of circular muscles.

The brain, or supraoeosphageal ganglion, is nearly spherical, suspended within the head coelom and surrounded by mesothelium and a thin layer of connective tissue.[17] Circumoesophageal connectives link the brain to the ventral nerve cord, which begins in the first segment and extends through the last segment of the body as a single structure of uniform diameter, part of the ventral body wall, between the two ventrolateral muscle bundles, and surrounded by a collagenous sheath. The nuchal organs are innervated by the brain.

Anatomy of the Larva

The trochophore of *Armandia brevis* is small with well-developed prototroch, neurotroch, apical tuft, stomodaeum, stomach, and intestine (Fig. 3). The pygidium bears cells that secrete mucous strands. These trail behind the swimming larva and entangle particular matter. The mouth opens beneath the ventral margin of the prototroch and the anus opens dorsally, anterior to the pygidial mucous glands. As the larva grows, it lengthens by adding segments, and the three prostomial ocelli, characteristic of the adult, appear within the brain rudiments of the episphere (Fig. 4). A pair of neuropodial and notopodial chaetal sacs form and begin to produce chaetae as each segment forms, or becomes distinguishable from the pygidial growth zone. Muscles attach to the chaetal

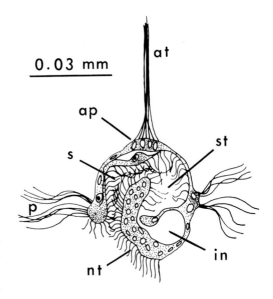

Fig. 3. Optical sagittal section of trochophore 39 hours after fertilization showing: prototroch (p), stomodaeum (s), apical plate (ap), apical tuft (at), stomach (st), intestine (in), and neurotroch (nt).

sacs. The chaetae are normally directed posteriorly along the sides of the body as the larva swims, but they can be erected to fan out in all directions away from the sides of the larva. Like those of the adults, these chaetae are simple, slender, tapering capillaries.

As the first larval segments form, the metatroch differentiates from a band of approximately ten ciliated cells surrounding the head behind the prototroch. A narrow groove separates these

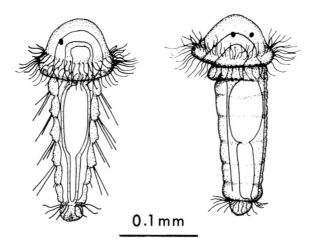

Fig. 4. Sketches of dorsal (left) and lateral (right) aspects of 5-segmented larva. One dorsal and a pair of ventral eyes have developed in the prostomium, the stomach and intestine extend into the segmental part of the body, and the telotroch encircles the pygidium posterior to the anus.

two bands and widens ventrally so that the prototroch forms the upper lip and the metatroch forms the lower lip of the mouth. The prototroch is the major propulsive organ of the fully formed larva. It is about 150 to 210 μm in diameter and consists of four ciliary bands. The second is the largest, produced by eight large trochocytes (Fig. 5). The cells forming the other three bands are smaller and somewhat more numerous. Whereas the apices of the trochocytes are smooth and contiguous, their bases are irregular and project forward into the episphere, between the prototroch and the brain rudiments. In the pygidium, four large cells form the telotroch, a band of cilia encircling the larva posterior to the anus and anterior to the pygidial mucous cells. On each segment patches of cilia form transverse bands on the dorsal and ventral surfaces at the level of the chaetal sacs (Fig. 6). On the prostomium, the apical tuft is reduced as the larva segments, and the nuchal organs appear as two small patches of cilia dorsolaterally between the brain rudiments and the prototroch.

Longitudinal muscles develop in the body wall enabling the larva to bend and undulate. Transverse muscle bundles appear toward the end of the larval period. Circular muscles do not form in the larval segments. In the head of the larva, a complicated system of small muscles connect nuchal organs, prototroch, metatroch, brain rudiments, and stomodaeum, and encircle the head beneath the prototroch and metatroch (Figs. 7, 8, 9).

The three major subdivisions of the trochophore's digestive tract, stomodaeum (oesophagus), stomach, and intestine, elongate and extend into the segmental parts of the body as the segments form (Fig. 10). A short proctodaeum also develops. In the fully formed larva the heavily ciliated oesophagus is not eversible. It extends from the mouth, between the prototroch and metatroch, to the third segment. The transition from oesophagus to stomach is distinct. The oesophageal cells are heavily ciliated and probably myoepithelial. Gastric cells are moderately ciliated and often filled with lipid spherules. The boundary between stomach and intestine is not distinct. The fact that the stomach is always somewhat distended with fluid and the intestine is only distended by particulate matter is the only basis for distinguishing the two regions in the fully formed larva. The proctodaeum, which develops at the end of the larval period, is sharply differentiated from the intestine. It is about 50 μm long, heavily ciliated, and connects the anus, at the anterior dorsal margin of the telotroch, with the intestine.

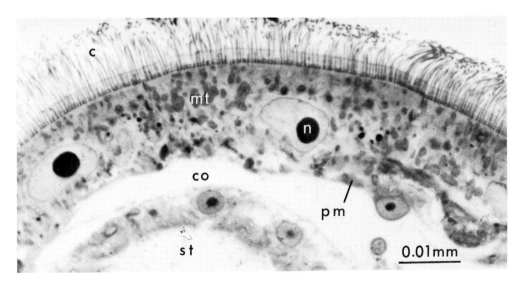

Fig. 5. Transverse section through the prototroch of a 6-segmented larva showing parts of three of the eight cells of the second prototroch row, their nuclei, nucleoli (n), mitochondria (mt), and cilia (c). Parts of the head coelom (co), stomach (st) and prototroch muscle (pm) are also shown. One μm thick section.

119

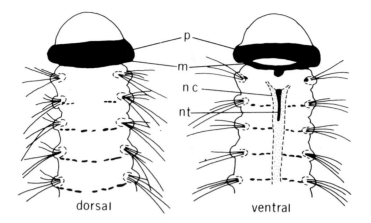

dorsal   ventral

Fig. 6. Camera lucida sketch of whole mount of 18-segmented larva superficially stained with Chlorazol black E to show distribution of epidermal cilia in the prototroch (p), metatroch (m), neurotroch (nt), and in transverse rows of patches on the dorsal and ventral surfaces of the segments; ventral nerve cord (nc), bases of circumoesophageal connectives, and chaetal sacs outlined by dashes. Nuchal organs not shown.

The coelom contains very little fluid, although it extends from the prostomium to the pygidium. A blood sinus surrounds the stomach and intestine of the fully formed larva, but other elements of the circulatory system, intersegmental septa, and larval tubular organs have not been discovered.

Metamorphosis

Larvae settle when they have 20 segments and the process of metamorphosis is initiated. If worms raised in culture dishes are not presented with mud from the adult environment they may advance to a 21-segmented condition without acquiring morphological attributes normally associated with 21 segments. Otherwise the number of segments strictly corresponds to other morphological parameters.

The diameter of the head is reduced in the 20- to 21-segmented stage through the contraction of circular muscles beneath the metatroch and prototroch muscles (Fig. 11). Cilia of the prototroch, metatroch, and telotroch are shed and the trochocytes degenerate. The cytoplasms of the prototrochal cells appear to be squeezed anteriorly into the intermediate zone between the prototroch and the developing brain, while the nuchal organ shifts posteriorly and enlarges

Fig. 7. Camera lucida sketch of head and first segment (1) of 19-segmented larva in dorsal view showing relative positions of the dorsal lobes of the brain (dlb), internuchal organ muscle (inom), metatroch (m), mid-neuropile muscle (mnm), nuchal organ (no), nuchal organ-stomodaeal muscle (nostm), the prototroch (p) and the stomodaeum (st).

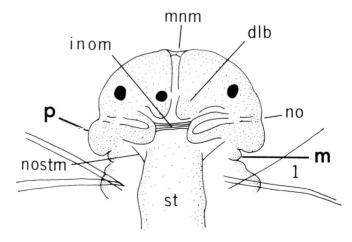

120

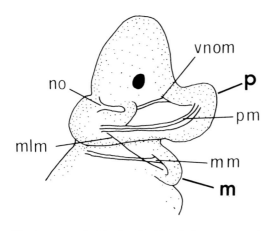

Fig. 8. Camera lucida sketch of head of 18-segmented larva in lateral view showing relative positions of the metatroch (m), metatroch muscle (mm), metatroch levator muscle (mlm), nuchal organ (no), prototroch (p), prototroch muscle (pm), and ventral nuchal organ muscle (vnom).

markedly, leading to the superficial impression that one has been transformed into the other. The mouth is constricted by circumoral muscles, and the stomodaeum is transformed into an eversible proboscis (Fig. 12). No larvae have been observed to evert the stomodaeum while even the youngest juveniles have been observed to do so. A pair of midgut diverticula develop from the anterior wall of the stomach, the midgut gland develops in the floor of the stomach, and the boundaries between stomodaeum, stomach and intestine also become obscure. The pygidium begins to form the anal funnel. A median tubercle at the posterior end of the 20-segmented larva begins to differentiate into the anal cirrus (Fig. 14).

Subsequent changes strictly correlate with the addition of segments. The prostomium becomes conical and the palpode forms at its tip (Fig. 13). The brain separates from the prostomial epidermis and is surrounded by and suspended within the head coelom. The prostomial mucous glands develop rapidly. The respiratory cirri form, enlarging as the worm segments. The nuchal organs enlarge further, become eversible, and shift posteriorly to occupy a position on either side of the head behind the brain and dorsal to the mouth. The eleven pairs of segmental ocelli become pigmented abruptly during the 26-segmented stage. The proctodaeum

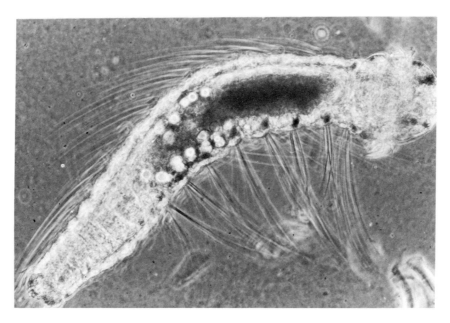

Fig. 9. Phase contrast photomicrograph of living, 16-segmented larva restrained in methyl cellulose-sea water. Thus restrained, the larva has reduced the diameter of the prototroch, erected the chaetae on one side of the body while attempting to undulate its way through the viscous medium. Note opacity of stomach due to food contents and refractibility of lipid spherules stored within gut epithelial cells.

Fig. 10. Ventral view of whole mount of 20-segmented larva ready to meta-morphose, stained with Chlorazol black E, showing three prostomial eyes (e), prototroch (p), metatroch (m), chaetal sacs from the first (l) to the twentieth segments (chaetae were not stained in this preparation and thus do not appear in the photograph), stomodaeum (st), stomach (s) and intestine (in) containing a *Didylum* and several other diatoms, and the telotroch (t) on the pygidium.

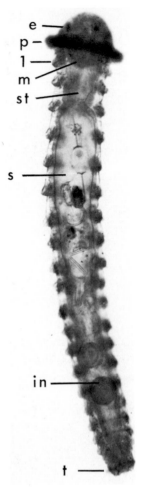

deepens by the posterior growth of the pygidium forming the dorsal and lateral margins of the anus and producing a pair of dorsal and a pair of ventral anal papillae (Fig. 14). More papillae are formed on the margin of the anal funnel during more advanced stages of development. The dorsolateral and ventrolateral longitudinal muscle bundles of the body wall enlarge markedly and extend forward into the prostomium. The transverse muscles and the coelom also enlarge. The major components of the circulatory system develop with the formation of the respiratory cirri. Differentiation of the latter lags behind the formation of the most posterior body segments. The full complement of 29 segments is formed when the worms are 5 mm long. From this stage onward the general proportions of the body remain unchanged and growth occurs without segmentation.

Behavior

Locomotion of the larva is produced by continuous metachronal beating of the prototrochal, telotrochal, and segmental ciliary bands. Throughout the segmental period of larval development, stimulation of the larva through encounter with nearly any solid surface causes the chaetae to be erected and the body to be flexed ventrally so that the larva is transformed into a ball of spines. The larva usually resumes normal ciliary swimming immediately. The larva may also be stimulated to swim rapidly for a short distance using the chaetae. The body undulates laterally with waves of undulation passing forward along the body; and the parapodia on alternate sides of the body are drawn sharply to the rear with the passage of the crest of each undulatory wave. This is the typical undulatory locomotor pattern of many errant polychaetes.[18] After settlement the worm no longer moves by means of cilia but crawls through the surface sediments using the parapodia. Pressing the worm against the bottom of a Petri dish with an eyelash probe causes it to swim rapidly with its chaetae. During the differentiation of the twenty-fifth and twenty-sixth segments, when the worm is about 2.5 mm long, this escape response changes to a short burst of smooth undulatory swimming, nematode- or amphioxus-like swimming characteristic of adult *Armandia*.[14] The chaetae remain at the sides of the body, the transverse muscles contract, compressing the body, and the large longitudinal muscles contract alternately. The coelom acts as a compressional strut. The interval hydrostatic pressure of the coelom prevents the body from shortening as the muscles contract. With further development, it burrows feebly with the parapodia and chaetae, or rapidly by means of nematode-like writhing. It glides through a mucous sheath secreted by the prostomial mucous glands. Development of the anal funnel assists this method of locomotion by pumping water, peristaltically, into the posterior end of the gut. This increases the turgidity of the body.

Feeding behavior likewise changes markedly during metamorphosis. During larval development the prototroch propels water past the head and edible material that is encountered is swept into the mouth by the metatroch. The shape of the head can undergo extensive changes

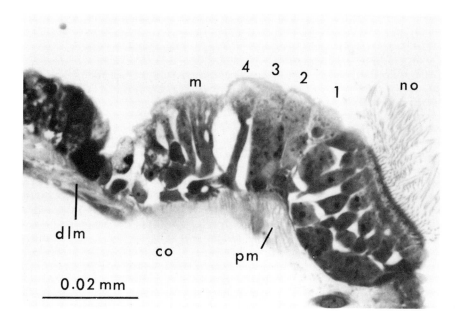

dlm co pm 0.02 mm

Fig. 11. Frontal section (anterior to the right, posterior to the left) through a 20-segmented specimen beginning to metamorphose showing one of the two ciliated nuchal organs (no), anterior to the four rows of prototroch cells (1-4), and the metatroch cells (m) that have recently lost their cilia. The prototroch muscle (pm), dorsal longitudinal muscle (dlm) within the first segment, and the coelom (co) are also shown. Section is 1 μm thick.

facilitating the ingestion of large, irregularly shaped material, such as tintinnids and *Didylum*, that would otherwise escape predation. The larva does not evert the pharynx. With the loss of the metatroch and prototroch at metamorphosis the pharynx becomes eversible. Food is no longer selected from suspended particulate matter propelled by and striking the prototroch but from the substratum by the tongue-like, ciliated proboscis that is protruded and appears to taste the substratum. To ingest food material, the proboscis simply envelops it and draws it into the mouth.

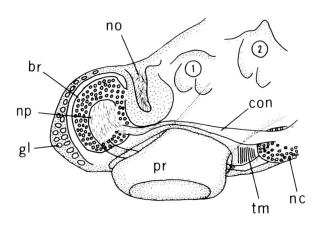

Fig. 12. Camera lucida sketch showing lateral view of whole mounted, 22-segmented specimen illustrating the changing relationships between structures in the head and the first two segments of the body. Brain (br), circumoesophageal connectives (con), glandular epidermis of the prostomium (gl), ventral nerve cord (nc), nuchal organ (no), neuropile (np), the partially everted proboscis (pr), transverse muscles behind the mouth (tm) and the parapodia of the first two segments (1, 2). Note that if the prototroch and metatroch were still present they would encircle the head between the nuchal organs and the first parapodia, passing either anterior or posterior to the mouth respectively. Also note the close connection between the nuchal organ and the brain and the separation of the brain from the prostomial epidermis.

Fig. 13. Superimposed camera lucida sketches of prostomia of whole mounted specimens with 21 to 28 segments. The brain and head coelom are outlined by dots in the 28-segmented stage.

DISCUSSION

The onset of metamorphosis in *Armandia brevis* is clearly marked by loss of the prototroch in the 20-segmented stage. But at what stage can one say that metamorphosis is complete? One could say that the metamorphosis is completed during the 20-segmented stage when the prototroch and telotroch are lost and the proboscis becomes eversible, or one could say that metamorphosis is not complete until 29 segments have formed and the 5 mm juveniles have developed an essentially adult morphology. Neither definition seems entirely satisfactory, and yet I can suggest none better. This reveals a fundamental property of metamorphoses: that they begin abruptly at settlement but end so gradually that a general definition cannot encompass both ends of the process.

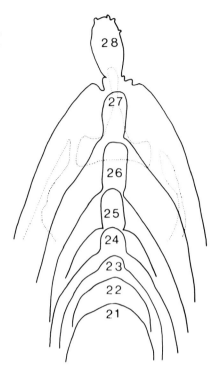

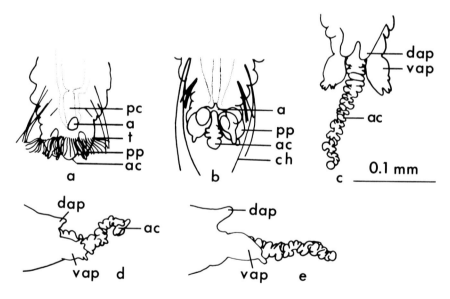

Fig. 14. Camera lucida sketches of the posterior ends of specimens showing the transformation of the pygidium during metamorphosis. a. Dorsal view of 20-segmented larva showing proctodaeum (pc), anus (a), telotroch (t), pygidial papillae (pp), and rudiments of the anal cirrus (ac). b. Dorsal view of recently settled, 20-segmented specimen showing absence of telotroch, enlargement of the pygidial papillae (pp), and the rudiment of the anal cirrus (ac), elongation of the chaetae (ch) on the last two segments, and posterior growth of the dorsal margin of the anus (a). c. Dorsal view of 24-segmented juvenile showing development of the anal funnel by further development of the pygidial papillae to form a ventral pair of anal papillae (dap), and the anal cirrus (ac). d. and e. Lateral views of 22- and 24-segmented juveniles showing development of the anal cirrus and the dorsal and ventral anal papillae to form the margins of the anal funnel.

What can be said about the role of settlement and metamorphosis in the life history of this species? *Armandia brevis* clearly has what one can call an opportunistic life history. It has a broad geographical range and appears to have few restrictions on its distribution within that range in the plankton or benthos. Population densities are highly variable[19] and they must have a very high growth potential that would come from the release of a great number of tiny gametes by adults that provide no parental care and die after spawning. The eggs are just large enough to produce tiny trochophores that must feed on nanoplankton to develop further. Presumably the advantages of planktotrophic development outweigh those of benthotrophic development until lengths of 1 mm and 20 segments are attained; metamorphosis thus occurs at the stage in development when the net efficiency of swimming and planktotrophy (in terms of growth and survival) is no longer greater than that of rapid burrowing and ingestion of benthic particulate matter. Epitoky, in which adults spawn once and die, insures that adult populations are unstable, but the production of large numbers of planktotrophic larvae insures a high rate of larval recruitment over broad areas of substratum.

The broad distribution and extended planktotrophic larval phase of *A. brevis* constrast sharply with the narrow distribution and short lecithotrophic larval phase in *Ophelia bicornis*.[3,4] The larvae of *O. bicornis* are highly selective in their settlement and they develop rapidly to a stage where they are competent to metamorphose as soon as an appropriate substratum is found. The role of a suitable substratum, restricted in its distribution, is so important for the ultimate survival of members of this species that substratum selection is the dominant factor in larval development and metamorphosis. The inherent instability and broad distribution of *A. brevis* populations, however, appear to make rapid growth the dominant factor in larval development and metamorphosis.

It is interesting to note that *Polyophthalmus pictus* and probably *Armandia cirrosa* are also epitokous species and yet they have larvae that settle with 5 segments and lengths of only 300 μm.[9,10] Why do the members of these species not develop more fully in the plankton, and why, if the adult's substratum is so important, does *O. bicornis* not undergo direct development like *Travisia forbesii*?[8] These are teleological questions which cannot be answered directly, but perhaps it would be illuminating to study the pattern of reproduction, larval development and metamorphosis in *A. brevis* at the two extremes of its geographic range (in the Kurile Islands[11] and in Acapulco[12]) in order to see if there are significant variations in reproductive and developmental patterns.

ACKNOWLEDGMENTS

This paper is a condensation of parts of theses[20,21] written under the supervision of Prof. R. L. Fernald at the Zoology Department and the Friday Harbor Laboratories of the University of Washington and supported by a National Science Foundation Marine Sciences Training Grant and a Public Health Services Predoctoral Fellowship. Advice, encouragement, and inspiration from R. L. Fernald, A. H. Whiteley, Paul Illg, R. A. Cloney, H. E. Potswald, R. P. Dales, and Gunnar Thorson during the course of the original investigation are gratefully acknowledged. Support for the preparation of this manuscript and for additional field work came from Public Health Service Research Grant GM 10292.

REFERENCES

1. Woltereck, R. (1902) Zoologica, Stuttgart, 34, 1-71.
2. Wilson, D. P. (1932) Phil. Trans. Roy. Soc., London, Ser. B, 221, 231-334.
3. Wilson, D. P. (1948) J. mar. biol. Ass. U.K., 27, 540-553.
4. Wilson D. P. (1958) in Perspectives in Marine Biology, Buzzati-Traverso, A. A. ed., Univ. California Press, Berkeley, pp. 87-103.
5. Dales, R. P. (1952) Biol. Bull., 102, 232-242.

6. Parke, S. R. (1975) M.S. Thesis, Univ. Pacific, Stockton, California, pp. 1-69.
7. Hartmann-Schröder, G. (1956) Zool. Anz., 157, 92-101.
8. Retière, C. (1971) C. R. Acad. Sci., Paris, 272, 3075-3078.
9. Guérin, J.-P. (1971) Vie Milieu, Ser. A, 22, 143-152.
10. Guérin, J.-P. (1973) Tethys, 4, 969-974.
11. Khlebovich, V. V. (1961) Issledovaniya Dalnevostochnykh Morei, Akad. Nauk SSSR, 7, 151-260.
12. Rioja, E. (1941) Ann. Inst. Biol. Mexico, 12, 669-746.
13. Hermans, C. O. (1964) Amer. Zool., 4, 292.
14. Clark, R. B. and Hermans, C. O. (1976) J. Zool., London, 178, 147-159.
15. Hermans, C. O. and Cloney, R. A. (1966) Z. Zellforsch., 72, 583-596.
16. Hermans, C. O. (1969) Z. Zellforsch., 96, 361-371.
17. Hermans, C. O. (1970) J. Ultrastr. Res., 30, 255-261.
18. Clark, R. B. and Tritton, D. J. (1970) J. Zool., London, 161, 257-271.
19. Woodin, S. A. (1974) Ecol. Mono., 44, 171-187.
20. Hermans, C. O. (1964) M.S. Thesis, Univ. Washington, Seattle, Washington, pp. 1-131.
21. Hermans, C. O. (1966) Ph.D. Thesis, Univ. Washington, Seattle, Washington, pp. 1-175.

METAMORPHOSIS IN *SPIRORBIS* (POLYCHAETA)

Herbert E. Potswald

Department of Zoology, University of Massachusetts, Amherst, Massachusetts 01003

Larval anatomy, subsequent settling, and metamorphosis are described primarily in the sinistral species *Spirorbis moerchi*, but comparative information is drawn from observations on the dextral species *S. spirillum* and *S. vitreus*. Initial tube formation in *S. moerchi* is due to the release of a calcareous secretion from the hind-gut, resulting in an opaque tube. The primary tube in *S. spirillum* and *S. vitreus* forms mainly from secretions of the ventral thoracic mucous glands; the resulting initial tube is transparent. Nevertheless, the hind-gut in *S. moerchi* and in the dextral species is obviously homologous. Within 24 hours of settling, the left thoracic muscle bundles grow into the opercular peduncle in *S. moerchi,* and upon contraction cause the body to turn to the left during secretion of the definitive tube. In both sinistral and dextral species, secondary tube formation is due largely to the secretory activity of the major subcollar glands. Gross asymmetry, characteristic of the adult, arises from differential growth of the larval abdominal region as it is transformed into the non-segmental achaetous zone of the adult. The peristomium in *Spirorbis* is clearly pre-segmental.

INTRODUCTION

Newly released *Spirorbis* larvae, given a suitable substrate, are usually planktonic only for a matter of hours. The behavior of recently hatched *Spirorbis* larvae has been extensively studied with special reference to choice of substrate and gregariousness.[1-7] The fact that some *Spirorbis* larvae have been demonstrated to settle preferentially on particular algal species[3,4,5,6,8,9] is especially interesting because it raises the possibility that speciation of the Spirorbinae has been largely accomplished by larval behavioral isolation.

In keeping with D. P. Wilson's[10] concept, but certainly not a principle that can be extended to all polychaete larvae,[11] settlement of *Spirorbis* larvae can be considered an event that precedes metamorphosis and results in the transition from larval to adult environment. The morphological events associated with settlement and secretion of the initial tube, in which the worm is destined to live for the rest of its life, have been described for *Spirorbis spirorbis* (*S. borealis*) in some detail by Nott,[12] and Nott and Parkes.[13] However, metamorphosis *per se,* again keeping with D. P. Wilson's[10] concept of the latter process as the change from larval to pre-adult morphology, has not been extensively studied in the Spirorbinae in a detailed, *i.e.,* histological manner. A detailed study of metamorphosis in *Spirorbis*, which necessitates delving into internal changes, is clearly warranted, not only for its intrinsic value, but at least for two other reasons. First the adults of the subfamily Spirorbinae are unique among their class in being grossly asymmetrical. Second, many of the existing textbook concepts of head segmentation, the origin and meaning of body segmentation, and the concomitant origin of germ cells in polychaetes have been based, to a large extent, on early studies of serpulid development. The fact that the subfamily Spirorbinae has been proposed to deserve separate family status, Spirorbidae,[15] is fully acknowledged. Nevertheless, Pillai[15] concedes that his family Spirorbidae is closely related to the family Serpulidae.

The present study deals primarily with an analysis of the histological events attending the metamorphosis of *Spirorbis moerchi* larvae. Comparative observations, although not described in detail, have been made on the dextral species *S. spirillum* and *S. vitreus.*

MATERIALS AND METHODS

The animals used in the present study were collected intertidally on San Juan Island, state of Washington. *Spirorbis moerchi*, often in association with *S. vitreus,* was found in only one location, Argyle Creek. *Spirorbis vitreus* was also found intertidally on San Juan at several other locations, often in association with *S. spirillum.* The worms were kept in the circulating sea water system at the Friday Harbor Laboratories, or were brought to the Seattle campus where they were kept in the Zoology Department's 10°C cold room.

In order to obtain large numbers of free swimming larvae, brood pouches containing actively moving larvae, whether in the operculum (*S. moerchi*) or in the tube *(S. vitreus* and *S. spirillum*), were dissected with #5 watchmaker's forceps or size 0 insect pins. The released larvae were transferred via pipette to containers previously kept in sea water so that they developed "films" of micro-flora and fauna.[1] Artificially released larvae showed the same behavioral responses and settled in about the same length of time as naturally released larvae.

Larvae, newly emerged, and at various stages following settlement, were fixed in aqueous Bouin's, dehydrated through ethyl and tertiary butyl alcohols and embedded in a paraffin-piccolyte mixture modified after Cloney.[16] Larvae and post-settled worms were individually oriented with warm needles in drops of embedding mixture under a dissecting microscope. Blocks, chilled to 4°C, were sectioned 4 to 6 μm in three planes, affixed to slides, and were stained with Harris' or Ehrlich's haematoxylin and counterstained with eosin. The same stages were fixed in cold buffered O_sO_4[17] and embedded in Epon 812.[18] Thick sections, $\frac{1}{2}$ to 1 μm, were cut on a Porter-Blum ultramicrotome, affixed to slides and stained with Richardson's stain.[19] Simple line drawings were made with the aid of a Zeiss drawing attachment.

RESULTS

1. The Fully Formed Larva—External Anatomy

Hatching is a mechanical process and is effected by a tearing of the fertilization envelope by means of the larval chaetae. At a constant temperature of 10°C, hatching in *S. moerchi* occurs at about six weeks. A chronology for the tube brooders has not been worked out.

At hatching, a fully formed *S. moerchi* larva has an overall length of about 320 μm and a diameter, measured across the collar, of about 150 μm. The larva can be conveniently divided into three regions: head, thorax, and abdomen (Figs. 1, 2). With a 40x objective one can observe the branchial rudiments tucked under the dorso-lateral sides of the head; however, the latter are never as prominent as described by Höglund[20] until the larva has started to undergo metamorphosis. The most characteristic feature of the *S. moerchi* larva is the large white sac of secretion in the hind-gut. The latter has been referred to as the "attachment gland"[1] and the "primary shell gland."[20] The latter term will be used in this paper because I feel it more accurately describes the function of the "gland." The external anatomy of *S. vitreus* and *S. spirillum* is similar to that of *S. moerchi* except that the larvae of the latter species lack a fully formed primary shell gland.

2. The Fully Formed Larva—Internal Anatomy

A. Ectoderm. The ventral portion of the larval head or prostomium contains massive mucous glands (Figs. 3, 4), undoubtedly playing some role in the initial searching and settling behavior of the larva. Lateral to the prostomial mucous glands there is a single eye and a pigment spot on each side of the head. Bordered ventrally by the prostomial mucous glands and laterally by the eyes is the supraesophageal ganglion (Fig. 4). The apical organ consists of three elongate cells which extend from the surface of the head downward where they taper sharply and become confluent with the neuropile of the brain (Fig. 4). At about the level of the eyes the portion of the prostomium just dorsal to the brain is composed of two lobes of thickened ectoderm which

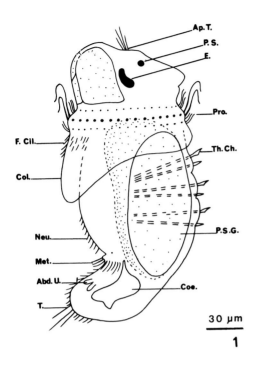

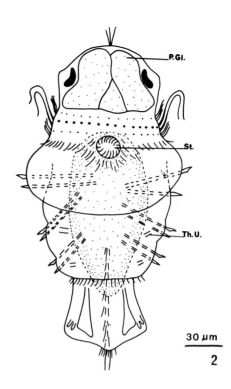

Fig. 1. Side view of a fully formed *Spirorbis moerchi* larva. Abd. U., abdominal uncini; Apt. T., apical tuft; Coe., coelom; Col., collar; E., larval eye; F. Cil., cilia of the stomodaeal region; Met., metatroch; Neu., neurotroch; P.S., pigment spot; P.S.G., primary shell gland; Pro., prototroch; T., telotroch; Th. Ch., thoracic chaetae.

Fig. 2. Ventral view of a fully formed *S. moerchi* larva. P.G., prostomial glands; St., stomadaeum; Th. U., thoracic uncini.

overhang the branchial rudiments, giving the latter the appearance of being tucked into the dorsal side of the head. On the left side there are three rudiments; the middle one is the largest and is the rudiment of the operculum (Fig. 3). In the larvae of *S. vitreus* and *S. spirillum,* the opercular rudiment develops on the right. The large size of the opercular rudiment imposes on the larva the *only element of asymmetry*. There are three tentacle rudiments on the right side (*S. moerchi*).

The prototroch consists of a complete circlet of cilia arising from eight ectodermal trochoblasts (Figs. 3, 4). In the midline of the ventral surface, the prototroch overhangs the deeply invaginated stomodaeum like a large lip (Fig. 4). The ectoderm of the stomodaeum makes contact with the endoderm of the gut, but the lumina of the stomodaeum and gut are not continuous.

The characteristic collar extends from below the stomodaeum posteriorly to the level of the third thoracic segment. Folding of the lateral edges of the collar occurs during larval development, but because the folds in at least some specimens do not extend to the midline, the median portion of the collar remains singular in appearance (Figs. 3, 4).

A striking feature of *Spirorbis* larvae is the complex of gland cells in the ventral and lateral ectoderm of the thorax. Similar glandular regions have been described in the larvae of *S. spirorbis* (*S. borealis*) by Nott.[12] On each side, just below the origin of the collar, there is an invagination of the lateral ectoderm, the primordium of the major subcollar gland, which secretes the adult tube (Fig. 5).

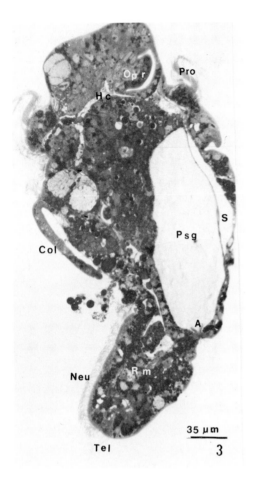

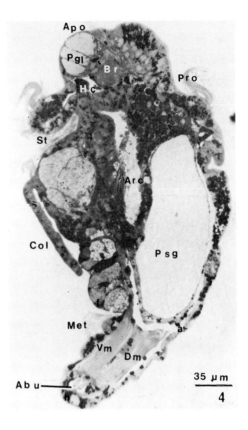

Fig. 3. Oblique sagittal section of a *S. moerchi* larva showing the opercular rudiment (Op r), prototroch (Pro), head coelom (H c), collar (Col), primary shell gland (P s g), anus (A), residual mesoderm (R m), neurotroch (Neu), and telotroch (Tel). The space (S) between dorsal ectoderm and primary shell gland is artifact.

Fig. 4. Oblique sagittal section of a *S. moerchi* larva showing prostomial glands (P gl), apical organ (Ap o), supraesophageal ganglion (Br), head coelom (H c), prototroch (Pro), stomodaeum (St), archenteron (Arc), collar (Col), primary shell gland (P s g), metatroch (Met), dorsal longitudinal muscle bundle (D m), ventral longitudinal muscle bundle (V m), and abdominal uncini (Ab u).

It should be emphasized that *S. moerchi* larvae lack posterior abdominal vesicles; however, *S. vitreus* and *S. spirillum* larvae possess the latter.

The ectoderm of the thorax and abdomen is relatively thick ventrally and laterally, and thins out considerably dorsally. Aside from the numerous glands, the ectoderm is heavily laden with yolk. The notochaetae of the three pairs of dorsally situated chaetal sacs have increased in length but not in number when compared with their appearance during larval development. Two thoracic uncini are found on each ventro-lateral side of segments two and three. Two uncini are also found in lateral pits on each side at the posterior end of the abdomen (Fig. 4). As in the adult, uncini are absent on segment one. The paired intra-epidermal nerve cords are composed of three pairs of ganglia, one pair per segment, which are connected longitudinally by connectives and transversely by commissures. The ventral nerve cords remain independent of the supraesophageal ganglion. Nervous tissue cannot be demonstrated posterior to segment three.

On the dorsal side of the abdomen, the anus can be observed to be fully formed (Fig. 3). At the opening of the anus, the ectoderm meets the endoderm, but is not strongly invaginated as a proctodaeum. Cilia surround the anus, both internally and externally.

B. Mesoderm.   The head coelom is a small unpaired cavity found below the brain and surrounding the anterior portions of the fore-gut and stomodaeum. The mesoderm of the head arises independently of the mesodermal bands and is believed to be of ectodermal origin, arising from cells of the embryonic animal pole.[14]

At the posterior end of the abdomen, where they had their origin, the longitudinal muscle bundles are found as a perfectly symmetrical, bilateral pair (Figs. 4, 6). They insert on the "basement membranes" of the lateral ectodermal pits bearing the abdominal uncini. Anteriorly, at about the level of the metatroch, each lateral muscle bundle can be seen to divide into two components: a ventro-lateral and a dorso-lateral muscle bundle. The ventro-lateral muscle bundles extend anteriorly beneath the ectoderm, reducing greatly in thickness at the level of the first segment, and attach at the base of the prostomium. The dorso-lateral muscle bundles, which are smaller than the ventro-lateral muscle bundles, also extend anteriorly beneath the ectoderm and attach at the base of the prostomium (Figs. 7, 8). Of considerable importance is the fact that the longitudinal muscle bundles are perfectly symmetrical throughout their length.

Three paired, small coelomic cavities, corresponding to the three larval segments,[14] are found between the ventro-lateral and dorso-lateral muscle bundles. A septum, composed of two epithelial layers, separates the head coelom from somite one, somite one from somite two, and somite two from somite three. The cavity of somite three is continuous posteriorly with that of the larval abdomen which in the adult corresponds to the achaetous zone. The splanchnic mesoderm of the paired cavities surrounds the endoderm of the gut laterally and meets in the ventral midline to form a double-layered ventral mesentery (Fig. 7). A dorsal mesentery has not been observed. The ventral mesentery extends posteriorly and ends in a mass of undifferentiated mesoderm.

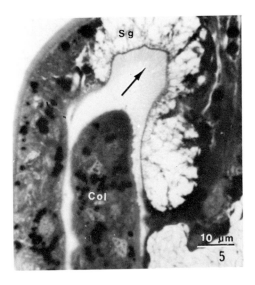

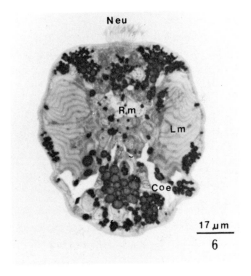

Fig. 5. Frontal section through the major subcollar gland rudiment (S g) of a S. moerchi larva. Note folded collar (Col) and fine spicules (arrow), probably calcareous, issuing from the surface of the gland.

Fig. 6. Cross section through the abdomen of a S. moerchi larva just above the abominal uncini showing the neurotroch (Neu), convergence of ventral and dorsal muscle bundles to form a single pair of longitudinal muscle bundles (L M), residual mesoderm (R m), and coelom (Coe).

131

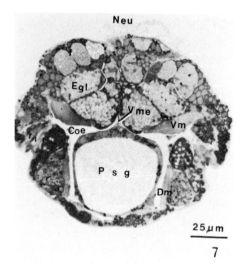

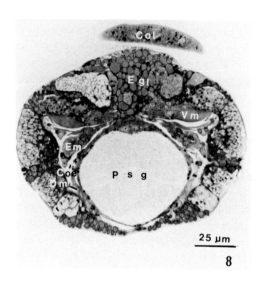

Fig. 7. Cross section taken just below the third thoracic segment of a *S. moerchi* larva showing the neurotroch (Neu), ventral longitudinal muscle bundles (V m), coelom (Coe), dorsal longitudinal muscle bundles (D m), ventral mesentery (V me), primary shell gland (P s g), and ectodermal glands (E gl).

Fig. 8. Cross section through the second thoracic segment of a *S. moerchi* larva showing the posterior end of the collar (Col), ventral longitudinal muscle bundles (V m), extensor muscles (E m) of the chaetal sacs, dorsal longitudinal muscle bundles (D m), coelom (Coe), primary shell gland (P s g), and various types of ectodermal glands (E gl).

C. Endoderm. The primary shell gland in *S. moerchi* is a dorsally situated spacious sac, which extends from the anterior portion of the larval abdomen forward to just below the level of the prototroch (Figs. 1, 2, 3, 4, 7, 8). If a living larva is pressed between a slide and a cover slip with sufficient pressure to rupture the gland, its contents are observed to be composed of tiny spicules suspended in a semi-fluid matrix which, after a few minutes in contact with sea water, solidifies. The shell gland gives a positive test for calcium when the chloranilic acid technique is applied.[21] The epithelium composing the dorsal lining of the gland is exceedingly thin and has the appearance of being stretched. Generally, the latter pulls away from the dorsal ectoderm during fixation leaving a space as an artifact (Fig. 3).

An anterior archenteron starts at a level just behind the blind stomodaeal invagination and extends posterioraly, ventral to, and parallel with, the anterior half of the primary shell gland (Fig. 4). A portion of undifferentiated endoderm separates the anterior archenteron from the lumen of the shell gland.

3. Larval Settlement

Just before actual settlement occurs in *S. moerchi*, creeping movements become slower, the larva turns more frequently and crisscrosses its path many times trailing a mucous thread. Finally, forward movement ceases altogether and the ventral surface of the larva appears to be firmly attached to the substrate, presumably the result of evacuation of ventral mucous glands. The larval thorax and abdomen contract violently, the contraction due largely to shortening of the paired ventro-lateral and dorso-lateral muscle bundles. As a result of the strong contractions, the contents of the primary shell gland empty via the anus and spread over the thorax and abdomen of the larva. The larva rocks back and forth on its antero-posterior axis and wiggles from side to side, aiding in the uniform dispersal of the gland contents and insuring firm adhesion to the substrate. The semi-fluid contents of the gland solidify after a few minutes in contact with sea water; therefore, the larva must effectively mold its initial tube during this time.

Within ten minutes after formation of the initial tube, the opercular and branchial rudiments make their appearance at a rapid rate. At this time the larva is on its side and the prototrochal region appears to shrink somewhat. About 30 minutes after formation of the initial tube, the trochoblasts become detached and are completely sloughed off. It is certain that *S. moerchi* larvae do not ingest prototrochal cells, as is thought to be the case in *S. spirorbis* (*S. borealis*) larvae.[12]

Two hours after formation of the initial tube, the post-settled larva has turned through 180° so that the dorsal surface faces the substrate. The collar, which in the free-swimming larva had been, for most of its extent, folded and pressed against the thorax, extends forward, unfolds, and then extends posteriorly to encompass the anterior rim of the primary tube. Lifting and the resulting unfolding of the collar are due, at least in part, to forward extension of the first pair of thoracic chaetae. The branchial and opercular rudiments are about twice the length they were in the fully formed larva and the opercular rudiment is much larger than any of the simple branchial rudiments. The pigmented eye cups have migrated to the midline and remain there for a day or so and then disappear. Also in the midline, ventral to the rapidly developing branchial crown, there is a proboscis-like structure forming. The post-settled larval abdomen becomes shorter and broader, and therefore is not clearly delimited from the thorax as in the free-swimming larva. The abdominal uncini serve to anchor and also allow the larva to "walk" up and down within the tube.

The initial or primary tube is rectilinear, chalky white, and opaque. In cross section, the primary tube is hemicircular, and the flat ventral portion, *i.e.*, the portion attached to the substratum and apposed to the dorsal surface of the post-settled larva, is transparent. The ventral portion of the tube is undoubtedly secreted by the ventral thoracic glands prior to turning over of the larva to assume the adult position. Soon after the collar has folded back to enclose the lip of the primary tube, deposition of the secondary tube begins. Calcium secretions originating from the major subcollar glands, as demonstrated by Swan,[22] and Hedley,[23,24] and most recently by Nott and Parkes,[13] are added to the anterior lip of the tube and molded into place by the encompassing collar fold. Twenty-four hours after settlement, the anterior end of the tube has started to turn in a clockwise fashion and the direction of coil is determined.

Upon hatching, larval behavior of *S. spirillum* and *S. vitreus* resembles that described for *S. moerchi*. During searching, *S. spirillum* and *S. vitreus* larvae, although they trail mucous threads, retain distinct paired terminal vesicles that do not appear to decrease in size. In the case of both species, the primary tube is a transparent sheath and is most likely formed from the secretions of the ventral thoracic mucous glands. It is possible that the hyaline contents of the posterior gut cavity contribute to formation of the initial tube, but judging from the size of the gut cavity, this would have to be a minor component. A complete time-sequence study has not been possible but at the end of 24 hours, secondary tube formation, marked by the presence of opaque calcareous material deposited on the primary tube, is visible and the tube has started to turn in a counterclockwise direction.

4. Changes Occurring During the First Day Following Settlement

A. Head. The earliest post-settled stages to be examined in section were fixed within an hour of attachment after the larvae had sloughed their prototrochs. Examination of newly released prototroch fragments reveals that the prototroch is cast off in quadrants. Loss of the prototroch results in a deep circumferential constriction separating the head or prostomium from collar and stomodaeum (Fig. 9). The ventral prostomial mucous glands are as large and as conspicuous as they were prior to settlement. Apical tuft cilia are lost at the time of settlement but the apical cells are retained and apparently are incorporated into the brain.

At the end of 24 hours the prostomium has undergone a drastic reorganization. The branchial rudiments and operculum are well formed and the ventral prostomial glands have become consolidated into a proboscis-like structure, which is situated in the ventral midline (Fig. 10). In the region previously occupied by the prototroch, a shortening has occurred obliterating the circumferential groove separating prostomium from peristomium. The branchial crown in *S. moerchi* is composed on the left side, starting at the dorsal midline, of a small branchial rudiment, a large opercular rudiment, and then ventrally a bilobed branchial rudiment; on the right side, also starting at the dorsal midline, there is a small branchial rudiment followed ventrally by a larger bilobed branchial rudiment. The "proboscis" separates the two halves of the crown ventrally, and a gap, produced by lateral migration of the rudiments during consolidation of the ventral prostomial glands, separates the halves of the crown dorsally.

At the end of 24 hours, the opercular rudiment is no longer proportionately larger than the other branchial rudiments. Within the central mesodermal core of the opercular peduncle are muscle fibres, extensions of the longitudinal muscle bundles, which are absent on the opposite side (Fig. 11).

The peristomium, consisting of stomodaeum and collar, shows few changes during the first day of post-larval development. The stomodaeum remains closed and consists of a deep invagination which extends from the ventral surface inward and somewhat caudad against the anterior end of the endodermal mid-gut. Muscle fibers extend into the collar at its origin and upon contraction apparently draw the collar forward so that it encompasses the branchial crown when the worm withdraws into its tube. The major subcollar glands are more deeply invaginated than in the larva.

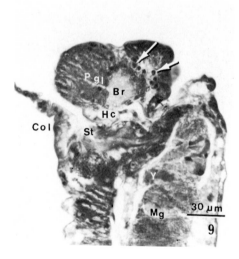

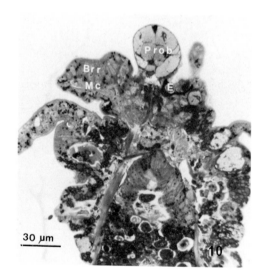

Fig. 9. Sagittal section through a *S. moerchi* larva, fixed within one hour after settling, showing mitotic figures (arrows) in the dorsal part of the prostomium, prostomial glands (P gl), supraesophageal ganglion (Br), collar (Col), stomodaeum (St), head coelom (H c), and yolk (Y) in the mid-gut (M g). Note the absence of the prototroch.

Fig. 10. Frontal section through a *S. moerchi* larva 24 hours after settlement showing the "proboscis" (Prob), branchial rudiments (Br r), mesodermal core (M c) of the branchial rudiments, and a pigment cup of one of the eyes (E).

B. Thorax. The most striking changes to occur within the thorax concern those associated with the further differentiation of the yolky mid-gut. Because this portion of the gut differentiates more rapidly in the larvae of *S. spirillum* and *S. vitreus*, apparently in contrast to larvae which develop a fully formed abdominal shell gland, the changes following initial settlement are more dramatically illustrated in *S. moerchi* than in the latter two species. With evacuation of the abdominal primary shell gland in *S. moerchi*, the mid-gut is displaced both dorsally, to occupy the space previously held by the gland, and posteriorly to fill the anterior portion of the abdominal region. As in the fully formed larva, the anterior portion of the mid-gut (that portion which makes contact with the ectodermal stomodaeum) is differentiated into a layer of columnar epithelium bearing cilia at its surface and a brush border. There is very little yolk in the latter region and a number of the columnar cells have started to accumulate a mucous secretion. Posterior to the differentiated portion of the mid-gut, the abundant yolk inclusions have started to break down.

Most of the proteid yolk of the undifferentiated portion of the mid-gut breaks up into discrete rod-shaped bodies of about 1.5 to 2.5 μm in length. Proteid yolk breakdown takes place within the cytoplasm of the mid-gut cells and then is extruded into the enteric lumen (Fig. 12).

C. Achaetous Zone. The larval abdomen is actually equivalent to the achaetous zone of the adult. Shortening and broadening of this zone is due to contraction of the paired longitudinal muscle bundles. As a result, the anus, although still displaced to the dorsal side, becomes more terminal in position.

With emptying and subsequent collapse of the abdominal shell gland, the latter becomes restricted to the posterior end of the achaetous zone where it communicates to the outside via the anus (Fig. 12). Upon discharge, then, the abdominal shell gland in *S. moerchi* becomes the hind-gut. The latter is directly comparable, not only in position and size but also in structure,

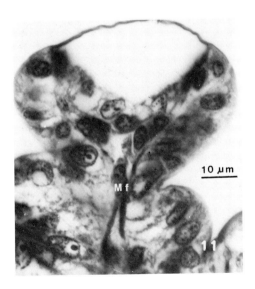 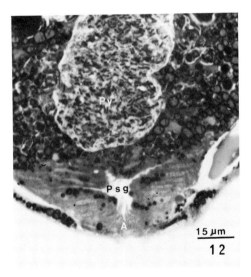

Fig. 11. Sagittal section through the operculum (Op) of a *S. moerchi* larva 24 hours after settlement. Note muscle fibers (M f) in the opercular peduncle.

Fig. 12. Frontal section through the posterior end of a *S. moerchi* larva 24 hours after settlement showing the rod-shaped proteid yolk bodies (P y) within the lumen of the mid-gut, the empty primary shell gland (P s g), or hind-gut, and anus (A).

to that in *S. spirillum* and *S. vitreus*. In all three species, the lumen of the hind-gut is separated from that of the mid-gut by a partition of undifferentiated endoderm.

As in the thorax, the muscle bundles of the achaetous zone retain their perfect bilateral symmetry. The short region in back of the posterior uncini is the pygidial growth region which is followed by the pygidium bearing the anal opening.

5. Changes Occurring Between the Second and Fourth Day

A. Head. At the end of the second day, the constriction at the base of the "proboscis" reaches a maximum and the structure is cast off; consequently, the ventral prostomial glands together with the apical tuft and various trochs are the only larval structures lost during metamorphosis. The circumesophageal connectives make contact with the first pair of thoracic ganglia, thus connecting the supraesophageal ganglion with the ventral nerve cords (Fig. 14); the connectives are the only nervous structures in the peristomium. In *S. moerchi* the pigmented larval eye cups start to break down (Fig. 14). The larval eyes are retained by *S. spirillum*.

By the fourth day, the stomodaeum moves anteriorly and opens terminally and ventrally within the branchial crown; it is surrounded by the branchial crown dorsally and laterally at approximately the same site previously occupied by the "proboscis." Examination of sagittal and transverse sections (Figs. 15, 16) clearly shows the topographical relationship of the dorsally situated supraesophageal ganglion to the stomodaeum. Concomitant with its anterior migration, the stomodaeum fuses with the endodermal midgut and the lumina of the two become confluent (Fig. 15). The point of fusion between ectoderm and endoderm is distinct and remains so even in the adult; *i.e.*, the mid-gut retains its much wider proportions as compared with the narrower ectodermal component or esophagus.

B. Thorax. The rod-shaped proteid yolk fragments are no longer visible in the lumen of the mid-gut at the end of the second day of settlement and presumably have been absorbed after extracellular digestion. A large amount of lipid yolk is retained by the cytoplasm and can also be observed in the lumen (Fig. 13).

C. Achaetous Zone. As can be anticipated from adult morphology, it is in the achaetous zone where gross asymmetry of the body will arise.

By the end of the second day of settlement, the pygidium and pygidial growth region have started to grow posteriorly, but the achaetous zone as a whole remains symmetrical at this time. The hind-gut lumen remains separate from that

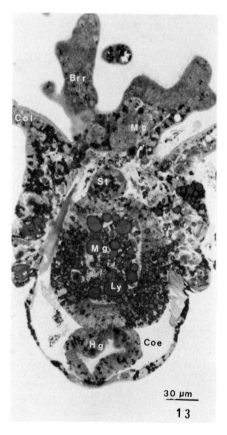

13

Fig. 13. Frontal section through *S. moerchi* two days after settlement showing branchial rudiments (Br r), mesodermal core (M c) of a branchial rudiment, collar (Col), stomodaeum (St), mid-gut (M g), coelom (Coe), and hind-gut (H g). Note that the rod-shaped proteid yolk bodies are absent and the mid-gut lumen now contains lipid yolk (L y).

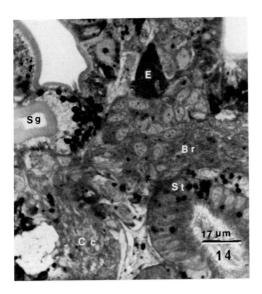

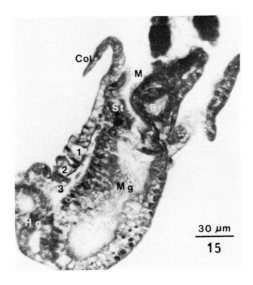

Fig. 14. Frontal section through the head of *S. moerchi* two days after settlement showing breakdown of one of the pigmented eye cups (E), supraesophageal ganglion (Br), major subcollar gland (S g), stomodaeum (St), and circumesophageal connective (C c).

Fig. 15. Sagittal section through *S. moerchi* four days after settlement showing the supraesophageal ganglion (Br), collar (Col), stomodaeum (St) which now opens terminally as the mouth (M), mid-gut, (M g), hind-gut (H g), and the coelomic cavities of the three thoracic segments (1, 2, 3). Note that the ectodermal stomodaeum, now the esophagus, is confluent with the endodermal mid-gut.

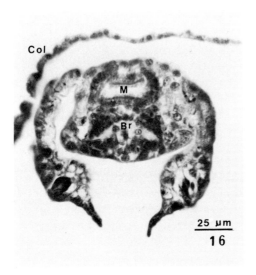

Fig. 16. Cross section through the head of *S. moerchi* four days after settlement showing the collar (Col), terminal mouth (M), and supraesophageal ganglion (Br).

of the mid-gut, owing to the persistence of the endodermal partition. The mesothelial layers of the ventral mesentery separate, and within the lumen thus formed a granular substance can be observed (Fig. 17). This marks the first appearance of the ventral blood vessel.

During the third and fourth day, the achaetous zone more than doubles its original length at settlement, and initial signs of asymmetrical development become evident. Growth occurs mainly in the pygidial region; little or no growth occurs anterior to the posterior uncini. As a result, the longitudinal muscle bundles remain about the same length as in the larva. Growth is greater on the right side than on the left and, in effect, the posterior end of the body turns to the left. It is also evident at this time that the achaetous zone starts to twist on its longitudinal axis to the left and, as a result, the longitudinal muscles of the right side start to be pulled over to the left side (Fig. 18). The longitudinal muscle bundles of the left side remain in position.

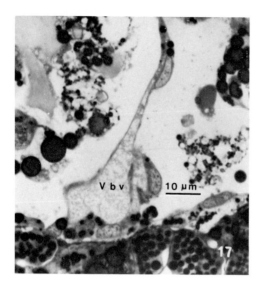

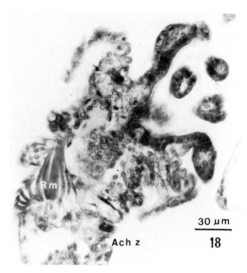

Fig. 17. Oblique frontal section through the achae-tous region of *S. moerchi* two days after settlement showing the ventral blood vessel (V b v) forming in the ventral mesentery.

Fig. 18. Frontal section through *S. moerchi* four days after settlement showing that the longitudinal muscle bundles of the right side (R m) are being pulled over to the left or concave side as a result of differential growth in the achaetous zone (Ach z).

The same situation applies to dextral species of (*S. spirillum* and *S. vitreus*) except for the fact that the morphology is a mirror image of that of *S. moerchi* and, therefore, the relationship between right and left sides is reversed. Unlike *S. moerchi*, the lumina of the mid-gut and hind-gut in *S. spirillum* and *S. vitreus* are confluent.

6. Changes Occurring Between the Sixth and Eleventh Day

A. Head. Between the sixth and eleventh day of settlement the branchial rudiments reach a length equal to about half that of the entire body. Although the crown is not yet complete, in that the fourth branchial rudiment is not distinguishable on the right side, pinnule formation has commenced. The pinnules grow out from the tentacle shaft at an angle of 45° and appear successively in pairs from the base towards the tip. When forming, the pinnules are composed of ciliated cuboidal epithelium surrounding a solid mesodermal core, but the core soon hollows out and the central cavity thus formed becomes continuous with that of the main shaft.

One change of interest occurring just before or on about the sixth day, is the enlargement of two distinctive cells in the dorsal portion of the supraesophageal ganglion. The cells are located bilaterally on each side of the midline and have somewhat irregularly shaped nuclei (Fig. 19). These cells persist in the adult where it appears that they give rise to giant axons, characteristic of sedentary polychaetes.

B. Thorax. On the sixth day of post-larval development the thorax is still symmetrical, both externally and internally, but by the eleventh day the thorax has an asymmetrical appearance characteristic of the adult. The number of uncini is greater on the concave side (left side in *S. moerchi*; right side in *S. spirillum* and *S. vitreus*) than on the opposite side.

A series of cross sections taken through the thorax of an eleven day post-settled animal correspond to those taken through an adult, at least as far as the arrangement of the longitudinal muscle bundles is concerned. In *S. moerchi*, a section taken through the first segment reveals a longitudinal muscle bundle in both the dorsal and ventral body wall on the right side (convex) as well as on the left side (concave); a section through the second segment shows the presence of a longitudinal muscle bundle on the right ventral side but lacking on the right dorsal side; and finally, a section taken through the third segment reveals that the longitudinal muscle bundles are concentrated on the left side (concave) and are completely absent from the right side. The situation is reversed in *S. spirillum* and *S. vitreus*. The arrangement of the longitudinal muscle bundles within the thorax will be made clearer below when further development of the achaetous zone is considered.

Just before, or on about the sixth day, the partition between mid-gut and hind-gut opens so that the gut is complete from mouth to anus for the first time. The mid-gut retains its central position within the thorax and anterior part of the achaetous zone prior to the sixth day; however, between the sixth and eleventh day the mid-gut shifts to the convex side, where it forms a conspicuous bulge in the anterior portion of the achaetous zone. During this period the remaining yolk inclusions are absorbed and the animal starts to feed, as evidenced by the presence of detritus within the gut lumen. The planchnic mesoderm separates from the ventral portion of the mid-gut wall, thus initiating the formation of the gut sinus. A pair of vessels is given off from the forming sinus at the level of the first segment (Fig. 20). The latter are the branchial vessels which grow around the esophagus and then extend anteriorly and dorsally to the base of the branchial crown where, on about the eleventh day, they branch sending a single vessel into the hollow center of each branchial tentacle.

The paired ventral nerve cords retain their intraepidermal position and show no change (Fig. 20). Nervous tissue has not been demonstrated posterior to the third segment.

C. Achaetous Zone. By the sixth day of settlement, the twisting of this zone to the left in *S. moerchi* has gone through about 45°, but by the eleventh day the twist has gone through a full 90°. Longitudinal muscle bundles and posterior uncini now lie entirely on the left or concave side. The torsion of the achaetous zone, in effect, brings about the asymmetrical arrangement of the muscle bundles in the thorax. Not only is the ventral longitudinal muscle bundle of

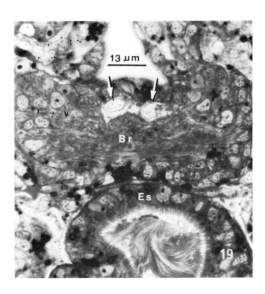

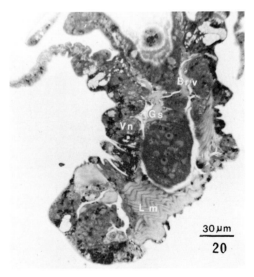

Fig. 19. Frontal section through the head of *S. moerchi* six days after settlement showing the supraesophageal ganglion (Br), a pair of giant nerve cells (arrows), and esophagus (Es).

Fig. 20. Frontal section through *S. moerchi* six days after settlement showing the branchial vessels (Br v), given off from the gut sinus (G s), and left ventral nerve cord (V n). Note that the longitudinal muscle bundles (L m) in the achaetous zone are concentrated on the concave side.

the right side pulled over to the left, but the right dorsal bundle is twisted ventrally and there merges with the right ventral bundle. The arrangement of the muscle bundles can be best understood by examining camera lucida drawings of an eleven-day post-settled animal (Figs. 21, 22).

It can be said that by the eleventh day of post-larval development the major features of the adult, with the important exception of abdominal or secondary segments, are either present or have started to form. Furthermore, the gross asymmetry characteristic of the genus has been shown to develop as a result of differential growth occurring primarily in the achaetous zone which corresponds to the larval abdomen. This change from an essentially bilaterally symmetrical larva to an asymmetrical pre-adult is considered to constitute metamorphosis.

DISCUSSION

Höglund's[20] recognition of three morphological types of *Spirorbis* larvae is based on the appearance of the primary shell gland: (1) larvae with one abdominal primary shell gland, (2) larvae which lack a primary shell gland, and (3) larvae with two thoracic primary shell glands. In the present study, an investigation of the first two types of larvae has shown that the abdominal primary shell gland is a modification of the hind-gut and is homologous to the posterior gut cavity of those species lacking the abdominal gland. According to Nott[12] *S. spirorbis* (*S. borealis*) larvae have a small amount of opaque material, presumably calcareous, in the posterior gut cavity. He refers to the hind-gut as the "attachment gland." It is clear, however, that the "gland" in *S. spirorbis*, although perhaps somewhat more developed than in *S. spirillum* and *S. vitreus*, is in no way comparable to the massive development of the primary shell gland in *S. moerchi*. Nott[12] has characterized a number of mucous glands in *S. spirorbis* which correspond closely to those in *S. moerchi* and which probably play an important role in larval settling and secretion of the larval tube. The secretions of the latter glands undoubtedly play a more important role in the initial tube formation of those larvae lacking a fully formed abdominal gland, as compared with those that possess a maximally distended hind-gut filled with calcareous secretion, as in *S. moerchi*.

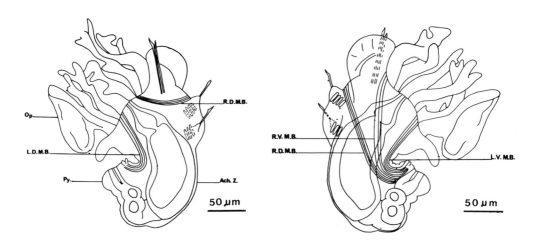

Fig. 21. Dorsal view of *S. moerchi* eleve   ays after settlement showing the operculum (Op.), right dorsal longitudinal muscle bundle (R.D.M.B.), left dorsal longitudinal muscle bundle (L.D.M.B.), achaetous zone (Ach. Z.) and pygidium (Py.).

Fig. 22. Ventral view of *S. moerchi* eleven days after settlement showing the right ventral longitudinal muscle bundle (R.V.M.B.), right dorsal longitudinal muscle bundle (R.D.M.B.), and left ventral longitudinal muscle bundle (L.V.M.B.).

A pair of anal vesicles develop in larvae belonging to the second larval type and are absent in *S. moerchi*. Whether the presence of anal vesicles is correlated with the absence of a fully formed primary shell gland or is simply fortuitous is not known. Nott[12] discusses this problem but does not reach a definitive conclusion except to suggest that they may be the remnants of discharged mucous glands.

Quiévreux[25] and Potswald[14] have attempted to compare larval types with adult characters and mode of brood protection in an effort to discern evolutionary trends within the subfamily. Although some trends are indicated, it appears that there are still too few detailed comparative data available to make this a meaningful approach.

The events associated with settling in *S. moerchi* differ from those described by Nott[12] for *S. spirorbis* in two respects: (1) the nature of primary tube formation, and (2) the time it takes for the larva to turn through 180° to assume the adult position. In *S. moerchi*, when conditions for settlement are optimal, the contents of the primary shell gland are extruded via the anus in an explosive fashion. The calcareous secretion is molded by the movements of the larva into a tube capable of housing the entire settled larva in less than five minutes. In *S. spirorbis* the initial tube is transparent and accommodates only the posterior end of the larva. According to Nott,[12] after secretion of the "attachment gland" in *S. spirorbis*, the hind-gut appears as a collapsed structure having no connection with the anus. It is presumed that the anus reopens some time later during metamorphosis. The lumen of the hind-gut in *S. moerchi* remains confluent with the anal opening following settlement. *Spirorbis spirorbis* larvae roll 180° within a minute after release of the contents of the "attachment gland" so that the settled larvae rapidly assume the adult position—*i.e.*, dorsal side facing the substrate. Nott believes that the rotation is due to ciliary movement of the prototroch. *Spirorbis moerchi* do not rotate through a full 180° until approximately two hours after discharge of the primary shell gland. The prototroch may be responsible for the 90° rotation in *S. moerchi*, but because the prototroch is lost 30 minutes after primary tube formation, it cannot be operating during the final stage of rotation. Too few observations have been made to comment on the time course of events during settling in *S. spirillum* and *S. vitreus*. Rapid rotation of the larvae of the latter two species would be advantageous because the initial tube in both species is primarily a mucous secretion and laying down of the definitive calcareous tube does not begin until the larvae have assumed the adult position.

The question arises as to whether there is a distinct survival advantage conferred upon larvae which possess a fully developed primary shell gland. The present study cannot provide an answer to this question; however, in the one area where *S. moerchi* is found in abundance *S. vitreus* is also found in at least equal numbers. The two species occur on the same type of substratum (rock and shell) and often in close association. On the average, brood size is greater in *S. vitreus* than in *S. moerchi*. It may be that larval mortality is greater in *S. vitreus* during settling because of the absence of a primary shell gland; however, this possibility would have to be tested by quantitative studies under conditions simulating those of the natural habitat, using larvae of both species.

Since neither the thorax nor the abdominal region shows any sign of asymmetry during the first day of post-larval development in *S. moerchi*, how then can the spiral determination of the tube by the end of 24 hours of settlement be explained? The answer to this question obviously must reside in the presence of muscle fibers in the opercular peduncle and their absence from the opposite side. The presence of muscle fibers in the peduncle effectively introduces an element of asymmetry and their contraction would cause the anterior end of the body to bend to the right in a dextral species such as *S. spirillum* (operculum on right) and to the left in a sinistral species such as *S. moerchi* (operculum on left). This would account for the fact that the umbilicus of the tube is always located on the same side as the operculum.

Although settlement in *Spirorbis* is a rather dramatic event, metamorphosis of the essentially bilateral larva to an asymmetrical pre-adult is a surprisingly slow process. It has been shown in the present study that the key to understanding the asymmetry of the adult resides in changes occurring in the larval abdomen as it transforms into the adult achaetous zone. These changes seem to occur primarily as a result of differential growth so that the larval abdomen grows faster on the convex than the concave side and thus causes the developing achaetous zone to turn either to the right or to the left, depending on the species. During this differential growth, the achaetous zone also twists along its longitudinal axis to either the right or left and causes the right and left dorsal and ventral longitudinal muscles of the thorax to meet and coalesce as a single longitudinal muscle bundle on the concave side of the achaetous zone. The thoracic muscles penetrating the opercular peduncle by the end of 24 hours of settling may, in addition to causing the initial turn of the primary tube, play a role in shaping the differential growth of the achaetous zone. Growth of the thoracic longitudinal muscle bundles has not been detected during the first 11 days following settlement; consequently, the thoracic muscle bundles of the side opposite the operculum are believed to be displaced passively to the concave side of the achaetous zone.

The concept of heteronomy, as it pertains to polychaetes, has been reviewed recently by Schroeder and Hermans.[11] This paper presents evidence to refute Meyer's[26] report that the mouth (stomodaeum) in serpulids is surrounded by a pair of chaetal-bearing segments. Not only are there no chaetal sacs surrounding the stomodaeum during the metamorphosis of *Spirorbis,* but the only nervous tissues in this region are the circumesophageal connectives. The peristomium which bears the collar is obviously pre-segmental in *Spirorbis.*

SUMMARY

1. Upon hatching, *Spirorbis* larvae exhibit perfect bilateral symmetry, except for the presence of a large opercular rudiment situated on either the right or left side, depending on the species involved.
2. Settling in *S. moerchi* results in the explosive release of a calcium-containing secretion from the hind-gut which results in the formation of the primary tube. *Spirorbis spirillum* and *S. vitreus* lack an abdominal primary shell gland, and initial tube formation appears to be due to mucous secretions issuing from the ventral thoracic glands.
3. Metamorphosis involves little loss of larval tissue. The ventral prostomial mucous glands and trochoblasts pinch off and are shed. It is certain that *S. moerchi* post-settled larvae do not ingest prototrochal cells. Larval eyes are lost in *S. moerchi* but are retained by *S. spirillum* and appear in the adult.
4. Initial development of asymmetry following settlement results from longitudinal muscle bundles growing into the opercular peduncle. Contraction of the latter causes the definitive tube to be secreted either to the right or left, depending on the species.
5. The major metamorphic change in *Spirorbis* is the asymmetrical development of the larval abdomen as it transforms via differential growth into the non-segmental achaetous zone of the adult. Except for the thoracic muscles inserting into the opercular peduncle, the thoracic muscles on the opposite side probably play little role in this process.
6. Feeding in *S. moerchi* does not start until between the sixth and eleventh day following settlement.
7. The peristomium in *Spirorbis* is clearly pre-segmental.

ACKNOWLEDGMENTS

I wish to thank Professor Robert L. Fernald for introducing me to *Spirorbis* and for his persistent encouragement and friendship over the past many years.

REFERENCES

1. Knight-Jones, E. W. (1951) J. mar. biol. Ass. U.K., 30, 201-222.
2. Knight-Jones, E. W. (1953) J. mar. biol. Ass. U.K., 32, 337-345.
3. Gross, J. and Knight-Jones, E. W. (1957) Ann. Rep. Challenger Soc., 3, 18.
4. Wisely, B. (1960) Aust. J. Mar. Freshw. Res., 11, 55-72.
5. de Silva, P.H.D.H. (1962) J. Exp. Biol., 39, 483-490.
6. Gee, J. M. and Knight-Jones, E. W. (1962) J. mar. biol. Ass. U.K., 42, 641-654.
7. Gee., J. M. (1965) Anim. Behav., 13, 181-186.
8. de Silva, P.H.D.H. and Knight-Jones, E. W. (1962) J. mar. biol. Ass. U.K., 42, 601-608.
9. Williams, G. B. (1964) J. mar. biol. Ass. U.K., 44, 397-414.
10. Wilson, D. P. (1968) in Perspectives in Marine Biology, Buzzati-Traverso, A. A., ed., Univ. Calif. Press, Berkeley, California, pp. 87-103.
11. Schroeder, P. C. and Hermans, C. O. (1975) in Reproduction of Marine Invertebrates, vol. III, Giese, A. C. and Pearse, J. S. eds., Academic Press, New York, pp. 1-213.
12. Nott, J. A. (1973) J. mar. biol. Ass. U.K., 53, 437-453.
13. Nott, J. A. and Parkes, K. R. (1975) J. mar. biol. Ass. U.K., 55, 911-923.
14. Potswald, H. E. (1965) Ph.D. Thesis, Univ. of Wash., Seattle, pp. 1-330.
15. Pillai, T. G. (1970) Ceylon J. Sci., Biol. Sci., 8, 100-172.
16. Cloney, R. A. (1961) Amer. Zool., 1, 67-87.
17. Bennett, H. S. and Luft, J. H. (1959) J. Biophys. Biochem. Cytol., 6, 113-117.
18. Luft, J. H. (1961) J. Biophys. Biochem. Cytol., 9, 409-414.
19. Richardson, K. C., Jarett, L. and Finke, E. H.(1960) Stain Technol., 35, 313-323.
20. Höglund, L. B. (1951) Zool. Bidr. Uppsala, 29, 261-276.
21. Carr, L. B., Rambo, O. N. and Feichtmeir, T. V. (1961) J. Histochem. Cytochem., 9, 415-417.
22. Swan, E. F. (1950) J. Morph., 86, 285-314.
23. Hedley, R. H. (1956) Quart. J. Microscop. Sci., 97, 411-419.
24. Hedley, R. H. (1956) Quart. J. Microscop. Sci., 97, 421-427.
25. Quiévreux, C. (1962) Cahiers Biol. Marine, 3, 1-12.
26. Meyer, E. (1888) Mitt. Zool. Stat. Neapel, 8, 462-662.

METAMORPHOSIS AND SETTLEMENT IN THE SABELLARIIDAE

Kevin J. Eckelbarger*

Harbor Branch Foundation, Inc., RR 1, Box 196, Fort Pierce, Florida 33450

Settlement and metamorphosis are reviewed in the polychaete family Sabellariidae with emphasis on some ultrastructural features of the premetamorphosed larva of *Phragmatopoma lapidosa*, an intertidal reef-building species. A possible sensory structure is described from *P. lapidosa* larvae which might play a role in substrate selection. Among the factors influencing settlement and metamorphosis in sabellariid larvae are type of substrate, degree of tidal exposure, tolerance of wave energy, phototaxis, presence or absence of gregariousness in settling, and ability of larvae to detect the tube cement of their own species. The larvae of most intertidal reef-building species are induced to settle and metamorphose upon contact with the adult or larval tube cement of their own species.

INTRODUCTION

The family Sabellariidae is a relatively small group of tube-building, marine polychaetes with a cosmopolitan distribution from the intertidal to abyssal depths. Their pelagic larvae are seasonally prominent members of the nearshore plankton in many parts of the world[1-4] and have representatives in the teleplanic larval community in the open ocean.[5] Sabellariid polychaetes have played a key role in demonstrating substrate selectivity by marine invertebrate larvae through the more recent studies of Wilson[6-9] on *Sabellaria alveolata* and *S. spinulosa*.

The purpose of this paper is to review settlement and metamorphosis in the family Sabellariidae and to present new findings regarding larval metamorphosis in the reef-building species, *Phragmatopoma lapidosa*.

MATERIALS AND METHODS

Larval stages of *Phragmatopoma lapidosa* were obtained from artificially fertilized eggs using the procedures of Eckelbarger[10] and from plankton samples kindly provided in part by Dr. M. J. Youngbluth and Ms. P. I. Blades of the Harbor Branch Foundation, Inc. Specimens for scanning (SEM) and transmission electron microscopy (TEM) were prepared according to the procedures of Eckelbarger and Chia.[11,12]

SETTLING BEHAVIOR

The larval development of eight species of sabellariids has been described in the literature—three species from Europe: *Sabellaria alveolata* (Linne),[1,6,8,13] *Sabellaria spinulosa* Leukart[1,9] and *Lygdamis muratus* (Allen)[1,14-16]; three species from the east coast of North America: *Sabellaria vulgaris* Verrill,[10,17,18] *Sabellaria floridensis* Hartman,[19] *Phragmatopoma lapidosa* Kinberg[4,20]; a single species from California: *Phragmatopoma californica* (Fewkes)[19,21,22] and a single species from the Indian Ocean: *Lygdamis indicus* Kinberg.[23] Many of these reports provide some data on larval settlement behavior.

Figure 1 shows a portion of an intertidal reef on the east coast of Florida (St. Lucie County) built by *Phragmatopoma lapidosa* to illustrate the massive nature of sabellariid reefs.

Sabellariid larvae generally respond positively to a weak light source. *Sabellaria floridensis* larvae, however, seem to be neutral to photo stimuli and swim randomly in larval culture during the entire developmental period.[19] Eckelbarger[10] reported that most *Sabellaria vulgaris* larvae

---

*Contribution no. 82, Harbor Branch Foundation, Inc.

Fig. 1. Portion of *Phragmatopoma lapidosa* reef in St. Lucie County, Florida.

in laboratory cultures tended to aggregate at the water line nearest the window illumination whereas a smaller number of larvae characteristically aggregated furthest from the light source. As development proceeded and larvae approached settlement, the photopositive larvae migrated towards the bottom of the culture. The remaining larvae, presumably not as far along in development, behaved similarly by aggregating at the water line and repeating the same migratory behavior as they approached settlement. Wilson[14] reported a similar massing of *Lygdamis muratus* larvae on the light and dark sides of his laboratory cultures.

As sabellariid larvae approach the stage shown in Figures 2A, 14, and 20, they tend to swim near the bottom of the culture and periodically cease swimming and crawl over the substratum. The larva crawls on its ventral side with its mouth applied to the substratum by pushing back the building organ to allow the ciliated surface of the lips of the mouth to be applied to the bottom. The larval tentacles are turned anteriorly so that their ventral ciliated food grooves contact the substratum. The tentacles alternately touch the substrate and are lifted up in a manner suggesting feeling or testing. During crawling behavior, larvae tenaciously cling to the substrate and considerable effort is required to dislodge them with squirts of water from a fine pipette. After a few seconds or minutes of crawling, the larvae abruptly resume swimming.

Phragmatopoma lapidosa larvae approaching settlement frequently have been observed swimming nearly upside down, stopping to apply their anterior region (episphere or "hood," as defined by Wilson[1]) gently to the substratum and then swimming on (Eckelbarger, unpublished observations). Occasionally a number of larvae display this behavior when they encounter a metamorphosed larva, appearing to take special interest in its mucoid tube. Although this behavior has been observed on numerous occasions in *P. lapidosa* larvae, it has not been reported for other sabellariid species.

Fig 2. A. Dorsal view of late larval stage of *Sabellaria floridensis* just prior to metamorphosis. B. Primary opercular palea from setal bundle. C. Larval nuchal spines from different individuals (modified from Eckelbarger[19]).

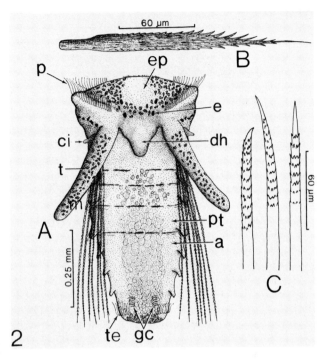

The larvae of *Sabellaria floridensis*[19] and *Lygdamis muratus*,[14] two non-rock-adherent species from Florida and the waters of the English Channel, respectively, behave differently from other sabellariids. In each species, the larvae crawl over solid surfaces less actively than other sabellariids and tend to burrow through the substratum, frequently burying themselves under sand grains and pebbles. Generally, they do not turn their tentacles forward to test the substrate, but use them frequently when on their sides or back.

LARVAL MORPHOLOGY

Before a discussion of sabellariid metamorphosis can be presented, it is necessary to describe the general external features of the larva just prior to this event. With the exception of species within the genus *Lygdamis*,[14,16,23] the larvae of *Sabellaria floridensis* and *Phragmatopoma lapidosa* are typical of most sabellariid larvae at settlement and will be used to illustrate general morphological features.

Figure 2A shows a late stage of a *Sabellaria floridensis* larvae with four crescent-shaped red eyespots, a pair of posteriorly projecting tentacles about one-half the body length, a well-developed prototroch and telotroch, two bundles of barbed, provisional setae, a dorsal hump posterior to the eyespots, three parathoracic segments with dorsal parapodial lobes each bearing four capillary setae and three abdominal segments with dorsal uncinigerous lobes. Yellow-green chromatophores and black pigment bands are present dorsally on the posterior borders of the parathoracic segments. Hidden among the provisional setae are broad, spiny primary paleae (Fig. 2B) and a pair of small nuchal spines (Fig. 2C). The latter figure shows variation in spine morphology from different individuals. In a study of the anterior region of *Phragmatopoma californica*, Dales[22] concluded that the larval tentacles were probably prostomial in origin because they are innervated from the brain mass in the adult. He added that the pigment patches which appear on the dorsal surface of the tentacles are additional evidence for a prostomial origin because such pigment is characteristically found on the asegmental regions of the larval body (i.e., pygidium and episphere).

Grasping Cilia

One peculiar feature common to sabellariid larvae is the presence of groups of tightly clus - tered cilia forming two dorsal longitudinal rows over the pygidium (Fig. 2A). The cilia are continuous with the telotrochal cilia, but when viewed by light microscopy, they appear to be fused and undergo flicking or vibratory movements as a unit rather than beating motions typical of the other telotrochal cilia. Wilson[1] referred to these cilia as "grasping cilia" and attributed their function to curling around and grasping the provisional setae when the larvae are swimming. Dales[22] referred to these cilia in the larvae of *Phragmatopoma californica* and attributed the same function to them. Cazaux[13] disputed this interpretation, but Wilson[6] reconfirmed his observations and the present author concurs with Wilson after extensive observations of the larvae of *Phragmatopoma lapidosa*. Figure 3 shows an SEM view of the grasping cilia which illustrates the close association of the cilia. The tips of the cilia in this figure and a TEM view of a longitudinal section through the cilia (Fig. 4) indicate that the cilia are not fused. Figure 4 reveals, however, that the basal bodies of the cilia appear to be fused, perhaps in order to lend greater support during the forceful flicking motions or to provide coordination of action.

Prototroch and Telotroch Cells

Figure 5 is a cross section through the prototroch cells illustrating the long, primary rootlets that penetrate deeply into the cell and appear to anchor to the nuclear membrane. Prototroch cells are large, cuboidal cells with spherical basal nuclei and numerous spherical or oblong mitochondria. Cell microvilli differ from those of other epidermal cells by their beaded appearance (Fig. 6) and appear to be less branching than those of adjacent cells. Telotroch cells are similar with long primary rootlets (Fig. 7) and branching secondary rootlets (Fig. 8). The cells are attached at their apicolateral margins by a junctional complex consisting of maculae adherens and septate desmosomes (Fig. 9).

Epidermal Gland Cells

At least two distinct types of gland cells, designated Types I and II, are present in the epidermis of premetamorphosed *Phragmatopoma lapidosa* larvae. Either of these cells may play a role in the secretion of the larval mucoid tube during or after metamorphosis. Each type is based solely upon the ultrastructure of its respective secretion droplets; no cytochemical tests were performed to characterizie or differentiate them.

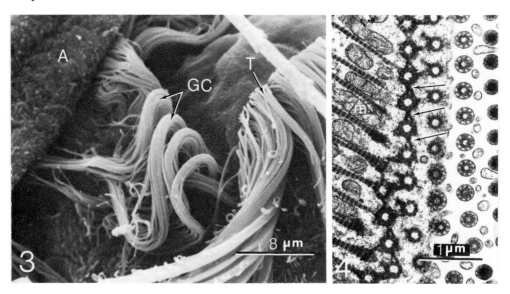

Fig. 3. SEM view of "grasping cilia" from pygidium of late stage larvae.
Fig. 4. TEM section through the base of the "grasping cilia" showing apparent fusion of basal bodies (arrows).

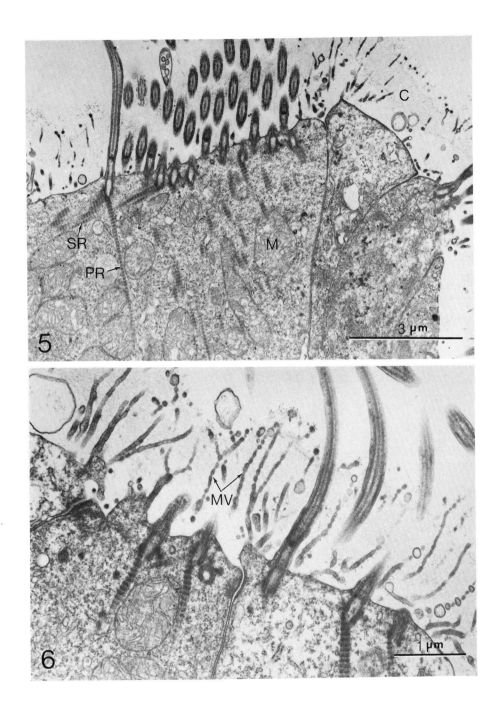

Fig. 5. Section through prototroch cell of *P. lapidosa* showing ciliary rootlets.

Fig. 6. Beaded microvilli of prototroch cell cuticle. Note unbeaded microvilli of adjacent epidermal cell to left.

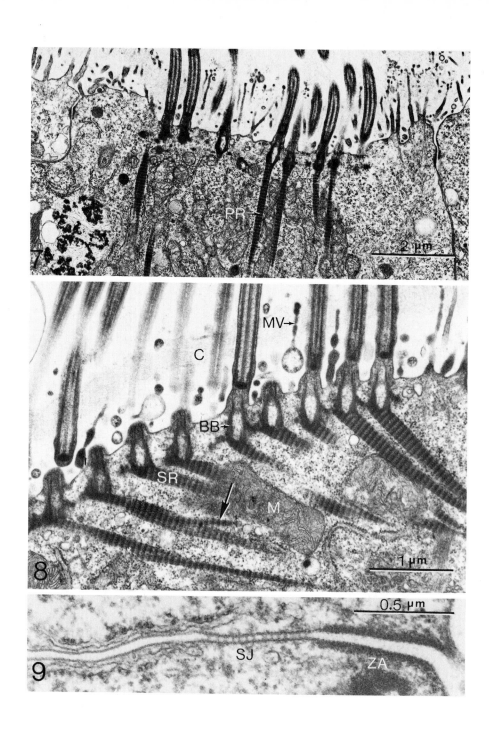

Fig. 7. TEM section through primary rootlets of telotrochal cilia.

Fig. 8. Section through secondary rootlets of telotrochal cilia. Note branching of rootlet at arrow.

Fig. 9. Junctional complex at apicolateral margins of the telotrochal cells.

Type I gland cells are limited to the ventral portion of the first parathoracic segment, and they are smaller and less numerous than Type II cells, and contain secretion products composed of whorls of moderately electron-dense fibrillar material (Fig. 10). The gland pores are recessed beneath the cuticle and are surrounded by short, nonbranching microvilli. An SEM view of the ventral portion of the parathoracic segments reveals a number of pores which are believed to represent the pores of Type I gland cells (Fig. 11).

Type II gland cells are limited to the pygidial region surrounding the anus and the episphere or hood. The cells are large and contain somewhat spherical secretions consisting of fibrillar strands aligned in randomly arranged parallel arrays (Fig. 12). The apex of the cells is elevated above the level of neighboring epidermal cells and the gland pore is wide and usually observed to be releasing the secretory product. SEM views of the surface of the episphere of the larva reveal these raised gland pores (Fig. 13).

Sensory Structures

In his detailed light microscopic investigation of larval development in *Sabellaria alveolata,* Wilson[1] referred to the presence of "sensory cilia" on a number of locations, including the cirri and tentacles, on the surface of the premetamorphosed larvae. More recently, scanning electron microscopy was utilized to examine the surface of *Phragmatopoma lapidosa* larvae nearing settlement in order to determine whether sensory-like structures existed which were not easily discernible with the light microscope.[11] Figure 14 shows an SEM photograph of a larva just prior to settlement and Figure 15 illustrates the dorsal surface of one larval tentacle having a number of sensory-like tufts composed of a variable number of circularly arranged, stiff, radiating cilia. A transverse section through a tuft reveals a single, small supportive cell from which the stiff cilia project through the larval cuticle (Fig. 16). The cells contain a basally situated nucleus, large oval or irregularly shaped mitochondria, slightly electron-dense vesicles and thick, blunt, branching and nonbranching microvilli with bundles of microfilaments (stereocilia) which project well into the cytoplasm of the cell. These microvilli differ significantly from the thin, highly branching microvilli of regular epidermal cells. The modified cilia are characterized by their stiff posture and blunt tips (Fig. 17), which contrast with the curved, pointed cilia observed in other epidermal cells. Occasionally microtubules are observed in the cells. Synaptic contacts with sensory nerve fibers have been observed at their base. Figure 18, based on a compilation of scanning electron micrographs taken from approximately 70 larvae, represents the distribution of presumed "sensory tufts" over the surface of a *Phragmatopoma lapidosa* larva in the swimming-crawling phase just prior to settlement. The greatest concentration of "sensory tufts" per unit area occurs on the dorsal surface of the tentacles where they usually form two or three rows. The dorsal hump, a raised area between the tentacle bases, is the only other region on the dorsal surface where tufts were observed. These structures are conspicuously absent from the dorsal regions of the thoracic and abdominal segments. On the episphere or anterior surface of the hood, a small number of tufts are distributed in a horseshoe-shaped pattern along with a number of gland pores. Ventrally, "sensory tufts" are scattered from the neurotroch, laterally to the neuropodia on the thoracic segments, over the surface of the building organ and the lips of the mouth. Tufts are restricted to the ventral lateral regions of the last one or two abdominal segments. The pygidium possesses a large concentration of tufts scattered around the anus in association with gland pores and in the dorsal gap of the telotroch.

It is noteworthy that in the juvenile worm, ciliary tufts are not observed on the dorsal surface of the numerous feeding tentacles formed after metamorphosis, but they are still present on the original pair of larval tentacles which are retained after metamorphosis. Dales[22] pointed out that the latter tentacles probably have a sensory function in most sabellariids except in the genus *Phalacrostemma* where they function in feeding.

151

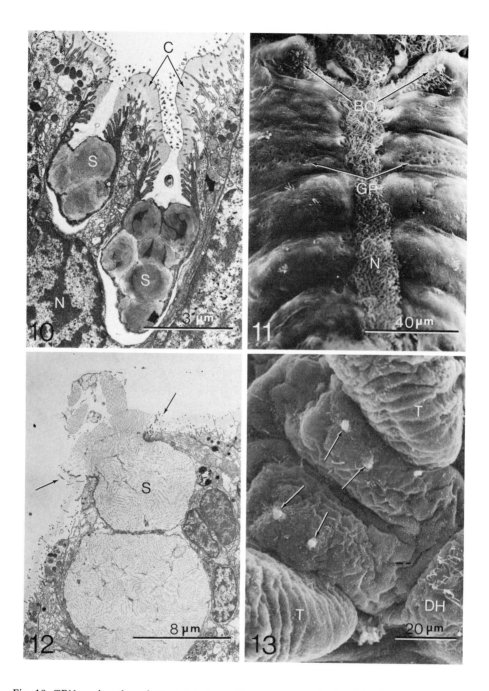

Fig. 10. TEM section through Type I ventral epidermal mucous glands of the first parathoracic segment of settling *P. lapidosa* larva. Note recessed gland pores.

Fig. 11. SEM view of ventral parathoracic segments of larva showing gland pores.

Fig. 12. TEM view of Type II epidermal gland cell from episphere of settling larva. Arrows indicate the elevation of the cuticle around the gland pore neck.

Fig. 13. SEM view of episphere of larva showing secretion droplets emerging from Type II gland pores (arrows).

Fig. 14. SEM view of right lateral side of *P. lapidosa* larva just prior to settlement.

Fig. 15. SEM view of sensory-like tufts over the dorsal surface of tentacles of larva in Fig. 14.

Fig. 16. TEM section through "sensory tuft" of *P. lapidosa* larva.

Fig. 17. Longitudinal section through cilium of "sensory tuft" showing blunt tip.

(Figs. 14 and 15 from Eckelbarger and Chia.[11] Reproduced by permission of the National Research Council of Canada from the *Canadian Journal of Zoology*, Volume 54, pp. 2082-2088, 1976.)

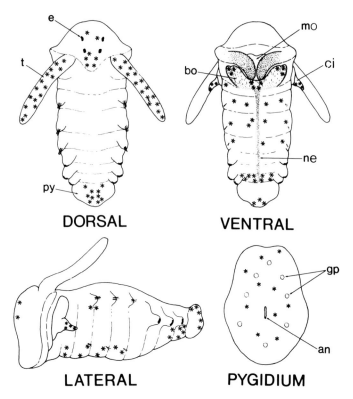

Fig. 18. Diagrammatic views of the pygidium and dorsal, ventral and lateral surfaces of a late stage *P. lapidosa* larva showing distribution of "sensory tufts " (∗).

It is unclear whether the tufts described from the larvae of *Phragmatopoma lapidosa* correspond to secondary sensory cell mechanoreceptors or chemoreceptors as defined by Welsch and Storch.[24] The cilia are closely grouped together and are modified from those of other ciliated cells. The microvilli are likewise modified, but do not encircle the cilia as in some mechanoreceptors.

Wilson[6] clearly established that *Sabellaria alveolata* larvae in the premetamorphosed searching phase can detect the adult tube cement or larval mucoid tubes of other sabellariids and that the attracting factor is not detected from a distance. During the searching phase, the larvae contact the substrate with their ventral surface, mouth region and tentacles. *Phragmatopoma lapidosa* larvae appear to be unique by "testing" the substratum with their head region as well. The distribution of tufts over these surfaces in *P. lapidosa* larvae suggests that they serve a sensory function during substratum selection although fine structural evidence alone is not conclusive. The function of the pygidial tufts is unclear as larvae do not appear to contact the substratum with this surface during the searching phase.

The tufts described here from *Phragmatopoma larvae* superficially resemble epidermal "diffuse sense organs" of adult *Nereis virens* described by Langdon[25] to consist of one or more distal processes penetrating the cuticle from basally located bipolar nerve cells. These structures were particularly numerous around the mouth and over the distal surfaces of cephalic appendages of *Nereis* such as the polyps, cirri and tentacles.

Wilson[6] compared the substrate selection behavior of *Sabellaria alveolata* to that of the barnacle cypris larva. Crisp and Meadows[26,27] reported that physical contact of the cyprid with the substrate containing the metamorphosis-inducing substance, to which they respond, was essential. Knight-Jones[28] had earlier concluded from studies of cyprid behavior that contact of the cyprid with settled barnacles or their bases was essential and that the perceived inducing substance was not water soluble. He concluded that the gregarious settling response in barnacles was probably due to contact with quinone-tanned proteins forming the epicuticle of settled animals. Crisp and Meadows[26] suggested that the cyprid is able to recognize a specific chemical structure or molecular configuration of substances attached to solid surfaces without the need for aqueous diffusion. This ability was referred to as a "tactile chemical sense."[27] Wilson[7] suggested that a similar mechanism is utilized by *S. alveolata* larvae during settlement, although the specific substance detected by sabellariid and cyprid larvae differs in that it is destroyed by cold concentrated HCl in the former case but not in the latter.

Crisp[29] questioned whether the surface would not be occluded by bacterial or other films which grow on it if larvae depend on close contact with the substratum and the settlement-inducing substance on its surface. Wilson[8] reported, however, that laboratory experiments on *Sabellaria alveolata* settlement behavior indicated that aged tube cement was less effective than fresh cement in inducing settlement, suggesting that indeed microbial growths or other surface films might prevent larvae from detecting these substances.

METAMORPHOSIS

The duration of larval development in sabellariids varies widely according to the literature (see Table 1): from 14-30 days for *Phragmatopoma lapidosa*[4] to over 32 weeks for *Sabellaria alveolata*.[8] It is always somewhat difficult to use laboratory development times to predict developmental rates in nature owing to the artificial conditions imposed on the organism in the laboratory setting. A thorough investigation of laboratory development times combined with field studies of spawning and settling rates is perhaps the most reasonable way to determine the true development time for a given species. After extensive laboratory and field observations over many years, Wilson[30] suggested that the normal development time for *S. alveolata* in nature (Duckpool, North Cornwall) probably ranged from 6 to 24 weeks, with the peak probably falling between 8 and 12 weeks.

The change in larval behavior from swimming to crawling is accompanied by a series of morphological changes which constitute metamorphosis. Metamorphosis in sabellariids follows a typical, distinctive pattern with only minor variations among species. Figures 19 through 22 photographically document metamorphosis in living *Phragmatopoma lapidosa*, giving a more life-like impression of the process. For comparative purposes, Figure 23 shows a juvenile worm two weeks after metamorphosis. Figures 24 through 26 represent similar stages in the larval metamorphosis of *Sabellaria floridensis* in order to more graphically demonstrate some of the detailed morphological changes which occur. Figure 27 represents a juvenile worm approximately the same age as the *P. lapidosa* juvenile shown in Figure 23.

Metamorphosis begins with a gradual shrinkage of the episphere, loss of the prototroch and enlargement of the building organ. The tentacles rotate anteriorly and become contractile (Fig. 20). The provisional barbed setae are lost (Fig. 21) and reveal a number of pairs of primary paleae projecting from the setal sacs (more easily observed in Fig. 24). The two setal sacs with their accompanying paleae and cirri, now collectively referred to as the opercular peduncles, rotate until the paleae project anteriorly (Fig. 25). The entire head region has now shrunk and 2 of the 4 red eyespots have migrated closer together. The entire larval body lengthens and the telotrochal swelling shrinks with the loss of the telotroch (Figs. 22 and 26).

At this stage, the larva initially constructs a mucoid tube which is attached to a stable substrate. Sand grains are seized by the tentacles, conveyed to the mouth by the ventral ciliated food grooves, and cemented to the outside of the mucoid tube by secretions from the building organ. The metamorphosed larvae of *Sabellaria floridensis*[19] and *Lygdamis muratus*[14] are unusual among sabellariids in failing to attach themselves to the substrate. Rather, both encase

155

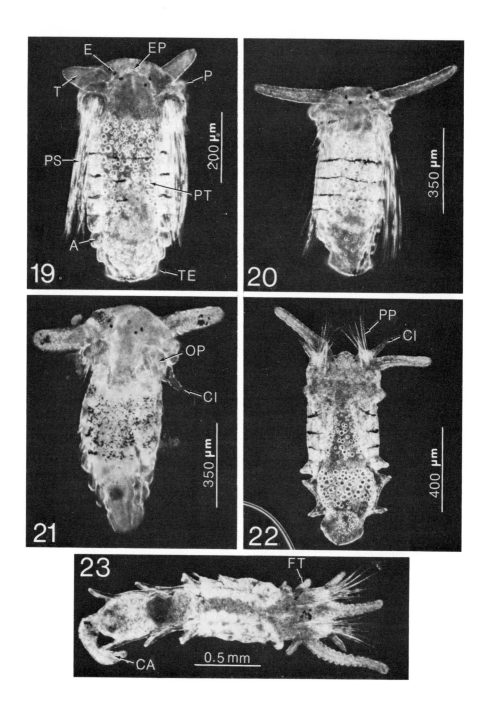

Figs. 19-23. Stages in the metamorphosis of living *Phragmatopoma lapidosa* larvae. 19. Larva about a week before metamorphosis; tentacles turned anteriorly during narcotization and do not represent the normal condition at this stage. 20. Larva in the crawling stage with tentacles rotated anteriorly. 21. Larva has lost provisional setae and telotroch. 22. Larva has rotated opercular peduncles with primary paleae and opercular papillae; episphere has shrunk, eyespots have migrated together, prototroch has been lost and body has elongated; larva has secreted a mucoid tube. 23. Juvenile worm removed from tube two weeks after metamorphosis; note appearance of new feeding tentacles and cauda.

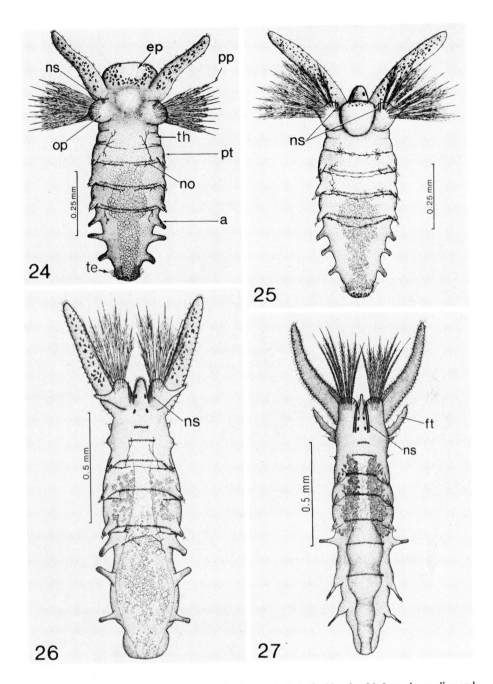

Figs. 24-27. Stages in the larval metamorphosis of *Sabellaria floridensis*. 24. Larva is crawling and has rotated tentacles anteriorly and lost provisional setae. 25. Opercular peduncles bearing primary paleae have rotated anteriorly and episphere has shrunk. 26. Completely metamorphosed larva in tube. 27. Juvenile worm two weeks after metamorphosis (modified from Eckelbarger[19]).

TABLE 1

Sabellariid Ecology and Larval Behavior

Species	Development[a] Time	Temp. (°C)	Adult Habitat	Tube-Building Habits	Type of Settler	Metamorphosis Induced by Tube Cement	References
Sabellaria alveolata[b]	6-32½ weeks	±15	Intertidal[d]	Reef-building	Gregarious	Yes	1,6,8,13
Phragmatopoma lapidosa[c]	14-30 days	21-23	Intertidal[d]	Reef-building	Gregarious	Unknown (suspected)	4,20
Phragmatopoma californica[b]	18-25 days	21-23	Intertidal[d]	Reef-building	Gregarious	Unknown (suspected)	12,21
Sabellaria spinulosa[b]	5½-12 weeks	±15	Intertidal & subtidal[d]	Reef-building/small aggregations	Gregarious	Yes	1,9
Sabellaria vulgaris[b]	19-30 days	21-23	Intertidal & subtidal	Colonial/small aggregations	Gregarious	Unknown	10,17,18,31
Sabellaria floridensis[b]	18-27 days	21-23	Intertidal & subtidal	Small aggre-gations/solitary	Nongregarious	No	19
Lygdamis muratus[b]	about 6 weeks	15-20	Subtidal	Solitary	Nongregarious	No	14-16

[a]Period from fertilization to settlement and metamorphosis.

[b]Warm-temperate species.

[c]Tropical species.

[d]Prefer high wave energy for tube-building.

themselves in a covering of agglutinated sand grains to form a tube which frequently is free on the bottom. Young *L. muratus* worms, unlike most sabellariids, can leave their tubes and build others. Newly metamorphosed individuals of *S. floridensis,* can crawl around in culture dishes on their tentacles dragging the tubes with them.

Among the last events in metamorphosis is the growth of the pygidium into a long, achaetous appendage which reflexes ventrally to form the cauda characteristic of the adult worm (Fig. 23). Whereas metamorphosis up to the formation of the cauda has been estimated to occur within 24 hours (*Lygdamis muratus*[14]) to two days or longer (*Sabellaria alveolata*[1]), formation of the cauda can take a week or more (*Sabellaria alveolata*[1]; *S. vulgaris*[10]).

During early larval development in sabellariids, the parathoracic segments (Figs. 2 and 24) are the first to develop, with the abdominal segments behind this region becoming recognizable with increasing age. The thoracic segments, which develop anterior to the parathoracic segments, are peculiar in that they are not clearly defined until metamorphosis.

An additional number of minor external morphological changes are part of metamorphosis in each species of sabellariid, but will not be detailed here. The reader should refer to Eckelbarger[19] for a list of references.

With the exception of the larvae of the genus *Lygdamis*,[14,16,23] late stage larvae within the genera *Sabellaria* and *Phragmatopoma* appear markedly alike to even the trained eye. As a result, prominent chitinous structures of systematic importance such as the larval primary opercular paleae and opercular spines are of considerable importance to zooplanktologists wishing to identify sabellariid larvae to species level. Larval opercular paleae are formed in the late stages of development and are carried in bundles in each opercular peduncle, but are hidden from view by numerous barbed, provisional setae. They are homologous to the adult opercular paleae and following metamorphosis are periodically lost and replaced by new, more adult-like paleae during the succeeding juvenile stages. Figure 28 shows the variation in those larval opercular paleae that have been described from developmental studies of sabellariids.

Larval opercular spines are smaller, less numerous structures which are also found hidden within the provisional bundles in all sabellariids described except *Sabellaria floridensis*.[19] Opercular spines are formed in the late larval stages and lost at metamorphosis or soon after. It is not clear what, if any, adult structures are homologous to the spines. Figure 29 shows the variation in larval opercular spine morphology.

Although not strictly a part of metamorphosis, the development of the larval cuticle is a dynamic process spanning the entire period of larval development and continuing after metamorphosis. Eckelbarger and Chia[12] studied the morphogenesis of the larval cuticle in *Phragmatopoma lapidosa* from egg envelope formation to larval metamorphosis and determined that the egg envelope is retained as the larval cuticle through the trochophore stage and that it is then gradually lost and replaced by a new cuticle. Figure 30 diagrammatically summarizes the ultrastructural events occurring during the ontogenetic development of the egg envelope and larval cuticle in *Phragmatopoma*. Figures 31 through 34 show the variation in the morphology of the cuticle on the ventral surface of the tentacles, the ventral parathoracic segments, the dorsal parathoracic segments and the episphere of a late stage *P. lapidosa* larva (Fig. 20). Figure 35 shows the cuticle of a juvenile worm approximately 2 weeks after settlement (similar to Fig. 23).

FACTORS INFLUENCING SETTLEMENT AND METAMORPHOSIS

D. P. Wilson's pioneering studies on larval settlement behavior in the sabellariids *Sabellaria alveolata*,[6,8] *S. spinulosa*,[9] and *Lygdamis muratus*,[14] have provided most of our information on settlement behavior in this group. Arduous observations on the larvae of the reef-building species, *Sabellaria alveolata*,[6] demonstrated that purely physical factors have only

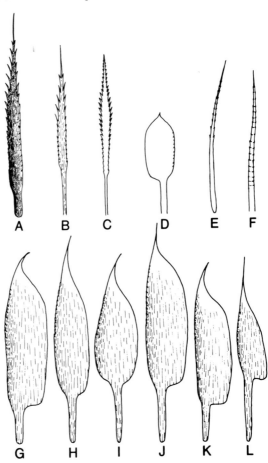

Fig. 28. Primary opercular paleae from late stage sabellariid larvae: A, *Sabellaria floridensis*[28]; B, *Sabellaria vulgaris*[19]; C, *Sabellaria spinulosa*[14]; D, *Sabellaria alveolata*[14]; E, *Lygdamis indicus*[32]; F, *Lygdamis muratus*[25]; G-I, *Phragmatopoma lapidosa*[17]; J-L, *Phragmatopoma californica*.[28]

159

Fig. 29. Larval opercular spines from late stage sabellariid larvae: A, *Sabellaria vulgaris*[19]; B, *Sabellaria spinulosa*[14]; C, *Sabellaria alveolata*[14]; D, *Lygdamis indicus*[32]; E, *Lygdamis muratus*[25]; F, *Phragmatopoma lapidosa*[17]; G, *Phragmatopoma californica*.[28]

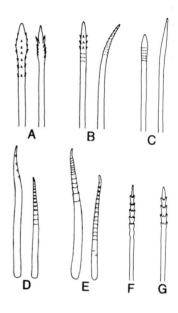

a minor influence on settlement. Such physical factors as clean, stable substrates washed over by sea water carrying suspended sand grains and a slight preference for shallow cracks or corners had some effect on settlement behavior. Wilson could find little preference for rough or smooth surfaces, little influence of surface bacterial slime films as a settlement attractant. In the laboratory, *Sabellaria alveolata* larvae are scarcely influenced by the mineralogical nature of adult tube walls, settling readily on tubes of their own species built of mineral grains, whereas in nature the adults build primarily with shell fragments if abundant.

Biochemical factors were found to have the most powerful influence on larval settlement patterns in *Sabellaria alveolata* larvae.[6] Some of the strongest stimuli to settlement and metamorphosis resulted from accidental contact with conspecific adult tubes, tube remnants, or with the tubes of recently settled young, whether simple mucoid tubes or newly constructed sandy tubes. Wilson found the specific factor triggering the settlement response to be the tube cement secreted by young or adult worms. Contact with the cement by larvae was essential because no evidence was found to suggest that it can be detected from a distance. Furthermore, the presence of newly metamorphosed worms on or in an attractive material makes the material even more attractive.

Wilson determined that the metamorphosis-inducing component of the cement is insoluble in water and unaffected by drying. The inducing material was destroyed by cold concentrated HCl without destroying the entire tube wall.

The settling larvae of *Sabellaria alveolata* are able to distinguish between the natural tubes of their own species and those of the sympatric species, *S. spinulosa*.[8] However, in laboratory tanks when adult worms of both species built tubes from the same shore sand, most *S. alveolata* chose *S. spinulosa* tubes in preference to their own, a phenomenon that could not be explained.

Wilson's[8] experiments on *Sabellaria spinulosa*, a non-colonial subtidal species in the Plymouth area, demonstrated that the larvae were strongly stimulated to metamorphose by the tube cement secretions of their own species and rarely failed to distinguish such secretions from those of *S. alveolata*. *S. spinulosa* larvae were not misled into showing preference for tube material from *S. alveolata*, and *S. alveolata* tubes were only slightly more effective in stimulating *S. spinulosa* larvae to settle and metamorphose than was sand from the seashore. Scallop shells, in particular, were shown to have some slight settlement-inducing properties especially when covered with silt and sand grains.

In larval development studies of the intertidal and subtidal non-reef building sabellariids *Sabellaria vulgaris*,[10] *S. floridensis*[19] and the intertidal reef-building species *Phragmatopoma lapidosa*[4] and *P. californica*,[19] no specific laboratory substrate experiments were undertaken. However, all these species will settle and metamorphose in the presence of sand from tubes of any of the other species.

Most sabellariid larvae are markedly gregarious in their settlement patterns and generally settle on top of or alongside the tubes of previously settled worms. However, the subtidal, non-reef building species, *Sabellaria floridensis*[19] and *Lygdamis muratus*[14] are exceptions. The

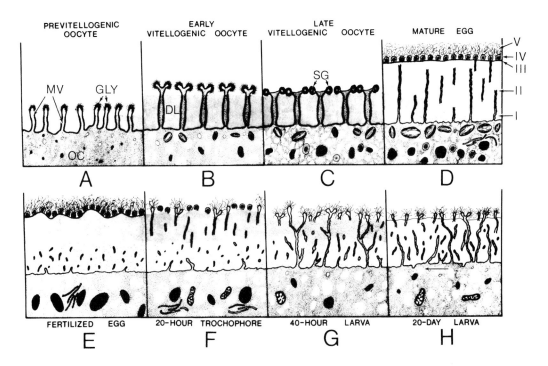

Fig. 30. A-H. Diagrammatic summary of the development of the egg envelope and its gradual replacement by the larval cuticle in *Phragmatopoma lapidosa*. A-C. Formation of surface granules by oocyte microvilli. D. mature egg showing 5 zones of the egg envelope including an outer jelly coat (Zone V), a granular layer (Zone IV), and underlying, thin, electron-dense layer (Zone III) and two inner layers (Zone I and II) of varying thickness and electron density. E. Withdrawal of microvilli into Zone I at fertilization and isolation of outer granular layer. F. Loss of jelly coat (Zone V) and Zone III and partial loss of surface granules (Zone IV); note penetration of new microvilli to surface of 20-hour trochophore cuticle. G. Complete loss of surface granules, formation of highly branching microvilli and partial loss of Zone II in 40-hour larva. H. Ventral cuticle of larva just prior to metamorphosis showing complete absence of original layers of egg envelope. Reproduced by permission of Springer-Verlag from *Cell and Tissue Research*, Volume 186, pp. 187-201, 1978.

larvae of both species crawl over and through the substratum for some period and finally build solitary tubes. It is interesting to note that although the larvae of *Sabellaria spinulosa*[9] and *S. vulgaris*[10] are gregarious settlers under laboratory conditions, they are not colonial reef-builders in nature except at some localities in the former species.[9] Substrate experiments with *L. muratus* larvae[14] demonstrated that they are not, unlike *S. alveolata* larvae, influenced by the cement of newly metamorphosed or adult worms. The larvae are capable of choosing between different kinds of deposits, favoring sandy sediments containing pebbles and mud and preferring non-sterile to sterile deposits.

Delay of metamorphosis for periods of weeks in the absence of a suitable substrate has been demonstrated in the laboratory for *Sabellaria alveolata*[6] and *Lygdamis muratus*.[14] Larvae remain in the swimming-crawling phase during this period and although some may eventually metamorphose in the absence of normal environmental stimuli, others die or metamorphose abnormally. The larvae of *Sabellaria alveolata*[8] and *S. vulgaris*[10] also survive periods of starvation by ceasing growth and resuming development when food is again available.

Agitation or circulation of larval cultures was noted by Wilson[6] to shorten larval settlement times. Eckelbarger[4,10] demonstrated that settlement times for the larvae of *Sabellaria vulgaris* and *Phragmatopoma lapidosa* were doubled in unstirred cultures, compared to stirred cultures. The stimulating effect of agitation on larval settlement probably is reflected by the tendency of

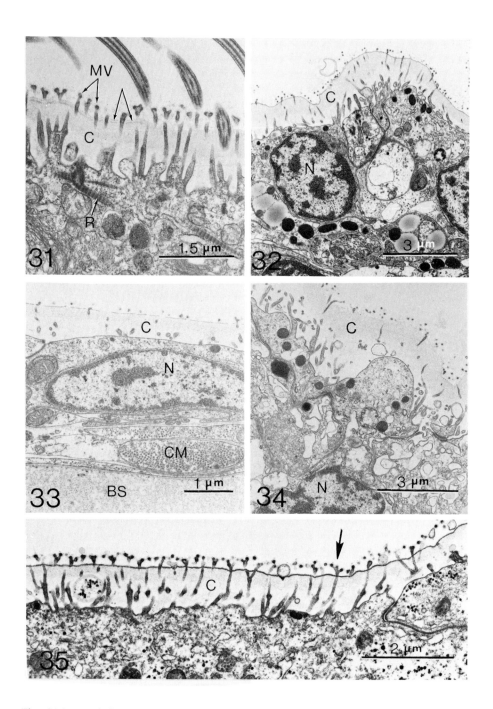

Figs. 31-35. Variation in cuticle morphology from various regions of late stage *Phragmatopoma lapidosa* larva. 31. Ventral ciliated cuticle of larval tentacle showing closely spaced, straight microvilli with bifurcated tips with glycocalyx and two-layered, faintly electron-dense outer region (arrows). 32. Ventral cuticle of parathoracic segment with branching microvilli and thin, single-layered electron-dense outer region. 33. Thin dorsal cuticle from parathoracic segment with sparse, branching microvilli and faint, electron-dense outer layer. 34. Cuticle of episphere showing irregular thickness, highly branching microvilli, noticeable absence of outer electron-dense boundary and highly vacuolated underlying epidermal cells. 35. Dorsal cuticle of juvenile worm 2 weeks after metamorphosis showing closely spaced microvilli with branching tips (arrow) and distinct, electron-dense outer boundary.

these species to select habitats in nature where some degree of wave action is present to provide tube-building materials. Agitation might serve as an additional cue to larvae in selecting an optimum habitat, particularly in intertidal reef-building species.

Additional factors apparently play a role in larval settlement in nature. Wilson[6] includes availability of clean, firmly anchored substrates continually washed with suspended sand grains. Depth and substrate preference and gregarious settlement behavior could also potentially determine the distribution of populations. Curtis[31] studied the intertidal, vertical distribution of *Sabellaria vulgaris* in Delaware Bay and found optimum colony growth was influenced by length of exposure time at low tides and availability of tube-building materials as reflected by wave action. He was not certain whether exposure affected the larval or adult stage but concluded it likely affected both. Colony formation is also limited by the strength of the wave action with such reef-building species as *Sabellaria alveolata* and *Phragmatopoma lapidosa* building strong colonies on the open coast where wave action is intense while *S. vulgaris* builds more fragile colonies that cannot withstand high wave energy.[19,31] All of the factors described above including larval substrate and tidal preferences, phototaxis, presence or absence of the ability of larvae to detect the cement of their own species, tolerance of wave action and presence or absence of gregariousness, undoubtedly contribute to ecological and reproductive isolation of natural populations.

ABBREVIATIONS

A, abdomen; AN, anus; BB, basal body; BO, building organ; C, cuticle; CA,cauda; CI, opercular cirrus; CM, circular muscle; DH, dorsal hump; E, eyespot; EP, episphere; FT, feeding tentacle; GC, grasping cilia; GP, gland pores; GLY, glycocalyx; M, mitochondrion; MO, mouth; MV, microvilli; N, nucleus; NE, neurotroch; NO, nototroch; NS, nuchal spines; OC,oocyte; OPP, opercular peduncle; P, prototroch; PP, primary paleae; PR, primary rootlet; PS, provisional setae; PT, parathoracic segment; PY, pygidium; R, rootlet; S, secretion droplet; SC, sensory cell; SG, surface granule; SR, secondary rootlet; ST, "sensory tuft"; SJ, septate junction; T, tentacle; TE, telotroch; TH, thoracic segment; ZA, zonula adherens.

ACKNOWLEDGMENTS

The author wishes to thank Mrs. P. A. Linley for her invaluable assistance in larval culturing and specimen preparation and particularly to thank Dr. Douglas P. Wilson who through mutual correspondence over the last five years, stimulated much of my interest in sabellariid biology. Special thanks to typist Mrs. J. McKay.

REFERENCES

1. Wilson, D. P. (1929) J. mar. biol. Ass. U.K., 15, 221-269.
2. Bhaud, M. (1972) Mar. Biol., 17, 115-136.
3. Guerin, J. P. (1972) Tethys, 4, 859-880.
4. Eckelbarger, K. J. (1976) Bull. Mar. Sci., 26, 117-132.
5. Scheltema, R. S. (1971) in Fourth European Marine Biology Symposium, Crisp, D. J. ed., Cambridge University Press, Cambridge, pp. 7-28.
6. Wilson, D. P. (1968) J. mar. biol. Ass. U.K., 48, 387-435.
7. Wilson, D. P. (1968) J. mar. biol. Ass. U.K., 48, 367-386.
8. Wilson, D. P. (1970) J. mar. biol. Ass. U.K., 50, 1-31.
9. Wilson, D. P. (1970) J. mar. biol. Ass. U.K., 50, 33-52.
10. Eckelbarger, K. J. (1975) Mar. Biol., 30, 137-149.
11. Eckelbarger, K. J. and Chia, F. S. (1976) Can. J. Zool., 54, 2082-2088.
12. Eckelbarger, K. J. and Chia, F. S. (1978) Cell and Tissue Res., 186, 187-201.
13. Cazaux, C. (1964) Bull. Inst. Oceanogr. Monaco, 62, 1-15.
14. Wilson, D. P. (1977) J. mar. biol. Ass. U.K., 57, 761-792.
15. Bhaud, M. (1969) Vie et milieu, ser. A., 20, 543-557.

16. Bhaud, M. (1975) Annales de l'Institut oceanogr., 51, 155-172.
17. Novikoff, A. B. (1957) in Methods for Obtaining and Handling Marine Eggs and Embryos, Costello, D. P. et al. eds., Marine Biological Laboratory, Woods Hole, Mass., pp. 93-97.
18. Curtis, L. (1973) Aspects of the Life Cycle of *Sabellaria vulgaris* Verrill (Polychaeta: Sabellariidae) in Delaware Bay, Doctoral Dissertation, Univ. of Delaware.
19. Eckelbarger, K. J. (1977) Bull. Mar. Sci., 27, 241-255.
20. Mauro, N. A. (1975) Bull. Mar. Sci., 25, 387-392.
21. Hartman, O. (1944) Allan Hancock Pacif. Exped., 10, 311-389.
22. Dales, R. P. (1952) Q.J. Microsco. Sci., 93, 435-452.
23. Bhaud, M. (1975) Cah. O.R.S.T.O.M., Ser. Oceanogr., 13, 69-77.
24. Welsch, U. and Storch, V. (1976) Comparative Animal Cytology and Histology, Univ. of Washington Press, Seattle, pp. 1-343.
25. Langdon, F. E. (1900) J. Comp. Neurol. 10, 1-78.
26. Crisp, D. J. and Meadows, P. S. (1962) Proc. Roy. Soc., B156, 500-520.
27. Crisp, D. J. and Meadows, P. S. (1963) Proc. Roy. Soc., B158, 364-387.
28. Knight-Jones, E. W. (1953) J. Exp. Biol., 30, 584-598.
29. Crisp, D. J. (1974) in Chemoreception in Marine Organisms, Grant, P. T. and Mackie, A. M. eds., Academic Press, New York, p. 177-265.
30. Wilson, D. P. (1971) J. mar. biol. Ass. U.K., 51, 509-580.
31. Curtis, L. A. (1975) Chesap. Sci., 16, 14-19.

METAMORPHOSIS IN MARINE MOLLUSCAN LARVAE: AN ANALYSIS OF STIMULUS AND RESPONSE

Michael G. Hadfield

University of Hawaii, Pacific Biomedical Research Center, Kewalo Marine Laboratory,
41 Ahui Street, Honolulu, Hawaii 96813

Molluscan metamorphosis is compared with that of Anura to expose similarities or differences in the mechanisms which control metamorphic events and thus to clarify the events in molluscs. One partially purified molluscan metamorphosis-stimulating substance resembles that of anurans (thyroxine) in molecular size and potency. The active substance of molluscan biology is, however, produced in the environment and not by an endocrine gland. Molluscs and anurans are similar in developmental-age specificity of the ability (competence) of larval tissues to respond to the metamorphic stimuli. The mode of action of the metamorphic inducers appears very different in the two groups of organisms. Thyroxine acts to stimulate new macromolecular syntheses; the rapid larval response to molluscan inducers resembles that of organisms to neurostimulants. Good comparisons can be made of the morphological responses in the two groups: larval organs are rapidly lost and adult structures emerge. New data are presented on artificial inducers of metamorphosis in the nudibranch gastropod *Phestilla sibogae*.

INTRODUCTION

Settlement and metamorphosis of marine invertebrate larvae have been thoroughly and excellently reviewed several times in recent years.[1,2,3,4] The viewpoint of the reviewers, like that of many of the researchers, has been to approach the problem of site-specific larval settling as a means of understanding the spatial distribution of animals in nature. Although much valuable embryological information has been gained, there has yet to appear a truly developmental analysis of the interactions which occur when larvae choose, or are induced, to settle in a particular spot. In this paper I will attempt such an analysis of the circumstances of molluscan metamorphosis.

Two well-known metamorphic sequences are those of insects and amphibians.[5] Both are under hormonal control, and the events preceding, during, and following the hormonally mediated metamorphic events have been described in elegant detail. Insects and amphibians differ mainly in the manner in which metamorphosis is stimulated, or released, by hormones. In amphibians, thyroxine progressively stimulates development of adult structures and culminates, through massive secretion, in inducing resorption of specific larval structures. In insects the process is in many ways reversed. Insect juvenile hormone serves to retain the larval anatomy, but simultaneously "allows" preformation of much adult structure in the imaginal discs. Thus juvenile hormone has an effect opposite to that of thyroxine: it stimulates larval formation and inhibits complete expression of adult structures. Metamorphosis occurs when secretion of juvenile hormone fails; the adult expression is effectively released. There is increasing evidence that prolactin, present in developing tadpoles, has a role analogous to that of juvenile hormone.[5]

The molluscan larval form is not, so far as we know, maintained through hormonal action. Rather, it exists until triggered by certain environmental stimuli to perish. When larval organs are lost, the adult form emerges. Thus, of the two discussed above, the better model for molluscan metamorphosis appears to be the frog. In the discussion that follows, an analogy will be drawn between thyroxine as the anuran metamorphic stimulator and the environmental cues which trigger molluscan metamorphosis. Hopefully, a better understanding of molluscan developmental biology will emerge.

Metamorphosis has yet to be thoroughly analyzed in any mollusc. Studies on morphological events are plentiful, but tend to be separate from those on stimuli. For only two species have fairly extensive data been gathered on both aspects, the oyster *Ostrea edulis*[6-9] and the aeolid nudibranch *Phestilla sibogae*.[10-15] *Phestilla* is used as the major molluscan example in the discussion below both because of my greater familiarity with the species and because metamorphosis is a more drastic and pervasive event in nudibranchs than it is in bivalves and thus exposes more details for analysis.

Metamorphosis, a highly accelerated series of the kinds of events which occur throughout the development of all animals, can be divided—albeit artificially—into several processes. In the discussion below, the following phenomena are examined: (1) the nature of the metamorphic stimulus (the "inducer"), (2) tissue competence to undergo metamorphic changes, (3) the primary mode of action of the stimulator, and (4) the morphological metamorphic responses. In each instance, an evaluation of whether or not molluscs agree with the anuran model is made.

Amphibian metamorphosis has been thoroughly described in many modern texts[5] and need not be presented again in detail. Analysis here will briefly outline events under the categories listed above.

METAMORPHIC STIMULUS

The Model

The action of the thyroid hormone thyroxine (tetraiodothyronine) in stimulating amphibian metamorphosis is so well known that it is often invoked as a model of hormone action. Thyroxine is synthesized in the thyroid gland from tyrosine and is the major iodine-requiring compound in vertebrate biology. The potency of thyroxine is great; some larval tissues respond to it in concentrations as low as one part per billion.

Thyroxine is a relatively small molecule (MW- 777) whose activity can be mimicked by its precursor, triiodothyronine, and other artificially synthesized compounds.[16] Hormonal events controlling synthesis of thyroxine are well known, but the ultimate control of the system, probably environmental, is yet uncertain.

The Mollusca

The nature of the stimulator of metamorphosis varies almost per molluscan species. For many bivalves, and probably many members of other groups as well, only a non-toxic surface is necessary to stimulate the behavioral settling events that probably trigger metamorphosis. Many other molluscs will not metamorphose in any numbers unless the available surface is at least roughened or otherwise textured. In the most interesting instances, specific biological-chemical stimuli are required. Species whose metamorphoses are surface triggered are doubtless the most numerous among molluscs. However, owing to lack of good experimental data they are not exclusively reviewed here.

Table 1 lists molluscan species for which there are some experimental data to establish a stimulatory effect on metamorphosis by another organism or (presumably) biogenic substance. In most of these instances, the metamorphic trigger arises from a more or less specific plant or animal on which adults of the metamorphosing species feed. In the best studied instances, a particular compound or series of compounds is implicated in the induction. In oysters, for instance, these are large proteinaceous substances, the specificity for which is not high.

In many, but not all, nudibranchs the presence of living adult prey is an absolute prerequisite for metamorphosis (see Table 1). In *Phestilla sibogae* the prey is the coral *Porites compressa*, although at least one other species, *P. lobata*, will also serve. The specificity of this coral has been experimentally established[15] and a water soluble coral product which induces metamorphosis in larvae of *P. sibogae* has been partially purified. The inducer has a molecular weight

TABLE 1

Specific Settlement of Molluscan Larvae on Plants, Animals and Extracts

Molluscan Species	Settling Substrate	Extracts	Source
Gastropoda			
Prosobranchia			
Nassarius obsoletus	Mud from adult habitat	+	34
Philippia radiata	*Porites lobata* (Cnidaria)		35
Opisthobranchia			
Nudibranchia			
Adalaria proxima	*Electra pilosa* (Bryozoa)		35
Doridella obscura	*Electra crustulenta* (Bryozoa)		27
Eubranchus exiguus	*Kirchenpauaria pinnata*	+	37
	echinulata (Cnidaria)		
Phestilla sibogae	*Porites compressa* (Cnidaria)	+	10,15
Rostanga pulchra	*Ophlitaspongia pennata* (Porifera)		24
Trinchesia aurantia	*Tubularia indivisa* (Cnidaria)		38
Tritonia hombergi	*Alcyonium digitatum* (Cnidaria)		39
Sacoglossa			
Elysia chlorotica	Primary film of microorganisms		19
	from adult habitat		
Cephalaspidea			
Haminoea solitaria	Primary film of microorganisms		19
	from adult habitat		
Anaspidea			
Aplysia californica	*Laurencia pacifica* (alga: Rhodophyta)		22
A. brasiliana	*Callithamnion*(?) *halliae* (alga:		25
	Rhodophyta)		
A. dactylomela	*Laurencia* sp. (alga: Rhodophyta)		23
A. juliana	*Ulva* spp. (alga: Chlorophyta)		23
A. parvula	*Chondrococcus hornemanni*		40
	(alga: Rhodophyta)		
Dolabella auricularia	Unidentified blue-green algae		23
	(Cyanophyta)		
Stylocheilus longicauda	*Lyngbya majuscula*		23
	(alga: Cyanophyta)		
Amphineura			
Tonicella lineata	*Lithophyllum* sp. & *Lithothamnion* sp.	+	41
	(alga: Rhodophyta)		
Lamellibranchia			
Teredo navalis	Pine wood; dissolved humic substances		18,28
Teredo sp.	Wood		42,43
Bankia gouldi	Wood		28
Mercenaria mercenaria	Clam liquor; sand	+	44
Placopecten magellanicus	Adult shell, sand, etc.		31
Mytilus edulis	Filamentous algae; other, non-biological		45
	silk material		
Crassostrea virginica	Shell liquor; body extract;	+	46-48
	"shellfish glycogen"		
Ostrea edulis	Muscle extract and extrapallial	+	49-51
	fluid of adult		

167

below 500 and is relatively stable over a wide pH range and at temperatures up to 100°C. Chemists at the University of Hawaii have succeeded in purifying the inducer molecule in quantities of about 5 mg per 600 grams of crude, lyophilized distilled water extract of coral. Since maximally effective concentrations of the crude inducer are in the range of 0.1%,[15] the inducer itself must be active in concentrations of a few parts per billion. Its potency is thus equal to that of thyroxine.

Crisp[2] suggests that all invertebrate settling substances are detected as adsorbed layers. Evidence for at least the larvae of *Phestilla sibogae*[15] runs counter to this notion. Larvae of *P. sibogae* exposed to dissolved coral extract while in swirling suspension for as little as six hours will begin metamorphosis hours later after having been passed through numerous changes of fresh filtered sea water.[17]

There are known artificial inducers of molluscan metamorphosis. Culliney[18] has published very interesting observations on the capacity of "dissolved humic substances" to stimulate inappropriate settling in shipworm larvae. He suggests that natural occurrences of such false signals may influence the distribution of teredinid species in estuaries.

Bonar,[13] while attempting to paralyze the muscular system of metamorphosing *Phestilla sibogae* larvae, found that succinyl choline chloride actually stimulated metamorphosis, even in the absence of natural inducer. Harrigan and Alkon[19] later found that the compound had a similar influence on larvae of the sacoglossan *Elysia chlorotica*. We have conducted a lengthy series of experiments in an attempt to analyze the role of succinyl choline in opisthobranch metamorphic induction. The results are presented in Table 2.

Three major characteristics of the inductive capacity of succinyl choline are now apparent. First, it is not the entire molecule which is critical, but simply the choline moiety. In fact, choline chloride alone is as effective as succinyl choline in inducing metamorphosis in *Phestilla* larvae. Secondly, the activity pattern of choline-induced metamorphosis is quite distinct from that of natural inducer in its extended latency. Third, the maximual effective concentrations of choline are many thousandfold greater than those of the natural inducer.

TABLE 2

Compounds Tested for Artificial Metamorphic Induction Activity

Compound	Concentration Tested—% Solutions	% Larvae Metamorphosing	Time To Response In Days
Natural inducer control	(0.1)	70-90	< 1
Succinyl choline Cl	0.1	70-90	3-4
Succinyl choline Cl	0.05	10	3-4
Choline Cl	0.1	60-85	3-4
Choline Cl	0.05	25	3-4
Choline Cl	0.0125	0	—
Succinyl dicholine diiodide	0.1	20-60	4-5
Carbonyl choline Cl	0.1	30	3-4
Acetyl B-methyl choline Cl	0.1	10-40	< 4
Acetyl choline Cl[a]	0.1	0	—
Taurobetaine[a]	0.1, 0.5	0	—
Acetyl thiocholine Cl	0.1	0	—
Betaine HCL	0.1	0	—
Phosphoryl choline Cl	0.1	0	—
Succinic acid	10 ppm	0	—
Succinyl dichloride	10 ppm	0	—

[a] Toxic after 1-2 days as tested.

At present we cannot explain the choline-induced metamorphosis in molluscs. The known biological roles of this molecule present many possibilities. The latency suggests that it serves as a precursor in the synthesis of another compound.

TISSUE COMPETENCE

The Model

In anurans, different tissues become capable of responding to the metamorphic inducer, thyroxine, at different moments in the developmental time scale. Each cell or tissue type becomes more mature during development; complete maturation, in the present sense, is metamorphic competence. Preliminary evidence indicates that metamorphic competence in amphibian larval tissues is the acquisition of thyroxine binding sites in the cells.[20] Characteristic of different tadpole tissues, in addition to the timing of their metamorphic competence, is the concentration of thyroxine to which they respond. Thus hind-limb growth is stimulated by low levels of thyroxine and tail resorption by only very high levels. Increasing thyroxine concentrations in the blood stream of the tadpole thus produce an ordered sequence of metamorphic events, an important point of which is that requisite adult structures are functionally present *before* larval organs are precipitously lost.

The Mollusca

Except for those molluscs with so-called direct development, few if any produce larvae which are capable of metamorphosis immediately upon hatching. Most have, instead, some mandatory planktonic period during which all of the exhaustively discussed events and advantages of pelagic dispersal occur. Thorson[21] established tables comparing the pelagic periods of many invertebrate species. These tables, insofar as the data are valid, give estimates of the length of an important embryological stage, the precompetent period. Thus, although newly hatched larvae of most molluscs cannot undergo metamorphosis, at a later stage and under the right circumstances they can. Metamorphosis, except in holopelagic forms, is preceded or accompanied by settling from the pelagic realm. Settling is best seen as a behavioral—as distinct from a morphogenetic—act.

The duration of the minimal pelagic period is thus the precompetent developmental period and its end marks the onset of metamorphic competence. In most if not all forms, the competent stage is morphologically distinct. In larvae of *Phestilla sibogae* and other gastropods, a prerequisite of metamorphic competence is the appearance of the propodium[14] (Figs. 1 and 2). The same feature appears to characterize the metamorphosable stage, the pediveliger, of oysters and other bivalves.[6]

Competent molluscan larvae with their well-formed foot are analogous to prometamorphic anuran larvae. With intact velum, the veliger can, like the tailed tadpole, continue to swim. With the complete development of the juvenile foot, the molluscan veliger can crawl, a trait analogous to limb development in tadpoles. However, while there are morphological analogues between metamorphosing frogs and metaphorically competent molluscs, important differences remain. The prometamorphic tadpole (i.e., one which is developing limbs) is already responding to increasing levels of the metamorphic inducer thyroxine. Development of structural concomitants of competence in molluscan larvae is a typically ordered developmental event and is, so far as we know, under no hormonal influence.

The minimal time from fertilization or hatching to the development of competence is known mainly for molluscan species whose larvae are lecithotropic or only briefly planktotrophic. In a series of opisthobranchs with planktotrophic larvae, the precompetent period is frequently around one month,[22-26] though others have shorter feeding periods.[12,19,27] Many planktotrophic bivalve larvae, including shipworms,[28] mussels,[29-30] scallops,[31] and oysters[32] have

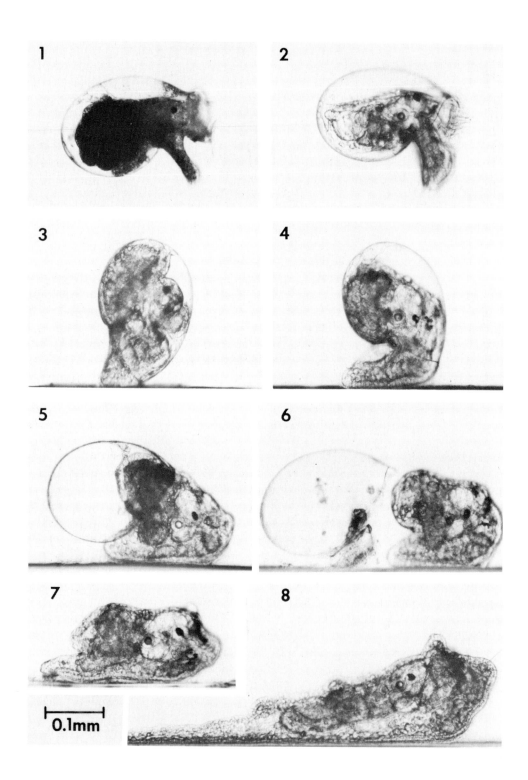

170

mandatory larval periods of 10 to 30 days. Only few data are available on the minimal period for development of metamorphic competence in most long-lived prosobranch larvae. Among sibling larvae of *Phestilla sibogae,* there is consistent variation in the age at which metamorphic competence is achieved.[15] At *ca.* 25°C the larvae hatch after eight days and by the end of the ninth day some are competent. However, maximal competence is usually not attained until they are 12 days old (*i.e.,* four days after hatching). The times from egg laying to hatching and from hatching to competence are extremely temperature sensitive in *Phestilla* and probably all other larvae.

Finally, important as are morphological criteria for competence, the true test is whether or not larvae can or will respond to appropriate environmental stimuli by undergoing metamorphosis. Anurans and molluscs, as exemplified by *P. sibogae,* are similar in this respect.

THE MECHANISMS OF METAMORPHIC INDUCTION

The Model

That the site of action of thyroxine in inducing amphibian metamorphosis lies deep within the cell is supported by autoradiographic evidence that it binds to chromatin in cells of responding tissues.[33] The latency of tissue response to thyroxine application suggests a rather lengthy response pathway, as confirmed by studies of the influence of inhibitors of macromolecular synthesis on tadpole metamorphosis.[5] In the tadpole liver, *de novo* protein synthesis, stimulated by thyroxine occurs some 100 hours after thyroxine application. Studies on RNA and protein synthesis suggest that, at least in the liver, thyroxine directly stimulates new ribosomal RNA synthesis and enhances the cytoplasmic events of protein synthesis. However, small amounts of mRNA are probably produced as well.[5] The appearance of particular new proteins defines liver events in metamorphosis. In excised tadpole tails, there is a 3 to 4 day latency between the application of thyroxine and the onset of involution.[16] Tail resorption is inhibited by actinomycin D, cycloheximide, and puromycin. Although the data for actinomycin D may be equivocal,[16] there are sufficient data to verify that new proteins are synthesized prior to tissue resorption.

Two other characteristics of thyroxine action are important: the effects are local and tissue specific, and intermediary compounds are not involved. Continued exposure to thyroxine is necessary for completion of metamorphosis. Tadpoles that are hypophysectomized midway through metamorphosis do not complete the process.[5]

The Mollusca

In larvae that respond to purely physical factors, the stimulus must be perceived strictly through surface mechanoreceptors. Vertical and horizontal surfaces may be distinguished by statocysts and lighted surfaces may be distinguished from shaded (underneath) ones by larval eyes. How such stimuli are translated into an internal trigger of rapid morphogenesis remains an enigma.

Figures 1-8 (*facing page*).

Fig. 1. Newly hatched, precompetent larva of *Phestilla sibogae*; larva is dense due to stored yolk.

Fig. 2. Competent larva of *P. sibogae*: note development of propodium.

Fig. 3. Metamorphosing larva of *P. sibogae* in position to pull loose the muscle-shell attachment.

Fig. 4. Metamorphosing larva of *P. sibogae*: body is loose in shell.

Fig. 5. Metamorphosing larva of *P. sibogae*: shedding shell and operculum.

Fig. 6. Juvenile of *P. sibogae* immediately after emergence from the shell.

Fig. 7. Juvenile of *P. sibogae* about 4 hours after leaving shell.

Fig. 8. Juvenile of *P. sibogae* 24 hours after the onset of metamorphosis.

(*All figures have the same magnification indicated by the scale at lower left.*)

171

Apparently only for *Phestilla sibogae* are there any analytical data on the mode of action of chemical/biological metamorphic stimuli, and here the findings are very preliminary.[17] As mentioned above, the inducer molecule is apparently perceived as a dissolved substance. Presumably the concentration of the substance is much greater around living corals in nature. As also stated above, successful exposure to the stimulus may be brief (2 to 6 hours). However, maximal response occurs after a group of larvae have been exposed to the inducer for about 18 hours.[15] Continuous exposure is not necessary and larvae exposed for 6 to 18 hours and then "washed" in many changes of sea water will undergo metamorphosis within the ensuing 24 hours.

In the rapidness of the response and the apparent triggering effect of the inducer (as opposed to the sustained maintenance requirement of thyroxine), the metamorphic stimulus in *P. sibogae* differs significantly from the thyroxine-induced metamorphosis of Anura. The rapid response suggests that the role of the inducer is not that of stimulating *de novo* macromolecular synthesis. Preliminary experimental data obtained in my laboratory indicate that metamorphic induction and response occur in the presence of relatively massive doses of actinomycin D, puromycin, and cycloheximide.*

Although analogy with hormonal metamorphic induction such as occurs in amphibians does not appear to be valid for molluscs, the comparison does tend to add clarity. The rapid and apparently triggering effect of molluscan inducers suggests a major role of the nervous system both in receiving the stimulus and transmitting its effect through the larval body. This hypothesis may subsequently be tested experimentally. It implies that the major event of metamorphic competence is the development of either specific receptors or pathways in the nervous system.

THE METAMORPHIC RESPONSE

The Model

Virtually every organ and tissue in a frog tadpole is altered at metamorphosis. Digestive enzyme systems are replaced, hemoglobin changes, visual pigments are altered, and new excretory products are produced.[5] The striking morphological changes, most important for our purposes, are the emergence of adult structures, most notably the limbs, and the loss of larval organs, most conspicuously the gills and tail.[5] In the developing frog, the limbs appear prior to tail and gill loss and other internal changes precede the final, irrevocable loss of the larval structures. Because of the drastic habitat and nutritional changes which accompany anuran metamorphosis, the physiological changes outlined above are requisite. All correlate with the switch from aquatic to terrestrial, and herbivorous to carnivorous life.

The Mollusca

Most molluscan veliger larvae respond to metamorphic stimuli by becoming benthic and losing the velum. The velum may be resorbed or it may fall apart with its component cells being lost or ingested. In many opisthobranch gastropods, the larval shell and operculum are cast as part of the metamorphic process, and in the remainder of the shelled molluscs there is apparently always some change in the pattern of shell growth (the chitons are probably an exception).

Larvae of *Phestilla sibogae* which have been stimulated to metamorphose settle to the bottom of a container and attach by the planar surface of the foot. They remain relatively motionless for a period of several hours during which the velum is lost. This occurs first by the detachment and ingestion of the larger ciliated cells and then more slowly by resorption of the remaining tissues.[14] Internally, the muscles and nerves which ran into the velar lobes degenerate *in situ*.

*Nine-day old larvae were simultaneously exposed to coral inducer and actinomycin D (5 and 10 ppm), puromycin (10 and 30 ppm), or cycloheximide (0.1, 1, 10, and 100 ppm), each with and without 1% DMSO added. Metamorphosis occurred in all cases in percentages equal to control groups.

The succeeding steps of metamorphosis, illustrated in Figures 3 through 8, are as follows. The larva attaches very securely to the substratum and raises the shell to a vertical position, pulling the anterior-dorsal beak of the shell against the substratum (Fig. 3). With this pivotal point established, the larval retractor muscle contracts very strongly until its posterior connection to the shell is ruptured. So violent is this action that cells actually seem to be torn free of the epidermis at this time (visible in Fig. 5).

The metamorphosing larva next relaxes from the vertical posture and proceeds to shed its shell and operculum (Figs. 4-6). This is apparently brought about by further contraction of the retractor muscle. The small, humped juvenile gradually becomes more compressed and elongate (Figs. 7 and 8). During the latter process, the shell epithelium is replaced by expanding epipoddial tissue.[14] Juveniles soon develop rhinophores and the buds of cerata (Fig. 8).

Details of metamorphosis of larvae of *P. sibogae* present many close analogies with processes in anurans. There is a precocious development of the foot (mollusc) and limbs (anuran). In metamorphic climax, major larval organs are precipitously discarded: the velum, shell and operculum of the veliger and tail, gills, and other larval organs of the tadpole. In both forms a swimming existence is replaced by a life on hard surfaces. Both have important changes in epithelial tissues. Finally, although not typical of the lecithotrophic larvae of *P. sibogae*, veligers of many gastropods are planktotrophic herbivores and the adults, like *Phestilla*, are carnivores. This switch, from larval herbivory to adult carnivory, is again similar to the changes which occur in anuran metamorphosis. Concomitant alterations in the digestive system, well described in frogs, have not, to my knowledge, been studied in Mollusca.

CONCLUSIONS

Comparison of molluscan metamorphosis with that of anurans is instructive both in the understanding it provides and in the research questions it defines. The metamorphic stimuli of molluscs, be they physical or chemical, do not appear to take good analogy with hormones, although final resolution of this question for all groups awaits chemical analysis of one or more inducers and further experimental examination of their mode of action. While these questions may soon be answered for the nudibranch *Phestilla sibogae*, the more telling results may come when we learn the chemistry of a second opisthobranch metamorphic inducer and can compare it with the first. A single class of compounds would suggest a closer analogy with hormonal activity; unrelated compounds will probably relegate the analogy to "smell." In potency, the *Phestilla* inducer is equal to that of thyroxine and other hormones.

Comparisons of the nature and temporal occurrence of metamorphic competence in tadpoles and veligers provide useful information in that they point to similar "developmental clock" events. Further experimentation on the timing and mechanisms of the acquisition of metamorphic competence in molluscs is in order. If the phenomenon in amphibians represents the establishment of hormone-binding sites, what is the comparable event in molluscs? In gastropod veligers it may simply be the morphological development of the juvenile foot, a requisite for benthic life. A better possibility, however, is that competence requires particular neural ontogeny; the evidence points to a neural role in the metamorphic response. Data on neurontogeny in gastropods are beginning to appear[52] and will add much to understanding of metamorphic events.

The morphogenetic patterns which typify molluscan metamorphoses, some of which are explored in more depth in other papers in this symposium, provide good analogy with anuran and other models. They appear no less complex and the precision of timing of particular events is equally critical. Examination of the anuran model suggests exciting research potential in studies of the gut and its enzymes in planktotrophic larvae and post-metamorphic molluscs, particularly the carnivorous gastropods. Development of laboratory culture techniques for a number of gastropod species[19,23,24,26,27] opens many possibilities for such studies.

It is obvious that data on site- or host-specific settling and other ecological aspects of marine invertebrate larval patterns have far outrun our understanding of embryological events. Nowhere is this more true than in molluscan studies. A new embryology, utilizing all of the modern analytical tools, is due for molluscan metamorphosis.

ACKNOWLEDGMENTS

Many of the ideas and hypotheses developed in this paper arose in discussions with pre- and postdoctoral colleagues in my laboratory. Comments of Drs. Marilyn Dunlap and Wm. Van Heukelem have been particularly stimulating. Data on choline analogues and inhibitors of macromolecular synthesis were gathered with the assistance of Kathy K. Milisen and Leo J. Hombach, S.J. The photographs in Figures 1-8 were taken by Wm. Van Heukelem. To all of the above-named scientists I extend my deepest appreciation. This work was partially supported by NIH Research Grant RR 01057.

REFERENCES

1. Meadows, P.S. and Campbell, J. I. (1972) Adv. mar. Biol., 10, 271-382.
2. Crisp, D.J. (1974) in Chemoreception in Marine Organisms, Grant, P.T. and Mackie, A.M. eds., Academic Press, London, pp. 177-265.
3. Scheltema, R.S. (1974) Thalassia Jugoslavica, 10, 263-296.
4. Crisp, D.J. (1977) in Adaptation to Environment. Essays on the Physiology of Marine Animals, Newell, R.C. ed., Butterworths, London, pp. 83-124.
5. Many texts present this information. The following two were useful in the preparation of this discussion: Graham, C.F. and Wareing, P.F. (1976) The Developmental Biology of Plants and Animals, W.B. Saunders Company, Philadelphia, 393 pp.; Grant, P. (1978) Biology of Developing Systems, Holt, Rinehart, and Winston, New York, 720 pp.
6. Hickman, R.W. and Gruffydd, L.D. (1971) in Fourth European Marine Biology Symposium, Crisp, D.J. ed., Cambridge, Univ. Press, Cambridge, pp. 281-294.
7. Cranfield, H.J. (1973a) Marine Biology, 22, 187-202.
8. Cranfield, H.J. (1973b) Marine Biology, 22, 203-209.
9. Cranfield, H.J. (1973c) Marine Biology, 22, 211-223.
10. Hadfield, M.G. and Karlson, R.H. (1969) Amer. Zool., 9, 1122.
11. Hadfield, M.G. (1972) Amer. Zool., 12, 721.
12. Harris, L.G. (1975) Biol. Bull., 149, 539-550.
13. Bonar, D.B. (1976) Amer. Zool., 16, 573-591.
14. Bonar, D.B. and Hadfield, M.G. (1974) J. exp. mar. Biol. Ecol., 6, 227-255.
15. Hadfield, M.G. (1977) in Marine Natural Products Chemistry, Faulkner, D.J. and Fenical, W.H. eds., Plenum Publishing Company, New York, pp. 403-413.
16. Weber, R. (1967) in The Biochemistry of Animal Development, Vol. II, Weber, R. ed., Academic Press, New York, pp. 227-301.
17. Unpublished observations made in the author's laboratory.
18. Culliney, J.L. (1973) in Proceedings of the Third International Congress on Marine Corrosion and Fouling, National Bureau of Standards, Gaithersburg, Maryland, Northwestern University Press, Evanston, Ill., pp. 822-829.
19. Harrigan, J.F. and Alkon, D.L. Veliger (in press).
20. Tata, J.R. (1971) Soc. Exp. Biol. Symp., 25, 163-181.
21. Thorson, G. (1950) Biol. Rev., 25, 1-45.
22. Kriegstein, A.R., Castelluci, V. and Kandell, E.R. (1974) Proc. Nat. Acad. Sci., 71, 3654-3658.
23. Switzer-Dunlap, M.F. and Hadfield, M.G. (1977) J. exp. mar. Biol. Ecol., 29, 245-261.
24. Chia, Fu-Shiang and Koss, R. Marine Biology (in press).
25. Strenth, N.E. and Blankenship, J.E. Veliger (in press).
26. Kempf, S.C. and Willows, A.O.D. (1977) J. exp. mar. Biol. Ecol., 30, 261-276.
27. Perron, F.E. and Turner, R.D. (1977) J. exp. mar. Biol. Ecol., 27, 171-185.
28. Culliney, J.L. (1975) Marine Biology, 29, 245-251.
29. Culliney, J.L. (1971) Bull. Mar. Science, 21, 591-602.
30. Costello, D.P. and Henley, C. (1971) Methods for Obtaining and Handling Marine Eggs and Embryos, Marine Biological Laboratory, Woods Hole, Mass., Second Edition, 247 pp.

31. Culliney, J.L. (1974) Biol. Bull., 147, 321-332.
32. Bardach, J.E., Ryther, J.H. and McLarney, W.O. (1972) Aquaculture, The Farming and Husbandry of Freshwater and Marine Organisms, Wiley-Interscience, New York, 868 pp.
33. Griswold, M.D., Fischer, M.S. and Cohen, P.P. (1972) Proc. Nat. Acad. Sci., 69, 1486-1489.
34. Scheltema, R.S. (1961) Biol. Bull., 120, 92-109.
35. Hadfield, M.G. (1976) Micronesica, 12, 133-148.
36. Thompson, T.E. (1958) Phil. Trans. Roy. Soc. Lond. B., 242, 1-58.
37. Tardy, J. (1962) C.r. hebd. Seanc. Acad. Sci., Paris, 254, 2242-2244.
38. Swennen, C. (1961) Neth. J. Sea Res., 1, 191-240.
39. Thompson, T.E. (1962) Phil. Trans. Roy. Soc. Lond. B., 245, 171-218.
40. Switzer-Dunlap, M.F. This volume.
41. Barnes, J.R. and Gonor, J.J. (1973) Marine Biology, 20, 259-264.
42. Isham, L.B. and Tierney, J.Q. (1953) Bull. Mar. Sci. Gulf Caribb., 1, 46-43.
43. Harington, C.F. (1921) Biochem, J. 15, 736-741.
44. Keck, R., Maurer, D. and Malouf, R. (1974) Proc. National Shellfisheries Assoc., 64, 59-67.
45. Bayne, B.L. (1965) Ophelia, 2, 1-47.
46. Crisp, D.J. (1967) J. Anim. Ecol., 48, 787-799.
47. Veitch, F.P. and Hidu, H. (1971) Chesapeake Science, 12, 173-178.
48. Keck, R., Maurer, D., Kauer, J.C. and Sheppard, W.A. (1971) Proc. National Shellfisheries Assoc., 61, 24-28.
49. Cole, H.A. and Knight-Jones, E.W. (1949) Fishery Invest. Lond., Ser. 2, 17, 39 pp.
50. Knight-Jones, E.W. (1952) Fishery Invest. Lond., Ser. 2, 18, 47 pp.
51. Bayne, B.L. (1969) J. mar. biol. Ass. U.K., 49, 327-356.
52. Kriegstein, A.R. (1977) Proc. Natl. Acad. Sci., 74, 375-378.

MORPHOGENESIS AT METAMORPHOSIS IN OPISTHOBRANCH MOLLUSCS

Dale B. Bonar

Department of Zoology, University of Maryland, College Park, Maryland 20742

The developmental patterns in the Opisthobranchia are briefly reviewed and a modification of Thompson's classification of development[5] is proposed which takes into account morphogenetic changes which occur at metamorphosis (=METAMORPHOGENESIS). The structure of a competent, pre-metamorphic veliger, relatively consistent for all opisthobranch groups, is presented. The subsequent metamorphogenetic changes are presented system by system (shell, velum, foot, mantle elements, digestive, circulatory, excretory, muscular, and nervous systems) and compared among different opisthobranch groups. The changes which occur at metamorphosis vary in degree depending on the adult morphology. The nudibranchs, which lose the shell, operculum, and mantle cavity, and become flattened and elongate, show the greatest degree of change and are discussed in the greatest detail, because more is known about the metamorphosis of this group than any other.

INTRODUCTION

The earliest detailed reports of opisthobranch metamorphosis are those of Schultze[1] and Nordmann,[2] who described external changes in two aeolid species. Since those reports, various aspects of the metamorphosis of more than 60 opisthobranch species representing all of the major opisthobranch groups have been described (Table 1). A substantial part of our understanding of opisthobranch metamorphosis has come from the work of Thompson,[3,4,5] who made extensive use of sectioned material to describe the morphogenetic events during the metamorphosis of several nudibranch species. Several other investigators, including Smith,[6] Tardy,[7] Thiriot-Quiévreux,[8] Bonar,[9,10,11] Bonar and Hadfield,[12] and Kriegstein[13,14] and his coworkers,[15] have subsequently reported histological changes in a variety of opisthobranch groups, including the Cephalaspidea, Anaspidea, Pteropoda, and Nudibranchia.

All of these reports, except for those of Thiriot-Quiévreux on the pteropod *Cymbulia* and Kriegstein on *Aplysia*, are on species which produce lecithotrophic larvae or have no free larval stage and pass through metamorphosis while still in the egg capsule. Until recently, the rearing of planktotrophic opisthobranch larvae has been a frustrating and usually unsuccessful venture, so that observations on the metamorphosis of species with planktotrophic development have been limited to those species for which mature, premetamorphic larvae could be collected from the plankton.[2,7,8,16,17-20] The major difficulties in rearing planktotrophic larvae have recently been overcome, however,[15,21,22] so that information on the metamorphosis of species exhibiting this type of development is rapidly advancing.[23,24,25] As can be seen in Table 1, more than one-half of the reports on opisthobranch metamorphosis are for nudibranch species. Except for the species studied by Smith,[6] Thiriot-Quiévreux,[8] and Kriegstein,[13,14] all of the most detailed studies of histological events occurring at metamorphosis are also on nudibranchs. The information now available allows us to make detailed developmental and phylogenetic comparisons of morphogenesis among the Cephalaspidea, the Anaspidea, the Nudibranchia, and to a lesser degree the thecosome Pteropoda, but does not allow us to make equally critical evaluations of the remaining groups. Now that planktotrophic larvae can be successfully reared through metamorphosis, it should be only a matter of time before we fill in the gaps in our knowledge for the other opisthobranch orders.

In this paper I will attempt to synthesize our current knowledge about morphogenesis occurring during opisthobranch metamorphosis. Since the information on metamorphic processes is

TABLE 1

Opisthobranch Metamorphosis: Developmental Patterns

Species	Development[a]	Ref.	Species	Development	Ref.
CEPHALASPIDEA			**NUDIBRANCHIA**		
Runcina setoensis	ameta	26	**DORIDACEA**		
Runcina coronata	ameta	27	*Cadlina laevis*	ameta	5
Acteocina senestra	ameta	28,29	*Casella obsoleta*	ameta	49
Retusa obtusa	cap	6	*Doridopsis limbata*	cap	40
Acteocina canaliculata	plank	50	*Chromodoris loringi*	cap	51
Philine aperta	plank	31	*Adalaria proxima*	leci	3
Philine denticulata	plank	32	*Discodoris erythroensis*	leci	52
			Trippa spongiosa	leci	53
PYRAMIDELLACEA			*Aegires punctilucens*	plank	16
Onchidella celtica	cap	33,34	*Doridella obscura*	plank	23
Brachystomia rissoides[b]	cap	35,36			
Brachystomia rissoides[b]	leci	37	**DENDRONOTACEA**		
			Tritonia hombergi	leci	4
ANASPIDEA (APLYSIACEA)			*Tritonia diomedea*	plank	54
Phyllaplysia taylori	cap	38	*Melibe* sp.	plank	54
Aplysia californica	plank	13,15			
Aplysia dactylomela	plank	22	**AEOLIDACEA**		
Aplysia juliana	plank	22	*Okadaia elegans*	ameta	55
Aplysia brasiliana	plank	25	*Coryphella stimpsoni*	ameta	56
Aplysia parvula	plank	25	*Aeolidiella alderi*	cap	7,17
Aplysia pulmonica	plank	22,25	*Cuthona pustulata*	cap	57
Dolabella auricularia	plank	25	*Aeolis* sp.	cap	58
Stylocheilus longicauda	plank	22,25	*Tenellia pallida*[b]	cap	64
			Trinchesia granosa	cap	59
NOTASPIDEA			*Amphorina doriae*	leci	17,19
Berthellina citrina	leci	39,40	*Cuthona adyarensis*	leci	60
			Cuthona nana[b]	leci	61
SACOGLOSSA (ASCOGLOSSA)			*Embletonia mediterranea*[c]	leci	62
Coenia cocksi	ameta	28,35	*Embletonia pallida*[c]	leci	37
Vayssieria caledonica	ameta	41	*Phestilla sibogae*	leci	10,12,63
Elysia timida	cap	42	*Tergipes lacinulatus*[c]	leci	1
Elysia chlorotica[b]	cap	43	*Tenellia pallida*[b]	leci	64
Elysia cauze[b]	cap	44	*Spurilla neapolitana*[b]	leci	83
Elysia cauze[b]	leci	44	*Spurilla neapolitana*[b]	plank	83
Tamanovalva limax	leci	45	*Aeolis* sp.	plank	19
Elysia cauze[b]	plank	44	*Cuthona nana*[b]	plank	61
Elysia chlorotica[b]	plank	24,46	*Facelina coronata*	plank	19
Alderia modesta	plank	47	*Phestilla melanobranchia*	plank	63
Limapontia capitata	plank	20	*Tergipes despectus*	plank	18
			Tergipes edwardsii	plank	2
PTEROPODA					
Cymbulia peroni	plank	8			
Limacina retroversa	plank	48			

[a]ameta = ametamorphic capsular embryogenesis; cap = metamorphic capsular veliger; leci = lecithotrophic veliger stage; plank = planktotrophic veliger stage.

[b]Species with more than one reported developmental pattern.

[c]Species synonymized as *Tenellia pallida*.[84]

best known for the Nudibranchia, and since most of my experience has been with nudibranch species, the bias in this paper will be toward a detailed evaluation of the changes in the various opisthobranch groups as they resemble or differ from the changes occurring in nudibranchs.

DEVELOPMENTAL PATTERNS

The function of metamorphosis in most opisthobranchs, as in other marine invertebrates which exhibit this phenomenon, is the conversion of an organism whose body plan is specialized for a pelagic larval existence to a form specialized for a benthic adult existence. The ecological demands of the adult existence typically involve mechanisms of feeding, locomotion, and predator avoidance different from those required in the larval phase, so that substantial morphological change is necessary in the transition between the two phases of the life cycle. This conversion involves both constructive and destructive elements, as adult morphology cannot be attained until the larval morphology is abolished. Obviously, the degree of morphogenetic change during metamorphosis will depend on the degree of difference between the larval and subsequent juvenile body plans. Among the more than 1100 species of opisthobranch molluscs can be found an extremely divergent array of adult body plans.[65] A comparison of the shelled Acteonidae (considered the most primitive opisthobranch group) and the unshelled Nudibranchia (the most advanced group) dramatically underlines the differences. While the acteonids have retained many of the "primitive" characters of the prosobranch ancestors (streptoneurous nervous system, retention of an external shell and operculum), the nudibranchs have lost many of the prosobranch features (shell, operculum, mantle cavity, primary gills), and for the most part have entirely reversed the effects of torsion. As a consequence of the widely divergent adult morphologies (more so than in any other molluscan group), the morphogenetic changes which occur at metamorphosis (hereafter termed METAMORPHOGENESIS) likewise vary considerably among species. In spite of these differences, it is possible to compare the metamorphic ontogeny of substantially dissimilar species, because the premetamorphic veligers of most opisthobranch species are morphologically quite similar. Any substantial variations in pre-metamorphic veliger morphology are usually the result of condensed, or reduced development, in which the larval phase has been completely suppressed and hatching occurs only after juvenile morphology has been attained. Those species which show an extremely condensed development (in which typically "larval" characters do not appear or are only vestigial) do not really undergo a true metamorphosis, but show a more consistently progressive nature of development. From a morphogenetic point of view, therefore, opisthobranchs can be considered to have either a *metamorphic* or an *ametamorphic* mode of development.

Much of the literature concerning opisthobranch development emphasizes the many variations of the life cycle which can be seen in different species of this group. These "ecological" descriptions of development have tended to mask the basic similarity of morphogenesis seen in most opisthobranchs. Thompson's classification of development into three types (planktotrophic, lecithotrophic, and direct) has been widely accepted by workers in the field; however, this classification treats ontogeny as an ecological entity rather than as a morphogenetic one.[5] A simple modification of Thompson's developmental scheme (presented below) produces a more workable framework in which to compare species of different developmental types by both ecological and morphogenetic parameters. This modification takes into account the fact that although details of the life cycle may vary (see Fig. 1), the morphogenetic events which ultimately occur in each of those cycles to produce a post-metamorphic juvenile can be virtually identical. For example, when a planktotrophic larva reaches the point of metamorphic competence it acquires a morphology virtually identical to that of a newly hatched, competent lecithotrophic larva. This situation is shown clearly by the related species *Phestilla melanobranchia* and *P. sibogae*.[63] The former species hatches as a planktotrophic larva which must feed for at

179

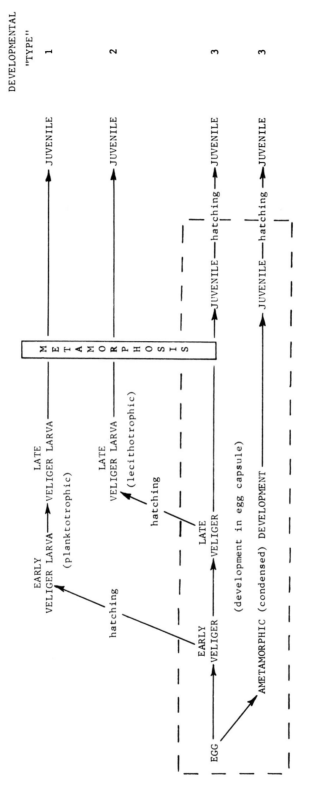

Fig. 1. Summary of the various patterns of early development through metamorphosis seen in the Opisthobranchia. The dashed rectangle represents development occurring within the egg capsule. The relationships of these patterns to the developmental "Types" described by Thompson[5] are shown on the right.

least 8 days before it completes larval morphogenesis and becomes competent to metamorphose. *Phestilla sibogae*, on the other hand, hatches as a lecithotrophic larva and can metamorphose without a feeding phase within one to two days.[66] The premetamorphic morphology of the competent larvae and the subsequent metamorphogenesis in the two species are virtually identical through the early juvenile stages. In other instances a single species has been reported to have more than one type of life cycle[24,36,37,43,44,46,63,64,67] (see fn *b* in Table 1). Variations in developmental "Type" have been reported for geographically separated populations of the pyramidellid *Brachystomia rissoides*[36,37] and the sacoglossan *Elysia chlorotica*,[24,43,46] and for temporally separated members of the same populations of the aeolid *Cuthona nana*[61] and the sacoglossan *Elysia cauze*.[44] Eyster and Stancyk[64] have noted that different individuals of the same population of the aeolid *Tenellia pallida* exhibit both larval and capsular metamorphic development (see Fig. 1), and Clark[83] has found that whereas the aeolid *Spurilla neapolitana* typically produces lecithotrophic larvae, starved adults will produce egg masses from which small planktotrophic larvae hatch.

Thompson's[5] developmental Type 3, direct development, actually includes a variety of morphogenetic patterns. This category includes species such as *Elysia cauze* and *E. chlorotica* which have a functional veliger with fully developed "larval" organs, as well as species such as *Coenia cocksi* or *Okadaia elegans* in which the "larval" organs (shell, operculum, velum) are completely absent or only transitorily present as extremely reduced vestiges. Although it has not been reported, it is likely that artificial liberation from the egg capsule of the veligers of capsular developing species such as *E. cauze*, *E. timida*, *Phyllaplysia taylori*, etc., would result in these veliger larvae exhibiting a behavior identical to that of species in which the veligers normally hatch as lecithotrophic larvae. There are reports of hatching occurring while metamorphosis is in progress. *Elysia timida*, *Trippa spongiosa*, and *Trinchesia granosa*, for example, have all been reported to crawl from the egg mass while the velum is regressing but prior to shell loss.[42,53,59] In addition, West has noted the simultaneous emergence of both swimming lecithotrophic veligers and crawling metamorphosing veligers from the same egg mass of *Elysia chlorotica*.[43]

It is clear therefore that substantial variation occurs in ecological patterns of development whereas the morphological pattern is basically unchanged. The difference in ecological patterns (reflected as time of hatching) in the foregoing examples seems to depend on the quantity of yolk stored in the egg during oogenesis. Planktotrophic larvae develop from smaller eggs than do lecithotrophic larvae.[20,36,37,67] Extrinsic factors which influence the adults (salinity, season, starvation) in some way influence the quantity of yolk laid down in the eggs.

In order to clarify the relationships between the various developmental patterns in the opisthobranchs, the following classification is proposed (Table 2). As noted earlier, this is a modification of Thompson's classification,[5] and involves the separation of his developmental Type 3 (direct development) into two groupings, depending on the degree of developmental condensation which is seen during embryogenesis.

TABLE 2

Patterns of Opisthobranch Development

Metamorphic Development	Ametamorphic Development (= Condensed or Reduced Development)
1. Planktotrophic veliger larva (Type 1 of Thompson) 2. Lecithotrophic veliger larva (Type 2 of Thompson) 3. Capsular veliger stage (Type 3 of Thompson)	1. Capsular embryogenesis (Type 3 of Thompson)

The terminology used in this scheme, as well as throughout this article, follows the guidelines defined by Giese and Pearse,[68] who use the term *embryo* to refer to all developmental stages which occur within the egg case or egg membrane, and *larva* to refer to free–swimming stages which pass through a metamorphosis to form the immature *juvenile* individual. If larvae are nourished by yolk reserves they are *lecithotrophic,* whereas if they must feed in the plankton they are *planktotrophic.* Juveniles are miniature adults resulting from larval metamorphosis or post-hatching stages which grow in size to attain adulthood without metamorphosis.[68]

As with most attempts to classify biological systems, this one has an inherent weakness. A survey of the species known to undergo "direct" development (Table 3) shows there is a gradation in developmental condensation from a typical "larval" morphology, as in *Elysia timida* or *Phyllaplysia taylori,* to the extremely condensed development of *Okadaia elegans* or *Runcina coronata,* where no trace of shell, operculum, or velum occurs during development.[27,55] For the species described to date, most show either a substantial condensation or very little condensation, so that they can easily be categorized as ametamorphic or metamorphic, respectively. Two species in particular, however, *Retusa obtusa* and *Doridopsis limbata,* seem to be intermediate forms which clearly show the "larval" characters, but in partially reduced stage. I have classed these species as metamorphic, however, since the recognizable larval structures are present almost up to hatching, when they are lost during processes recognizable as metamorphic. For purposes of morphogenetic comparisons I have designated capsular developing species as metamorphic if they appear to have sufficiently well-developed "larval" organs to function in the larval environment if artificially liberated from the egg capsule. Certainly, however, species such as *Retusa obtusa* and *Doridopsis limbata* are intermediate forms which are evolving toward an ametamorphic development and which are no longer well adapted to a free-swimming larval existence.

With this division of opsithobranch development into metamorphic and non-metamorphic classes as a basis, the rest of this paper will discuss the structure of the metamorphic veliger and the subsequent morphogenetic changes which occur during metamorphosis. A brief consideration of ametamorphic development is included at the end of the paper to demonstrate the effects of developmental condensation on metamorphic processes.

TABLE 3

Opisthobranchs with Capsular Development

Ametamorphic (= Condensed Development)	Ref.	Metamorphic	Ref.
Runcina setoensis	26	*Retusa obtusa*	6
Runcina coronata	27	*Brachystomia rissoides*	37
Acteocina senestra	28,29	*Onchidella celtica*	35,36
Coenia cocksi	28,35	*Phyllaplysia taylori*	38
Vayssieria calendonica	41	*Elysia timida*	42
Cadlina laevis	5	*Elysia chlorotica*	43
Casella obsoleta	49	*Elysia cauze*	44
Coryphella stimpsoni	56	*Chromodoris loringi*	51
Okadaia elegans	55	*Doridopsis limbata*	50
		Aeolidiella alderi	7,17
		Cuthona pustulata	57
		Aeolis sp.	58
		Trinchesia granosa	59
		Tenellia pallida	64

VELIGER MORPHOLOGY PRIOR TO METAMORPHOSIS

The information presented in this section is condensed from the papers cited in Table 1. Although metamorphosis has been reported for at least 61 species, the great majority of these reports contain only fragmentary information on the morphogenetic events which occur at this time. The most detailed investigations of opisthobranch metamorphosis, involving extensive use of sectioned material, are those of Thompson,[3,4,5] Smith,[6] Tardy,[7] Thiriot-Quiévreux,[8] Bonar and Hadfield,[12] Bonar,[9,10,11] and Kriegstein.[13,14] The only ultrastructural studies of opisthobranch metamorphosis to date are those of Bonar[9,10,11] and Bonar and Hadfield.[12]

The structure of a "typical" opisthobranch veliger just prior to metamorphosis is diagrammed in Figure 2. Although size, pigmentation, amount of yolk, depth of mantle cavity and extent of mantle edge vary with species (discussed below), opisthobranchs with metamorphic development exhibit this general morphology. The detailed structure and variability of the individual organ systems in premetamorphic veligers are discussed below.

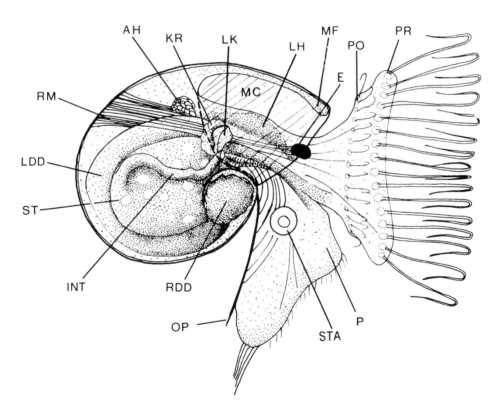

Fig. 2. Diagrammatic representation of a premetamorphic opsithobranch veliger. AH, adult heart rudiment; E, eye; INT, intestine; KR, adult kidney rudiment; LDD, left digestive diverticulum; LH, larval heart; LK, larval kidney; MC, mantle cavity; MF, mantle fold; OP, operculum; P, propodium; PO, postoral velar ridge; PR, preoral velar ridge; RDD, right digestive diverticulum; RM, retractor muscles; ST, stomach; STA, statocyst.

Shell

The shell is either inflated or coiled, with three-quarters to one and one-half whorls (shell types 2 and 1 respectively, according to Thompson[69]). Shell type has proved to be an effective taxonomic indicator for the Sacoglossa and Nudibranchia,[5,7] but cannot be used as an indicator of developmental type (*i.e.*, planktotrophic *vs.* lecithotrophic or capsular).[5] The shell rarely shows any evidence of sculpturing, although considerable growth of the shell occurs in at least some species during pelagic larval life.[23,24,54,70]

A curious feature of opisthobranch veligers is the sinistrality, or "left-handed," coiling of the veliger shell, whereas the asymmetric body plan follows a dextral, or "right-handed"organization. This condition, termed hyperstrophy, is usually reversed at metamorphosis in those species which retain the shell into the adult state. The mechanistic reason for hyperstrophy is not known, although Smith[6] suggests that for *Retusa obtusa* this condition exists as a result of the early hyperstrophy of the left digestive diverticulum during embryogenesis. The inherent tendency toward endogastric coiling is modified by the presence of the enlarged diverticulum, so that the developing shell gland conforms to the visceral topography and secretes a shell which is more pronounced on the left-hand side. This visceral "apex" imparts the sinistral direction of coiling to the fully formed shell.[6]

Velum

The velum is a bilobed structure, each lobe consisting of a single row of large marginal cells supported and attached to the head by upper and lower epithelial layers.[9,12,71] The densely ciliated marginal cells of the preoral ciliated band produce the locomotory force and primary feeding currents of the pelagic larva.Behind the preoral band, and separated from it by a sparsely ciliated feeding groove, is the sub-velum, or postoral ciliated band, which is active in feeding. The function of these ciliated bands in locomotion and feeding has been described in detail by Thompson.[72] Planktotrophic veligers often have larger velar lobes and longer preoral cilia than do lecithotrophic or capsular veligers, but only the veliger larvae of *Aegires punctilucens*[16] and certain gymnostome pteropods[73] show a hypertrophy of the velar lobes comparable to that common in prosobranch veligers.[74] The only opisthobranch larva which is reported to show any tendency toward increasing the lobation of the velar edge is *Philine denticulata,* which has shallow, paired indentations producing a four-lobed aspect.[32] The simple velar lobes of most opisthobranch veligers reflect the generally smaller size and shorter time spent in the plankton by these larvae (on the average) than by prosobranch veligers. In species having a capsular metamorphosis, the postoral band is often reduced or missing, as in the cephalaspid *Retusa obtusa.*[6]

Foot

The foot of an opisthobranch veliger is one of the best morphological indicators of metamorphic competence. A veliger which has developed to the point where it is competent to metamorphose invariably has a large, well-developed propodium, the result of hypertrophy of metapodial mucous glands within the cavity of the foot.[3,4,5,12,13,30] Thompson[3] has reported four gland types present in competent larvae of *Adalaria*: the propodial, accessory, metapodial, and lateral pedal mucous glands. Two distinct gland types have been reported for *Tritonia hombergi* and *Phestilla sibogae* veligers.[4,12] The pedal glands are presumed to secrete the tenacious mucus which anchors the larva to the substratum at metamorphosis. The metapodial mucous glands are reported to be reduced or lacking in species which have ametamorphic development.[5] Hypertrophy of pedal mucous glands generally occurs only a few days prior to acquisition of metamorphic competence and often signals the onset of "swim-crawl" searching behavior.[3,4,12] The ventral surface of the foot is covered with short cilia and may bear a dense mid-ventral band of cilia which produces a particle rejection current in planktotrophic larvae.[72] Around the margin of the foot, especially on the posterior tip, can be seen tufts of long, nonmotile cilia which presumably have a sensory function.[3,4,7,38] The metapodium bears a thin, usually transparent, operculum.

Mantle

The mantle cavity, located on the dorsal and right dorsolateral side of the body, varies in depth depending on species. The mantle fold extends as a roof over the mantle cavity to the shell margin where shell deposition occurs. Shell deposition typically ceases just prior to the acquisition of metamorphic competence and the mantle withdraws from the shell margin.[3,4,7,13] In nudibranchs the retraced mantle fold may remain as a thin roof of tissue covering much of the mantle cavity, as in *Phestilla sibogae*,[10] *Elysia chlorotica*,[46] or *Tenellia ventilabrum*,[7,10] or it may become thickened and withdraw to the back of the mantle cavity, as in *Adalaria*,[3] *Tritonia*,[4] and *Aeolidiella*,[7] where later it will be involved in forming the dorsal integument of the adult. In opisthobranchs which retain the shell into the juvenile stage, the mantle may remain attached to the shell margin, as in *Philine*,[32] or it may withdraw partially from the shell edge, as in the aplysiids.[13,22,25]

In *Philine aperta* and *Acteocina canaliculata* the rudiment of the gill can be seen on the roof of the mantle cavity just prior to metamorphosis.[30,32]

Digestive System

The premetamorphic veliger possesses a complete digestive tract, beginning with the food groove of the velum which leads into the lateral borders of the mouth. The short, cilitated esophagus leads back to the foregut, or buccal cavity, the floor of which is usually invaginated to form a muscular radular sac. At the onset of metamorphosis radular teeth are usually present.[3,4,12,38] In *Adalaria,* for example, six radular teeth, corresponding to the hooked lateral teeth of the adult radula, are present when metamorphosis begins and are capable of eversion through the mouth immediately following metamorphosis.[3] In certain species which show capsular metamorphosis, such as *Onchidella,* the radular teeth form late in metamorphosis, but are ready for feeding by the time hatching occurs.[34]

The stomach, or midgut, is typically flanked by two diverticula: a larger left diverticulum and a small right one. During development prior to metamorphosis the left diverticulum continues to enlarge while the right remains small or regresses.[3,4,12,13,38,65] The left diverticulum has an unciliated lumen which connects directly to the stomach lumen. The right diverticulum generally lacks a lumen. As the left diverticulum enlarges and extends backwards during development it causes a displacement of the stomach, so that the dorsal portion of the stomach rotates toward the left side of the larval body as the ventral portion rotates toward the right.[3,6,34] The intestine exits from the posterodorsal aspect of the stomach and curves anteriorly to the anal opening at the base of the mantle cavity on the right side of the body just adjacent to the shell aperture. Two large anal cells, a characteristic of early opisthobranch development,[75] may still be visible prior to metamorphosis,[38,70] but usually disappear during pre-metamoprhic veliger development.[3,4,32]

Excretory System

The embryonic and larval excretory systems of opisthobranchs have been the subject of much speculation owing to the variety of structures which may be present.[38,75,76] Mazzarelli[77] recognized two embryonic excretory systems: the first, a pair of unicellular "primitive kidneys," or nephrocysts, located in the neck region adjacent to the statocysts, and the second, a single (occasionally paired) "secondary kidney" located just dorsal or posterior to the anus. This structure is often darkly pigmented,[34,76] presumably from the stored excretory materials. The nephrocysts are occasionally lacking.[6] The ultrastructure of the nephrocysts has been described by Bonar and Hadfield.[12] The secondary kidney may have a definite excretory pore opening to the mantle cavity,[3,34,38,] but in many cases no communication with the exterior is visible.[7,12] The retention of waste materials by this structure and its subsequent loss or resorption at metamorphosis[3,4,6,7,12,34] suggest it does not actively excrete materials during embryonic or larval life. Horikoshi,[32] however, notes that in *Philine* the secondary larval kidney is not lost, as was previously reported,[31] but may be retained in the juvenile form.

Posterior and dorsal to the secondary kidney, a mass of undifferentiated cells in veligers of some species, such as *Onchidella, Adalaria,* and *Tritonia,* is the rudiment of the adult kidney.[3,4,34] This rudiment will develop following metamorphosis.

In addition to the nephrocysts and secondary embryonic kidneys, it has been suggested that the anal cells may play a part in excretion during early embryogenesis.[75,76] These structures, which appear early in embryogenesis and mark the future site of the anus, have been shown to accumulate trypan blue and other vital dyes[75]; they usually disappear about the time the secondary kidneys develop.

Heart

Although a larval heart is a structure typical of prosobranch veligers,[74] it is present only occasionally in opisthobranch veligers. The presence of a larval heart has been reported for all aplysiids studied to date,[13,25,38] for the cephalaspids *Philine*[31,32] and *Acteocina,*[30] and for a single nudibranch, *Adalaria.*[3] The small, vesicular, regularly pulsing larval heart is usually situated beneath the floor of the mantle cavity and although its function is not known, Thompson[3] has suggested that its regular pulsations accelerate the flow of water through the mantle cavity. In several species, an undifferentiated mass of cells can be seen prior to metamorphosis in the body some distance behind the larval heart. These cells are the rudiment of the adult heart, which develops independently of the larval heart during or following metamorphosis. According to Franz,[30] the position of larval and adult hearts in *Acteocina* are the reverse of that described above, although their ontogenetic order of appearance is the same.

Musculature

Two muscle systems are present in opisthobranch veligers: a system of velar and pedal retractor muscles and a diffuse network of sub-epithelial muscle fibers in the cephalopedal region.[3,4,12] The retractor muscles attach to the shell at its most posterior aspect on the shell midline or just to the left of the midline. In some species, such as *Adalaria*[3] or *Phyllaplysia,*[38] the retractors arise as a single large bundle of muscle fibers which pass anteriorly through the body beneath the floor of the mantle cavity and branch into the foot and velar lobes. Other species, such as *Phestilla,*[12] have paired retractor muscles, one of which serves the foot and operculum, while the other serves the velar lobes. Contraction of these muscles, which insert on the tissues of the velar lobes, foot, and operculum, causes retraction of those structures into the mantle cavity (or shell cavity if the mantle fold is retracted from the shell mouth.)[3,4,69] The insertions of the retractor muscles on the shell and operculum are mediated by modified squamous epidermal cells.[11] This adhesive epithelium contains bundles of intracellular tonofilaments which traverse the cells and embed in hemidesmosomes on apical and basal surfaces. These cytoplasmic anchors, in conjunction with some extracellular cement, firmly attach the body musculature to the external shell and operculum.

Nervous System

Comprehensive treatments of the development and organization of the opisthobranch nervous system are given by Kriegstein[13,14,15] for *Aplysia* and by Thompson[3] and Tardy[7] for the nudibranchs *Adalaria* and *Aeolidiella*, respectively. The embryological order of appearance of ganglia and their relative positions are similar in the different species, but as development proceeds notable differences occur between the anaspid and nudibranch nervous systems. Whereas the nudibranch nervous system has a strong tendency towards condensation of the ganglia (termed cephalization and telencephalization by Tardy), the aplysiid ganglia not only remain discrete, but also become progressively more so as connectives lengthen.[13,14] These changes reflect the adult conditions where extreme condensation is characteristic of nudibranch nervous systems and a more diffuse system is characteristic of the aplysiids.[65,78]

Sense Organs

The statocysts, with enclosed statoliths, are the earliest parts of the embryonic nervous system to develop during opisthobranch ontogeny[3,7] and remain relatively unchanged through metamorphosis. In virtually all gastropod species the statocysts are found situated next to, or embedded in, the pedal ganglia; however, innervation comes from the cerebral ganglia.[6,65]

The eyes, on the other hand, usually develop rather late in embryogenesis, and along with the appearance of an enlarged propodium signal the approach of metamorphic competence. Lecithotrophic species, such as *Phestilla sibogae*,[12] *Adalaria*,[3] or *Cuthona adyarensis*,[60] usually have eyes when they hatch, whereas planktotrophic species, such as the aplysiids, typically develop eyes at some time during the larval phase.[13,25] *Acteocina*, which has a short-term planktotrophic larva, hatches with only the right eye present; the left develops two days later.[30] Opisthobranch larval eyes typically form just behind the velar lobes as invaginations of the ectoderm.[65] This is not the case for the aberrant *Onchidella celtica*, however, as its eyes form in the intervelar area on short tentacles.[34] Hughes[79] has described the ultrastructure of the larval eyes of the aeolid *Trinchesia aurantia*, and has shown them to be simplified versions of the adult eyes.

Although Kriegstein has reported the presence of an osphradial ganglion in the premetamorphic larva of *Aplysia californica*,[14] the only report of a definitive osphradium or an osphradial rudiment being present prior to metamorphosis is for *Retusa obtusa*.[6] The osphradium in *Retusa* develops on the inner surface of the mantle fold just inside the mantle cavity on the right side.

A previously unreported sensory structure, the cephalic sense organ, has been described recently by Bonar[80] for *Phestilla sibogae*. This organ is located in the dorsal intervelar area and consists of several cells, each of which possesses a deeply invaginated lumen and is in intimate contact with the underlying cerebral ganglia and cerebral commissure. The lumen, which may retain its connection to the exterior, is filled with modified cilia arising from the cell surface. Although the origin of this organ has not been definitely ascertained, the cells involved appear to be derived from the original apical plate cells of the trochophore. I have recently examined the larvae of six other opisthobranch species (*Aplysia dactylomela*, *Stylocheilus longicaudata*, *Melibe* sp., *Hippselodoris vibrata*, *Aeolidia papillosa* and *Hermissenda crassicornis*) and several prosobranch species and have determined this same organ is ubiquitous, although reduced in some species.

METAMORPHOSIS

The degree of metamorphogenetic change in any species depends on the number of characteristics carried from the premetamorphic to the juvenile stage. The nudibranchs, for example, show much greater change than do the aplysiids, because the loss of the nudibranch shell and operculum results in a flattened elongated body form, reduced musculature, and loss of mantle cavity and related pallial elements such as gills, osphradium, and hypobranchial glands. Consequently, we cannot generalize a "typical" metamorphic sequence for all opisthobranchs, although this has been done for the nudibranchs.[12] As noted previously, we know more about the metamorphogenetic processes in nudibranchs than in any other group, partly because nudibranchs have a more cataclysmic metamorphosis than most other opisthobranchs, and partly because of the common occurrence of lecithotrophic and capsular development in the group. For these reasons the following system-by-system analysis of opisthobranch metamorphogenesis emphasizes the admittedly extreme changes in the nudibranchs, but relates these changes, where possible, to their equivalents in the shelled opsithobranchs.

Shell

The shell is lost at metamorphosis of all members of the Nudibranchia, Gymnosomata, and Acochlidiacea, and for certain species in the other opisthobranch groups.[65] In these groups the shell is cast usually after the velar lobes have been lost or are undergoing regression; however, several exceptions have been reported. Nordmann[2] suggested that shell loss in the planktotrophic larvae of *Tergipes edwardsii* occurred well before loss of the velar lobes. The fact that his drawings indicate cerata and rhinophores develop while functional velar lobes are retained, has led subsequent authors[3,7] to suggest that he was looking at abnormal individuals. Recent work by Thiriot-Quiévreux on the dorid *Aegires punctilucens* suggests, however, that Nordmann's observations may be correct.[16] *Aegires* has two sequential larval forms: the first is a typical planktotrophic veliger except for the hypertrophy of the bilobed velum. After an undetermined time in the plankton, the shell is lost and the larva continues to swim and feed by means of the enlarged velar lobes. During this second phase of larval life the flattened adult dorid body form becomes progressively more evident, eyes appear, papillae develop on the dorsum, and the larva eventually settles and completes metamorphosis with the reduction and loss of the velum. In another dorid, *Doridella obscura*, Perron and Turner[23] reported that shell loss preceeded loss of the velar lobes, although both of these events occur within a few hours of each other.

The mechanism by which the shell is actually cast has been reported in detail for the aeolid *Phestilla sibogae*.[10] In this species, the larval retractor muscles, once their attachment to the shell is severed, actively pull the visceral mass from the surrounding shell. During this process the ventral surface of the foot must adhere tightly to a surface (via secretions of the pedal mucous glands) so that the retractor muscles have an anchorage against which to exert a force. If the metamorphosing larva cannot gain purchase on a substratum, exit from the shell does not occur. For at least certain nudibranchs the shell apparently is pushed off the visceral mass by the progressive posterior reflexion of the mantle fold over the dorsal surface of the body (see section on **Foot** and **Mantle**).[3,7]

In those species which retain the shell through metamorphosis a variety of subsequent shapes are seen. In most cases the premetamorphic condition of hyperstrophy is reversed and the postmetamorphic shell forms in a dextral manner, consistent now with the dextral asymmetry of the body. This return to a dextral coiling from a hyperstrophic condition is termed heterostrophy, and is characteristic of all shelled opisthobranchs except for the Thecosomata, which retain the hyperstrophic condition as adults.[65]

The adult shell is usually small and quite fragile, and never becomes extensively calcified or elaborated as in prosobranchs.[65] In the Anaspidea the postlarval shell grows as a flattened shield and usually forms a convex plate-like structure in the adult which is almost entirely enclosed by the mantle folds or lateral extensions of the foot.[13,14,25,38] Details of shell formation in aplysiids can be found in the companion paper by Switzer-Dunlap in this volume.

The most fascinating shell form in postmetamorphic opisthobranchs is found in the curious bivalved sacoglossans, described by Kawaguti and Yamasu[45] for *Tamanovalva* (=*Berthelinia*) *limax*. Following metamorphosis the mantle edge secretes a shelf-like shell similar to that of the aplysiids[13]; however, a hinge line soon develops along the margin of the newly forming shell, so that two separate valves form and increase in size equally. The larval shell remains as the small apex of the right valve.

Velum

With the loss of a functional velar apparatus, a larva is irreversibly committed to metamorphogenesis. As noted in the previous section, loss of the velar lobes usually precedes other metamorphic events, such as shell loss, although exceptions do occur. The velar lobes usually are reported as being "lost, resorbed, reduced, or regressed"[3,7,30,45,60]; however, a detailed study of velar dissolution in *Phestilla sibogae*[9] has shown that all of the ciliated cells of the preoral band, postoral band and food groove selectively dissociate from the supporting tissues and are ingested as the first postlarval "meal." The mechanism by which only the ciliated cells dissociate from the velar apparatus is not known. The remnants of the velar lobes, consisting of the nonciliated supporting cells and subepidermal nerves and muscles, may be incorporated into the head epidermis[7,9,60] or may fuse and grow anteriorly to form the oral veil or oral hood, as in *Adalaria*,[3] *Melibe*,[54] or *Retusa*.[6] In other instances, the remnants may remain as discrete rounded lobes, sometimes fused medially at their bases, which subsequently grow out as precursors of the cephalic, or oral, tentacles, as in *Aplysia*,[13,22,25] *Phyllaplysia*[38] and *Philine*.[32]

Foot and Mantle

The thickened, triangular foot of the veliger lengthens and widens during metamorphosis. This is necessary, in part, because much of the visceral mass moves to a more ventral position within the cavity of the foot during the metamorphosis, especially in those forms which are shell-less as adults. The resulting dorsoventral flattening of the body leads to a further elongation and broadening of the foot to accommodate subsequent development of the digestive diverticulum. The lateral surfaces of the veliger foot may also contribute to the formation of the dorsal adult epidermis of certain species. This is discussed below in relation to the fate of the mantle fold. In some species, especially among the cephalaspids and anaspids, the lateral margins of the foot will extend outwards and dorsally as flattened parapodia which cover the shell and dorsal body surface.[6,15,25] Similar pedal extensions form the swimming "wings" of the Pteropoda during their metamorphosis.[8,48] Around the margin of the postmetamorphic foot stiff "sensory" bristles (actually bundles of cilia) can be seen projecting outwards from the body. Behavioral changes exhibited by crawling, postlarval forms when these bristles touch an object in the environment suggest the bristles have a tactile sensory function.

In species which retain the shell into the adult stage, the edge of the mantle fold continues as the site of shell formation and calcium deposition. As juvenile shelled opisthobranchs grow, the mantle edge may remain attached at the aperture lip, as in *Tamanovalva*,[45] or it may reflect back over the aperture margin and grow posteriorly to enclose the shell partially, as in *Aplysia californica*[13] and *Philine denticulata*,[32] or to enclose the shell totally, as in *Philine aperta* and *Philine scabra*.[32] For many of the nudibranchs, a similar reflexion of the mantle fold occurs prior to, or during metamorphosis.[3,4,7] The thickened edge of the mantle fold of *Adalaria*, *Tritonia*, and *Aeolidiella* undergoes reflexion within the shell cavity, however, and as the edge moves posteriorly it covers the visceral mass, simultaneously fusing with the underlying perivisceral membrane (see Fig. 3). Attachment to the shell is lost at about this time, so that the shell is pushed posteriorly off the visceral mass by the retreating mantle fold. This reflexion, or eversion, of the mantle fold is effected by the combination of differential growth processes and by contraction of muscle elements of the inner perivisceral membrane.[3] Whether the process of mantle eversion is the only force involved in removing the shell from the body in these species is not known. Action of the larval retractor muscles may aid in this process, as has been demonstrated for *Phestilla sibogae*.[10]

The result of mantle flexure in these species is the formation of dorsal postlarval epidermis from the tissues of the mantle edge[3,4,5,7] or from a thickening of the base of the mantle fold which subsequently spreads across the floor of the mantle cavity during metamorphosis.[7] Rapid proliferation of columnar epidermal cells is reported to aid the spread of the mantle over the dorsal surface of *Adalaria*.[3] In both *Adalaria* and *Tritonia*, the mantle tissues spread anteriorly as well, encircling the developing rhinophores and fusing in front of them so that the rhinophore sheaths are formed.[3,4] This origin of dorsal adult epidermis from tissues of the veliger mantle has been widely accepted[65]; however, it has recently been shown that mantle tissue does not contribute to the dorsal epidermis of the aeolid *Phestilla sibogae*.[10] Rather, the lateral surfaces of the veliger foot slide over the visceral mass to cover the dorsal surface as the metamorphosing larva exits from the shell and the visceral mass moves into the cavity of the foot (Fig. 3). Vacuolated epidermal cells, present only on the foot of the premetamorphic veliger, were seen to cover the dorsal body surface within 30 to 60 minutes of shell loss. The very thin mantle fold and epidermis of the mantle cavity floor could be seen as a crumpled remnant of tissue on the posterior aspect of the postlarval body at this time. Sections through metamorphosing individuals of *Elysia chlorotica* show the tissues of the mantle cavity to be almost idential to those of *Phestilla sibogae*, i.e., the epidermis of the mantle fold and floor of the mantle cavity is very thin.[46] Tardy[7] has reported this same structural organization for veligers of the aeolid *Tenellia ventilabrum*[7] (page 340, fig. 15a). It is likely that for both of these species the adult dorsal epidermis forms from veliger pedal epidermis. Several sacoglossan and nudibranch species are currently being studied in our laboratory to determine the origin of the dorsal adult epidermis. Tardy's clear demonstration of a mantle fold origin of dorsal epidermis for the aeolid *Aeolidiella*, coupled with my demonstration of pedal origin for *Phestilla*, suggests the origin of dorsal epidermis cannot be used as an indicator of polyphyletic origin of nudibranchs unless one wishes to claim polyphylogeny within the aeolid nudibranchs as well.

In those species which retain the shell (and thus the mantle cavity), the gill and osphradium develop shortly after metamorphosis.[13,30] In the aplysiids, the gill appears as an outpocketing of the mantle epithelium on the right side of the mantle cavity and folds repeatedly upon itself until it fills the mantle cavity.[13,38]

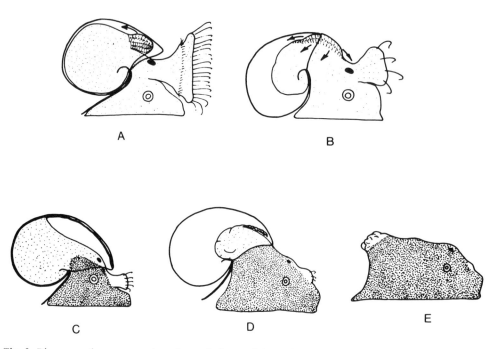

Fig. 3. Diagrammatic representation of mantle fate and formation of the dorsal adult epidermis at metamorphosis. A,B, eversion of the mantle fold to form the postlarval adult epidermis. C-E, loss of the mantle fold and formation of postlarval epidermis from larval epipodial epidermis. Figures A and B are redrawn from descriptions of Thompson[3,4,5] and Tardy.[7]

Digestive System

At metamorphosis the buccal sac is usually well developed and radular teeth are present, so that the postlarva is equipped to begin feeding relatively soon after settlement. In most species, feeding specializations of the mouth region, such as muscular sucking lips or labial palps, develop within a few days after metamorphosis, but in *Melibe leonina* the newly metamorphosed postlarvae already have an oral hood present.[54]

The stomach and digestive diverticulae undergo both positional and size changes during and following metamorphosis. The left diverticulum rapidly increases in size during late metamorphosis and early postlarval life and is soon much larger than the stomach,[3,7,12,19,28] which is often reduced to a vestibule at the anterior end of the diverticulum. The small right diverticulum is usually greatly reduced and may be missing entirely.[3,4,6] During these changes the displacement of the left diverticulum to a more ventral position causes a rotation of the stomach so that the opening between stomach and diverticulum becomes posterior and ventral.[3,12] The movement of the anus back along the right side of the body to a posterolateral position results in an effectively straight, un-torted digestive system, at least in early postlarval stages.[3,4,12]

Excretory System

Metamorphosis usually signals the loss of the larval kidney structures, both primary and secondary, which may be present and the rapid development of the adult kidney from the kidney rudiment. During the posteriorly directed translocation of the intestine along the right side of the metamorphosing body, these structures are carried along and retain their orientation dorsal, posterior or lateral to the anus.[3,4,7,12,38] The secondary larval kidneys may either degenerate *in situ*[3,38] or may be expulsed from the body during metamorphosis,[6,33] but most reports simply note they are "lost." Exceptions to the typically rapid disappearance of larval kidneys have been reported, however. Horikoshi[32] noted that for *Philine denticulata* the secondary larval kidney is visible well after metamorphosis as a small dark body adjacent to the anus, and suggests this may be true for other species as well. Kriegstein[13] reported that the larval kidney of *Aplysia californica* gives rise directly to the adult kidney, and Bridges has suggested a similar fate for the secondary larval kidney of *Phyllaplysia*.[38]

Heart

The adult heart usually becomes evident late in metamorphosis or at some time during postlarval life.[4,6,7,12,13,25,38] In *Retusa*, *Aplysia*, and *Phyllaplysia*, the adult heart develops above the intestine and adult kidney rudiment, beneath the floor of the mantle cavity, during the later stages of metamorphosis.[6,13,25,38] The aplysiid heart is two-chambered and is oriented with its long axis perpendicular to the anterior-posterior larval axis. The atrium is located on the right side of the larval midline and the larger ventricle is on the left side.[13,38] The adult heart of *Tritonia* begins beating two days after metamorphosis and within several more days it has enlarged and can be seen clearly as a single median ventricle with two posterolateral auricles.[4] The hearts of the aeolids *Phestilla* and *Aeolidiella* do not appear for several days after metamorphosis, and then only as small weakly beating rudiments on the dorsal midline behind the head.[7,12] A well-developed two-chambered heart is not seen in these species until later in the juvenile stage.

Musculature

While elements of the subepidermal cephalopedal muscle complex are retained through metamorphosis, the large pedal and velar retractor muscles are often lost, even in the shelled opisthobranchs.[3,6,10,34] In the nudibranch groups the retractor muscles begin degenerating immediately after shell loss, so that within 30 minutes the remnants of the larval retractor muscles

191

of *Phestilla sibogae* can be seen at the ultrastructural level as disorganized masses of filaments undergoing autolysis.[10]

In the shelled opisthobranchs the larval retractors may be maintained, although serving different functions in the adult. In the bivalved sacoglossan *Tamanovalva*, the velar and pedal retractor muscles of the larva become the head and foot retractors of the adult, but an entirely new muscle, which becomes the adductor muscle, appears two days after metamorphosis.[45] In the cephalaspid *Retusa obtusa*, on the other hand, both right and left larval retractor muscles degenerate at metamorphosis and a separate adult columellar muscle develops.[6] This is quite different from the muscle ontogeny of prosobranch gastropods whose post-torsional right larval retractor muscle typically becomes the adult columellar muscle.[74] Although the origin of the aplysiid adult retractor muscles is not specified, Bridges[38] has reported that juvenile *Phyllaplysia* have eleven muscles originating in the veliconch body which insert on the shell. As only two retractor muscles are evident in the larval body, either new muscles must develop or the original retractors must split to produce the numerous discrete muscle bands in the juvenile.

Nervous System and Sense Organs

As noted previously, the most complete descriptions of ontogenesis of opisthobranch nervous systems are those of Kriegstein[13,14] for *Aplysia californica* and Tardy[7] for *Aeolidiella alderi*. Kriegstein's work is reviewed in detail in this volume by Switzer-Dunlap.[25] The main changes which occur in the nervous system during metamorphosis and postlarval growth of *Aplysia* are the development of tentacular ganglia and the increased lengthening of virtually all ganglionic connectives. These changes result in a substantial spreading, or diffusion, of the nervous system of the adult. The nudibranchs, on the other hand, show an increasing condensation of the nervous system, reflected in the shortening or loss of many of the ganglionic connectives. Cerebral, optic, pedal and pleural ganglia usually fuse into a circumesophageal, or buccal ring, so that demarcation between ganglia is slight.[7]

The eyes are retained through metamorphosis and usually can be found in the adult at the base of the cephalic tentacles, although in the enigmatic *Onchidella celtica* the eyes are on the tips of the tentacles.[34] Many of the shelled opisthobranchs have both an osphradium and a hypobranchial gland in the adult mantle cavity; however, the only report of their development is that of Smith[6] for *Retusa obtusa*, noted earlier.

The cephalic sense organ of *Phestilla sibogae* can be detected in the early postlarval stages, but its eventual fate is unknown.[80] In the two-day postlarval stage the cells of this organ are still in contact with both the cerebral ganglia and the overlying epidermis medial to the eyes. This site is approximately where the cephalic tentacles will develop, and it is possible that these cells contribute to those structures.

Detorsion in the Opisthobranchs

The phenomenon of gastropod torsion has elicited substantial controversy over its origin and its relative importance to the larval and adult stages. In prosobranch gastropods, especially in diotocardians, torsion occurs early in development as a mechanical process in which the visceral mass undergoes a 180° rotation over the cephalopedal complex.[74] The presumed advantage of torsion is the movement of the "ancestrally"posterior mantle cavity to a position over the head. This relocation brings the gills and pallial sensory elements to an anterior position on the body where they receive a flow of water which has not been disturbed by turbulence created by the animal's forward movement.[74] Torsion also places the mantle cavity in a position to accommodate the head and foot when these structures are withdrawn into the protection of the shell.

The main controversy centers on whether this phenomenon originated as an advantage to the larval or adult phase of the life cycle. Recent summaries of the torsion controversy as it relates to opisthobranchs can be found in the papers by Thompson[3,4,81] Smith,[6] Tardy,[7] and

Kriegstein.[13] The embryogenesis and subsequent metamorphogenesis of opisthobranchs is especially germane to an understanding of the significance of torsion because the opisthobranchs generally are de-torted in the adult state.[65] The degree of detorsion varies in the opisthobranch groups from an almost total retention of the torted condition in the acteocinids to a totally detorted condition in the dorids. The torsional process in many opisthobranchs, and especially in the nudibranchs, has become incorporated so early in development that it is no longer seen as a mechanical process. Rather, the larval organs appear in the torted condition.[3,81] The effects of torsion on internal organs are often restricted to the digestive system, as the nervous system is condensed anteriorly to such a degree that no nervous elements have developed yet in the visceral mass when torsional effects become apparent. The intestinal loop and the anterior position of the anal complex in the mantle cavity is thus one of the few clear indications of torsion in these forms. Metamorphosis results in a posterior translocation of the mantle cavity complex for virtually all shelled species, so that the anal and kidney openings assume a posterior, or at least lateral position in the juvenile body. For the shell-less forms this posterior translocation is even more pronounced.

Although detailed studies of metamorphosis have suggested how detorsion occurs in opisthobranchs, they have not unequivocally answered *why* it occurs. The fact that the shell is lost or reduced in all opisthobranchs indicates that retention of the torted state in prosobranchs must be advantageous when a substantial shell is present. It has been argued that retention of torsion by the opisthobranch veliger reflects the advantage of torsion to the larval stage,[13,82] although these arguments are challenged by others.[81] The debate over adult versus larval advantage of the torted state will no doubt remain unresolved for a long time.

AMETAMORPHIC DEVELOPMENT

The progressive development of juvenile dorids from substantially reduced veligers has been described in detail by Thompson for *Cadlina laevis*.[5] Virtually all "larval" structures are extremely reduced or missing entirely (Fig. 4). The velar lobes are present as swollen ridges on the head which bear short, sparse cilia homologous with the long preoral cilia of veliger larvae. No subvelar ridges are present. The shell is transitorily present during embryogenesis as a thin, cup-

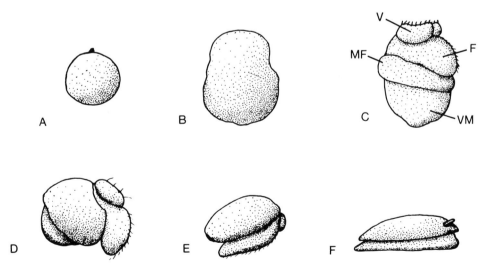

Fig. 4. Diagrammatic representation of ametamorphic development. A, zygote; B, late gastrula; C-E, progressive embryonic stages; F, hatching juvenile. F, foot; MF, mantle fold; V, velar ridge; VM, visceral mass. Redrawn from descriptions of Thompson.[5]

193

shaped structure which is later covered by eversion of the mantle fold and subsequently degenerates prior to hatching. No operculum ever forms. The thickened foot does not develop metapodial mucous glands and the "sensory" bristles are absent. Within the developing embryo there is no evidence of retractor muscles or nephrocysts, although a small secondary kidney is present.

The large quantity of yolk present in the embryo, coupled with the reduction or loss of larval structures, results in a slow, progressive development of adult body form without a rapid intermediate metamorphosis. Many of the morphogenetic events seen in *Cadlina*, although reduced, are basically similar to those of the metamorphosing larvae of *Adalaria*.[3,81] The thick mantle fold forms the dorsal adult epidermis in *Cadlina*, although its reflexion encompasses the shell in the process. The anus and "larval" kidney open on the anterior right side of the embryonic body, but move to a more posterior position as the juvenile form is attained. This is clearly homologous to the detorsion described for *Adalaria*.[3] The nervous system and radular sac develop by the time the juvenile hatches, but do not appear as early in development as they do in *Adalaria*.[3]

Thus, where vestiges of larval organs remain, their loss or transformation to adult structures can be homologized with similar events in metamorphic larvae, although the eventual loss of even these vestiges probably will result as evolution of the species continues. This extreme condensation of development has apparently evolved for the aolid *Okadaia elegans*. The diagrams of *Okadaia* development presented by Baba[55] show that virtually no "larval" characters are present. It is clear that the ametamorphic mode of development exhibited by *Cadlina* and *Okadaia* represent the evolutionary results of increasingly condensed development from an ancestral veliger larva.

ACKNOWLEDGMENTS

I would like to thank Dr. Gary Vermeij and Ms. Elizabeth Weber for their aid in translating articles in Portuguese and German. I am especially grateful to Dr. K. Clark for giving me manuscript preprints concerning nudibranch life cycles and to Dr. S. Stancyk, Ms. Hillary West, and Ms. Lindy Eyster for sharing with me their unpublished observations on the development of several Atlantic opisthobranchs. This research has been supported by National Science Foundation Grants BMS 75-0313, PCM 76-11479, and PCM 77-16262.

REFERENCES

1. Schultze, M. S. (1849) Arch. Naturgesch., 15, 268-279.
2. Nordmann, A. (1846) Ann. Sci. Nat. Zool., (3) 5, 109-160.
3. Thompson, T. E. (1958) Phil. Trans. Roy. Soc. London B, 242, 1-58.
4. Thompson, T. E. (1962) Phil. Trans. Roy. Soc. London B, 246, 171-218.
5. Thompson, T. E. (1967) J. mar. biol. Assn. U.K., 47, 1-22.
6. Smith, S. T. (1967) Canad. J. Zool., 45, 737-746.
7. Tardy, J. (1970) Ann. Sci. Nat. (Zool.), 12, 299-370.
8. Thiriot-Quiévreux, C. (1970) Zeit. Morph. Tiere, 67, 106-117.
9. Bonar, D. B. (1973) An Analysis of Metamorphosis in *Phestilla sibogae* Bergh 1905 (Gastropoda, Nudibranchia), Doctoral Dissertation, Univ. of Hawaii.
10. Bonar, D. B. (1976) Amer. Zool., 16, 573-591.
11. Bonar, D. B. (1978) Tissue and Cell, 16 (in press).
12. Bonar, D. B. and Hadfield, M. G. (1974) J. exp. mar. Biol. Ecol., 16, 227-265.
13. Kriegstein, A. R. (1977) J. Exp. Zool., 199, 275-288.
14. Kriegstein, A. R. (1977) Proc. Nat. Acad. Sci. U.S.A., 74, 375-378.
15. Kriegstein, A. R., Castellucci, V. and Kandel, E. R. (1974) Proc. Nat. Acad. Sci. U.S.A., 71, 3654-3658.
16. Thiriot-Quiévreux, C. (1977) J. exp. mar. Biol. Ecol., 26, 177-190.
17. Tardy, J. (1962) Comptes Rendues Acad. Sci. Paris, 254, 2242-2244.
18. Tardy, J. (1964) Comptes Rendues Acad. Sci. Paris, 258, 1635-1637.

19. Tardy, J. (1970) Bull. Soc. Zool. Fr., 96, 765-772.
20. Thorson, G. (1946) Medd. Komm. Danm. Fiskeri-og Havunders. Serie Plankton, 4, 1-358.
21. Franz, D. (1975) in Culture of Marine Invertebrate Animals, Smith, W. L. and Chanley, M. H. eds., Plenum Press, New York, pp. 245-256.
22. Switzer-Dunlap, M. and Hadfield, M. G. (1977) J. exp. mar. Biol. Ecol. 29 (in press).
23. Perron, F. and Turner, R. D. (1977) J. exp. mar. Biol. Ecol., 27, 171-185.
24. Harrigan, J. and Alkon, D. (1979) Veliger (in press).
25. Switzer-Dunlap, M. (1978) in Metamorphosis of Marine Invertebrate Larvae, Chia, F-S. and Rice, M. eds., Elsevier North Holland, New York (this volume).
26. Baba, K. and Hamatani, I. (1959) Publs. Seto Mar. Biol. Lab., 7, 281-290.
27. Vayssiere, A. (1900) Zool. Anz., 23, 286-288.
28. Pelseneer, P. (1899) Trav. Stn. Zool. Wimereux, 7, 513-521.
29. Colgan, N. (1912) Ir. Nat., 21, 225-231.
30. Franz, D. (1971) Trans. Am. Microsc. Soc., 90, 174-182.
31. Brown, H. H. (1934) Trans. Roy. Soc. Edinburgh, 58, 179-210.
32. Horikoshi, M. (1967) Ophelia, 4, 43-84.
33. Joyeux-Laffuie, J. (1882) Arch. Zool. Exptl. Gen., 10, 225-283.
34. Fretter, V. (1943) J. mar. biol. Assn. U.K., 25, 685-720.
35. Pelseneer, P. (1911) Mem. Acad. Roy. Belg. Cl. Sci. (2ieme Ser.), 3, 1-167.
36. Rasmussen, E. (1951) Vidensk. Meddel. Naturh. Foren., 113, 201-249.
37. Rasmussen, E. (1944) Vidensk. Meddel. Naturh. Foren., 107, 207-233.
38. Bridges, C. (1975) Ophelia, 14, 161-184.
39. Gohar, H.A.F. and Abul-Ela, I. A. (1957) Publs. Mar. Biol. Stn. Ghardaqa, 9, 69-84.
40. Usuki, I. (1969) Scient. Rep. Niigata Univ. Ser. D. (Biol.), 6, 107-127.
41. Risbec, J. (1928) Faune des Colonies Francaise, 2, 1-328.
42. Rahat, M. (1976) Israel J. Zool., 25, 186-193.
43. West, H. H. (1977) Personal communication.
44. Clark, K. B., Busacca, M. and Stires, H. (1978) in Reproductive Ecology of Marine Invertebrates, Stancyk, S. ed., Univ. of South Carolina Press, Columbia (in press).
45. Kawaguti, S. and Yamasu, T. (1960) Biol. J. Okayama Univ., 6, 150-159.
46. Bonar, D. B. Unpublished observations on the development of *Elysia chlorotica*.
47. Seelemann, U. (1967) Helgo. Wissen. Meer., 15, 128-234.
48. Lebour, M. (1932) J. mar. biol. Assn. U.K., 18, 123-128.
49. Gohar, H.A.F. and Soliman, G. N. (1967) Publs. Mar. Biol. Stn. Ghardaqa, 14, 149-166.
50. Tchang-Si, A. (1931) Comptes Rendues Hebd. Seanc. Acad. Sci., Paris, 192, 302-304.
51. Thompson, T. E. (1972) J. Zool. London, 166, 391-401.
52. Gohar, H.A.F. and Abul-Ela, I. A. (1959) Publs. Mar. Biol. Stn. Ghardaqa, 10, 41-62.
53. Gohar, H.A.F. and Soliman, G. N. (1967) Publs. Mar. Biol. Stn. Ghardaqa, 14, 269-293.
54. Kempf, S. (1976) Personal communication.
55. Baba, K. (1937) Jap. J. Zool., 7, 147-190.
56. Morse, M. P. (1971) Biol. Bull., 140, 84-94.
57. Roginski, I. S. (1962) Dokl. Akad. Nauk SSSR, 146, 488-491.
58. Fischer, H. (1892) Bull. Sci. Fr. Belg., 24, 260-346.
59. Schonenberger, N. (1969) Pubbl. Staz. Zool. Napoli, 37, 236-292.
60. Rao, K. V. (1961) J. mar. biol. Assn. India, 3, 186-197.
61. Harris, L. G., Wright, L. W. and Rivest, B. R. (1975) Veliger, 17, 264-268.
62. Vannucci, M. and Hosoe, K. (1953) Bolet. Inst. Oceanogr., Sao Paolo, 4, 103-120.
63. Harris, L. G. (1975) Biol. Bull., 149, 539-551.
64. Eyster, L. and Stancyk, S. (1977) Personal communication.
65. Hyman, L. H. (1967) The Invertebrates. Vol. VI. McGraw-Hill, New York.
66. Hadfield, M. G. (1972) Amer. Zool., 12, 271.
67. Mileikovsky, S. (1971) Mar. Biol., 10, 193-213.
68. Giese, A. C. and Pearse, J. S. (1974) in Reproduction of Marine Invertebrates, Vol. I, Giese, A. C. and Pearse, J. S. eds., Academic Press, New York.
69. Thompson, T. E. (1961) Proc. Mala. Soc. London, 34, 233-238.
70. Thompson, T. E. (1966) Malacologia, 5, 83-84.
71. Mackie, G. O., Singla, C. L. and Thiriot-Quiévreux, C. (1976) Biol. Bull., 151, 182-199.
72. Thompson, T. E. (1959) J. mar. biol. Assn. U.K., 38, 239-248.

73. Jagersten, G. (1972) Evolution of the Metzazoan Life Cycle: A Comprehensive Theory, Academic Press, London.
74. Fretter, V. and Graham, A. (1962) British Prosobranch Molluscs, Ray Society, London.
75. Raven, C. (1958) Morphogenesis: The Analysis of Molluscan Development, Pergamon Press, New York.
76. Saunders, M. and Poole, M. (1910) Quart. J. Microscop. Sci., 55, 497-540.
77. Mazzarelli, G. F. (1898) Biol. Centralbl., 18, 717. (Not seen in original. Cited from Raven.[75])
78. Kandel, E. R. (1976) Cellular Basis of Behavior, W. H. Freeman, San Francisco.
79. Hughes, H. (1970) Zeit. Zellforsch. Mikroscop. Anat., 109, 55-63.
80. Bonar, D. B. (1978) Tissue and Cell, 16 (in press).
81. Thompson, T. E. (1967) Malacologia, 5, 423-430.
82. Garstang, W. (1928) Brit. Ass. Rep. Glasgow, 77-98.
83. Clark, K. (1978) J. Molluscan Stud. (in press).
84. Roginskaya, I. S. (1970) Malacological Rev., 3, 167-174.

LARVAL BIOLOGY AND METAMORPHOSIS OF APLYSIID GASTROPODS

Marilyn Switzer-Dunlap

University of Hawaii, Pacific Biomedical Research Center, Kewalo Marine Laboratory,
41 Ahui Street, Honolulu, Hawaii 96813

Information on larval biology, metamorphosis, and early juvenile development of aplysiid gastropods is reviewed. *Stylocheilus longicauda* (Quoy and Gaimard) is used as a model to demonstrate the developmental stages of the larval and early postlarval phases of aplysiid ontogeny. Most aplysiid species develop planktotrophically and have a larval phase which includes an initial interval of shell growth to a species-specific size followed by an interval of arrested shell growth during which morphological developments prepare the organism for metamorphosis and postmetamorphic life. The larvae of each aplysiid species settle and metamorphose preferentially on one or a few species of benthic algae. Metamorphosis takes two to four days and transforms the pelagic, filter-feeding larva into a crawling, radular-feeding juvenile. Young juveniles commence feeding on algae, lose the operculum, resume shell growth, and develop parapodia, rhinophores, and structures of the mantle cavity. Larval and metamorphic events of aplysiids are briefly compared with those of other molluscan groups.

INTRODUCTION

Most members of the opisthobranch family Aplysiidae develop via planktotrophic larvae.[1,2,3,4] The only currently known exception to this pattern is *Phyllaplysia taylori* which has direct development.[2] Embryonic and cell-lineage studies of aplysiids were published early in this century,[5,6] but studies of later development were lacking until several years ago because of difficulties encountered in rearing planktotrophic larvae of these species.

Interest in aplysiid development has intensified recently because of the widespread use of these animals for studies by neurobiologists and behaviorists (e.g., Kandel[7]). Increasing numbers of species of Aplysiidae have been cultured through their entire life cycles in the laboratory. Kriegstein *et al.*[3] successfully reared *Aplysia californica.* In our laboratory four Hawaiian species have been cultured through metamorphosis[4] and, utilizing the same culture techniques, additional observations have been made on *Aplysia pulmonica* Gould and *Aplysia parvula* Guilding in Mörch. These studies have contributed substantially to our knowledge of larval biology and metamorphosis of aplysiids and form the basis of this report.

The purpose of this review is to: (1) summarize the available information on larval growth and development preparatory to metamorphosis in aplysiids; (2) describe the events of metamorphosis and early postlarval life in these animals; and (3) briefly compare these processes in aplysiids with those of other molluscan groups. *Stylocheilus longicauda* is used as a model to discuss the larval, metamorphic, and early juvenile phases of aplysiids.

APLYSIID LIFE CYCLES

With the exception of *Phyllaplysia taylori*,[2] known life cycles of aplysiids can be conveniently divided into five phases: embryonic, planktonic, metamorphic, juvenile, and adult. Kriegstein[8] further subdivided the four posthatching phases of the life cycle of *Aplysia californica* into 13 stages on the basis of morphological and behavioral characteristics identifiable in living specimens. Because of the great similarity of the life cycle of *A. californica* to those we have reported for *A. dactylomela, A. juliana, Dolabella auricularia* and *Stylocheilus longicauda*,[4,9] Kriegstein's staging is used in the following discussion.

LARVAL GROWTH AND DEVELOPMENT

Kriegstein[8] distinguished six stages of development during the planktonic phase of *Aplysia californica* and at least five of these are readily identifiable in other species. Newly hatched, Stage 1 veligers of aplysiids closely resemble each other except for shell size (see Table 1); shell length at hatching is a specific characteristic. Shells of all species examined are simple, unsculptured, amber to red in color, sinistrally coiled and consist of 1¼ to 1½ whorls (Type 1 of Thompson[10]). At hatching veligers of all species have a well-developed velum and prominent paired statocysts, but lack eyes and a propodium.

The sequence of developmental stages is similar in different species, but the duration of the stage varies, as does the total length of time needed to complete larval development. The larval phase includes a period of shell growth to a species-specific size, followed by a nongrowth period during which a number of morphological and behavioral changes occur and culminate in metamorphic competence. Larval shell growth and developmental stages of a representative culture of *Stylocheilus longicauda* are plotted in Figure 1; growth curves of other species are similar. Veligers commence feeding on phytoplankton and growing soon after hatching. Other visible changes accompany growth of the shell: in Stage 2, enlargement of the velar lobes and appearance of the eyes; and in Stage 3, development of the larval heart. The latter is a contractile sac located in the inner perivisceral membrane on the floor of the mantle cavity, dorsal to the esophagus. Although its contractions are readily evident in living larvae, the exact function of the larval heart is unclear. Possible functions of the larval heart are to increase the flow of fluid into and out of the mantle cavity[11] or to pump blood to the velum and foot.[12]

When maximal larval shell size has been attained, in Stage 4, larvae are not yet competent to metamorphose even though shell growth is arrested until after metamorphosis. Maximal larval shell size is a species-specific characteristic (Table 1). After the shell reaches maximal length, other visible changes occur in the larva. The mantle fold, which was attached to the inner lip of the shell previously and is responsible for shell deposition, becomes increasingly retracted from the shell aperture. At Stage 5 the propodium develops as a swelling in the middle of the foot, and both the eyes and the left digestive diverticulum enlarge and become darkly pigmented. The right digestive diverticulum does not enlarge or become pigmented during larval life. Clear

TABLE 1

Shell Lengths at Hatching and Metamorphosis and Lengths of Larval Phases for Planktotrophic Aplysiids Cultured Under Laboratory Conditions

Species	Average Shell Length at Hatching (μm)	Length of Larval Phase (Days)	Culture Temp. (°C)	Shell Length at Metamorphosis (μm)	Reference
Aplysia brasiliana	140	30-40	21-25	375-400	a
Aplysia californica	125	34	22	400	3
Aplysia dactylomela	144	30	24-26	310-315	4
Aplysia juliana	125	28	24-26	315-330	4
Aplysia parvula	105	no data		500[b]	This study
Aplysia pulmonica	128	24	24-26	330-340	This study
Dolabella auricularia	148	31	24-26	290-300	4
Stylocheilus longicauda	103	30	24-26	325-340	4

[a]Personal communication from Drs. Ned E. Strenth and James E. Blankenship, Marine Biomedical Institute, Galveston, Texas.
[b]Determined from larvae that were taken from plankton hauls and metamorphosed in the laboratory.

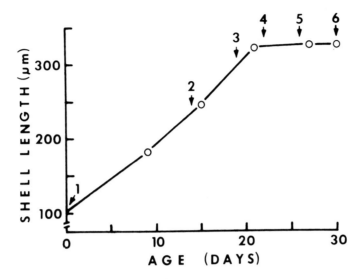

Fig. 1. Growth of the larval shell of *Stylocheilus longicauda* and age (days posthatching) and size at which the developmental stages occur (refer to text for further explanation of stages).

droplets, presumably lipid, accumulate in the left digestive diverticulum and the region surrounding the stomach. These are presumed to represent a food reserve to be used during the nonfeeding metamorphic phase.

Kriegstein described a Stage 6 for *Aplysia californica* characterized by the presence of six irregularly-shaped red spots located on the right side of the outer perivisceral membrane and by a red band overlying the mantle margin just beneath the shell. These pigmented areas are considered to be the most reliable indicators of competence in veligers of *A. californica*.[8] Similar pigmented areas have not been seen in late larvae of the species which I have observed and apparently are a specific character of *A. californica*. Competence in species without the pigmented areas must be determined empirically, that is, whether or not they metamorphose in response to the proper stimulus. A competent veliger of *Stylocheilus longicauda* is shown in Figure 2.

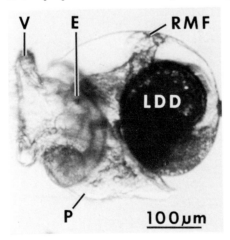

Fig. 2. Left lateral view of a competent larva of *S. longicauda*: E, eye; LDD, left digestive diverticulum; P, propodium; RMF, retracted mantle fold; V, velum.

Utilizing serial histological sections, Kriegstein[8,13] described the development of the major internal organ systems during the planktonic phase of *Aplysia californica*. His observations are summarized below. The central nervous system of an adult *Aplysia* consists of four pairs of ganglia in the head region, the cerebral, pedal, pleural, and buccal in addition to a single so-called abdominal (=parietovisceral) ganglion. Development of this nervous system is gradual. The cerebral and pedal ganglia develop late in the embyronic phase[6] and are the only ganglia present in newly hatched Stage 1 veligers. The eyes and optic ganglia develop from ectoderm overlying the cerebral ganglia about 14 dayas after hatching and characterize Stage 2 larvae. At Stage 3, about 21 days after hatching in *A. californica*, the osphradial, pleural and abdominal ganglia develop. The latter forms the fusion of three larval ganglia

199

which are probably homologues of the supraintestinal, subintestinal, and visceral ganglia of more primitive opisthobranchs.[13] The rudiment of the buccal mass forms during Stage 4 as a thickening of the ventral surface of the esophagus midway along its length. The buccal musculature develops at Stage 5. The paired buccal ganglia appear between Stages 5 and 6 on the posterodorsal surface of the developing buccal mass. By Stage 6 calcified teeth are present on the radula.[8]

When larvae are competent to metamorphose, they develop no further unless presented with the proper settling stimulus. Instead they continue to feed periodically on phytoplankton and spend more time near the bottom of their culture vessels swimming slowly about. Once competent, larvae retain the capacity to metamorphose for considerable periods of time. For example, in a culture of *S. longicauda* in which larvae were competent to metamorphose 30 days after hatching, over 90 percent of 50 larvae still surviving four weeks later metamorphosed when provided with the appropriate settling stimulus.

SETTLEMENT AND METAMORPHOSIS

Competent veligers of each aplysiid species settle and metamorphose preferentially on one or a few species of algae. Not surprisingly, the algal species triggering metamorphosis usually comprise a significant component of the adult diet of that aplysiid species in the field. The preferred algal settling substrata for a number of species are given in Table 2.

When presented with the proper algal stimulus, a competent larva usually begins crawling immediately upon contact with the alga. For up to an hour after settling, the larva will often alternate episodes of crawling with periods of sitting still. By this time if a larva is committed to metamorphose, it usually attaches to the alga by secretions from its metapodium and retracts almost completely into the shell with the operculum partially closed. Unless disturbed, the larva remains in its "attached" position during early metamorphosis. Depending on the species, metamorphosis, encompassing the period from loss of the velum to the beginning of radular feeding, takes two to four days in aplysiids. The description of metamorphosis below is for *Stylocheilus longicauda,* but the process is similar in other aplysiids. Differences will be noted for other species where they occur.

The onset of metamorphosis, Stage 7, is signalled by loss of the velum, a two-step process. Within several hours of settling, the velar cilia are lost. Ciliated cells of both the preoral and postoral ridges dissociate from their supporting cells and at least some of them are ingested, as evidenced by their presence   in the stomach. The remaining velar cells appear to undergo

TABLE 2

Preferred Algal Settling Substrata for Planktotrophic Species of Aplysiids

Species	Preferred Algal Substratum for Metamorphosis	Algal Class	Reference
Aplysia brasiliana	*Callithamnion balliae*	Rhodophyta	*a*
Aplysia californica	*Laurencia pacifica*	Rhodophyta	3
Aplysia dactylomela	*Laurenica* sp.	Rhodophyta	4
Aplysia juliana	*Ulva fasciata* and *U. reticulata*	Chlorophyta	4
Aplysia parvula	*Chondrococcus bornemanni*	Rhodophyta	This study
Dolabella auricularia	Unidentified blue-green	Cyanophyta	4
Stylocheilus longicauda	*Lyngbya majuscula*	Cyanophyta	4

[a] See note *a* in Table 1.

resorption over the next 24 hours to form two rounded lobes of tissue which are apparently rudiments of the oral tentacles (also called cephalic or anterior tentacles) (Fig. 3).

The definitive heart develops early in metamorphosis as a two-chambered organ in the right posterolateral region of the body beneath the larval shell. For a day both the larval heart and definitive heart beat, after which the larval heart ceases contractions and apparently is incorporated in the previsceral epidermis.[8] During the second day of metamorphosis, the animal begins crawling around and there is movement of the buccal mass although the animal does not actually feed. The rudiments of the oral tentacles fuse medially and begin growth laterally and anteriorly to assume their definitive shape (Fig. 4). Kriegstein[8] recognized a Stage 8 in the development of *A. californica* at this time characterized by the appearance of red pigment behind the eyes in the form of "eyebrows." Colored pigment is not present in other aplysiid species which have been studied at this stage. In *S. longicauda* longitudinal bands of purple pigment appear on the dorsal and lateral surfaces of the head region (Fig. 4). The metapodium begins growing posteriorly at this time but the operculum is retained. On about the third day after onset of metamorphosis, *S. longicauda* begin feeding on the blue-green alga *Lyngbya majuscula* signalling the end of metamorphosis, the completion of Stage 8, and the beginning of the juvenile phase.

During metamorphosis of *Aplysia californica*, the cerebral, pleural and pedal ganglia, which were anterior to the buccal mass around the esophagus at Stage 5, become positioned around the buccal mass at Stage 7. By postmetamorphic Stage 10 (10 days after onset of metamorphosis in *A. californica*), these ganglia lie around the esophagus behind the buccal mass in their definitive positions.[8]

POSTMETAMORPHIC DEVELOPMENT

Gradual postlarval modifications transform early juveniles into a recognizable adult form. Stage 9 is entered when animals first feed. Shell growth, which halted during late larval life, resumes soon after commencement of feeding. Prior to this, the mantle fold which retracted from the shell aperture during late larval life returns to a position at the shell lip. Whereas larval shell

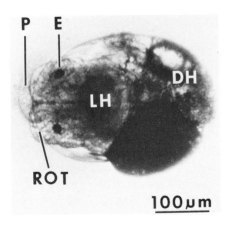

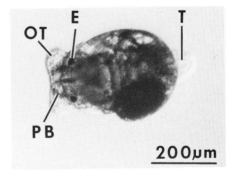

Fig. 3. Dorsal view of a metamorphosing, Stage 7, *S. longicauda*, 24 hours after loss of the velum: DH, region where definitive heart develops; E, eye; LH, region of larval heart; P, propodium; ROT, rudiments of the oral tentacles.

Fig. 4. Dorsal view of a metamorphosing, Stage 8, *S. longicauda*, 48 hours after velar loss: E, eye; OT, oral tentacle; PB, pigment band; T, tail.

growth was sinistral and helicoid, postlarval shell growth is dextral and forms a flattened, visor-like projection (Fig. 5). The parapodia develop as lateral outgrowths of the metapodium. The tail, which becomes particularly long in *Stylocheilus longicauda,* continues to grow posteriorly and the operculum is cast. The primordium of the anal siphon, a modification of the mantle margin, is visible next to the posterolateral edge of the growing shell on the right side. As the shell grows, the two-chambered definitive heart moves from a right posterior position to an anterolateral location beneath the left side of the shell. The mechanism by which the heart changes position is unknown but may relate to the movement of the mantle fold back to the shell aperture as well as to a change in direction of growth related to detorsion. During early juvenile life, the body and shell grow at similar rates so that most of the anterior portion of the animal is still covered by shell. The mantle gland develops in the mantle beneath the right side of the shell soon after juvenile shell growth commences. In species which secrete purple fluid from the mantle gland, the vesicles of this gland are dark (*e.g., A. californica* and *S. longicauda*), whereas in species like *A. juliana* which secrete only white fluid, the vesicles of the gland are translucent. As soon as feeding starts, the digestive gland increases in size and soon fills the entire larval shell. For species whose juveniles commence feeding on red algae (*A. californica, A. brasiliana, A. parvula*), the overall body color is pink initially and grows progressively darker with continued growth and feeding. In other species, specific pigmentation may also develop. In *S. longicauda* the longitudinal pigmented stripes on the head and oral tentacles and on the lateral surfaces of the foot increase in number as growth progresses (Fig. 6).

Kriegstein[8] recognized a Stage 10 in development of *A. californica* as characterized by the appearance of clustered white pigment spots on the tentacles, above the eyes, and on the siphon (Fig. 6). Similar pigment appears in *A. juliana*[4] and less obviously in *Stylocheilus longicauda.* The parapodia cover about half the shell at this stage and the anal siphon is very distinct at this time in most species of the genus *Aplysia* but is not prominent in *S. longicauda.* The digestive gland continues to enlarge and fills much of the area visible under the shell. A gill develops in the mantle cavity which is located under the shell on the right side. The rhinophores which characterize Stage 11 develop behind the eyes about eight days after onset of metamorphosis in *S. longicauda* (Fig. 7).

Further development includes additional growth of parapodia, head, and tail in relation to the shell. In all aplysiid species which retain their shells into adulthood, the anterior and lateral regions of the mantle margin reflect over the edges of the shell and eventually almost completely enclose it in a shell sac. In *S. longicauda* the mantle margin does not reflect over the shell but shell and body growth proceed as in other species for 10 to 12 days after metamorphosis at which time the parapodia overlap above the shell and the animal measures 6 to 8 mm in length. At this time the attachment of the shell to the underlying tissues becomes loose and the shell is shed. Adult *S. longicauda* do not have a shell. Two of these casts are shown in Figure 8.

Kriegstein[8] provides good diagrams of changes in internal organs of *Aplysia californica* during metamorphosis and postlarval development, although he mislabels the digestive gland (=midgut) as hindgut in several figures. The digestive system

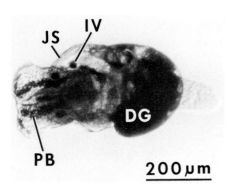

Fig. 5. Dorsal view of a young juvenile, Stage 9, *S. longicauda,* 72 hours after velar loss: DG, digestive gland; IV, ink vesicles; JS, juvenile shell; PB, pigment band.

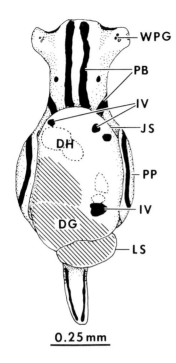

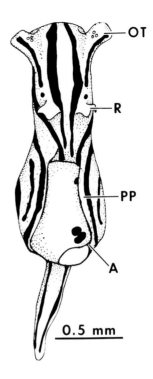

Fig. 6. Juvenile, Stage 10, *S. longicauda*, 6 days after velar loss: DG, digestive gland; DH, definitive heart; IV, ink vesicles; JS, juvenile shell; LS, larval shell; PB, pigment band; PP, parapodium; WPG, white pigment granules.

Fig. 7. Juvenile, Stage 11, *S. longicauda*, 9 days after velar loss: A, anus; OT, oral tentacles; PP, parapodium; R, rhinophore.

of *A. californica* undergoes partial detorsion during the early juvenile phase bringing the anus from its right lateral position in the larva to a position on the posterior dorsal surface of the adult. The nervous system of *A. californica* develops with all its ganglia in their posttorsional positions. The abdominal ganglion and the pleuro-abdominal connectives escape twisting related to detorsion because they move from the cephalic region, which is not involved in detorsion, into the visceral region only after detorsion is complete. The larval kidney, which is located above the anus on the right side of the larva, develops directly into the adult kidney in *A. californica*. Two further stages recognized by Kriegstein[8] which occur considerably later in ontogeny of *A. californica* are Stage 12, development of the genital groove, and Stage 13, the beginning of egg laying.

DIRECT DEVELOPMENT OF *PHYLLAPLYSIA TAYLORI*

Although development is direct in *Phyllaplysia taylori*, a well-developed "veliger" phase is passed through within the egg capsule during the embryonic period.[2] This "veliger" phase differs in some respects from its planktotrophic counterparts. Once the "larval" shell develops in *P. taylori*, its size (250 to 280 μm estimated from Bridges' figures) and structure remain constant until hatching. The larval heart apparently does not function in *P. taylori*. The definitive heart appears in its definitive location and begins to beat in the "late veliger" phase, but its contractions do not become steady until metamorphosis. Resorption of the velum and loss of the operculum occur in the capsule prior to hatching. In most respects a newly hatched *P. taylori*

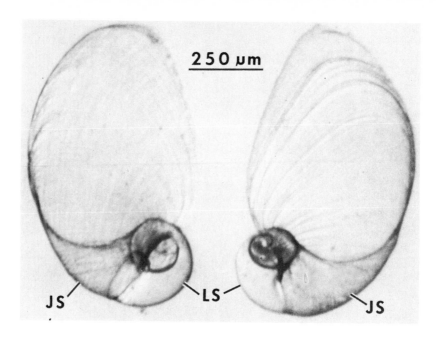

Fig. 8. Shells lost by *S. longicauda* juveniles 10-12 days after onset of metamorphosis: left, dorsal view; right, ventral view; larval shell (LS) and juvenile shell (JS); note change in direction of shell growth.

closely resembles early postmetamorphic stages of aplysiids with planktotrophic development. Growth of the juvenile shell of *P. taylori* occurs in the same manner but at a far slower rate than in other aplysiids. Bridges[2] reared juveniles of *P. taylori* for 60 days at which time the largest ones had shells 1.25 mm long which still covered most of the animal's body. Beeman[14] reported that the shell "pops off" the back of the limpet-like juvenile of *P. taylori* when they achieve a length of 2 mm. The process may be similar to shell loss in *Stylocheilus longicauda.*

FUNCTION OF THE PLANKTOTROPHIC PHASE

The majority of aplysiids and, in fact, opisthobranchs as a group, develop via planktotrophic larvae.[1,2] Veligers of planktotrophic species are usually smaller and at a less advanced stage at hatching than are larvae of species with lecithotrophic development. The pelagic larval phase of planktotrophic species, therefore, provides for both increase in size, and completion of "embryonic" development preparatory to metamorphosis. Both functions are accomplished at the expense of an external food source.

To some extent the phases of larval life of aplysiids concerned with growth on tne one hand, and development to metamorphosis on the other, are temporally separated (Fig. 2). Energy goes primarily to growth during the first two-thirds of the larval phase after which growth (*i.e.*, shell growth) ceases. Thereafter, energy is channeled to development of definitive organs (nervous system, buccal mass, and radula) and presumably to accumulation of energy reserves for the nonfeeding metamorphic phase. However, development of these definitive organs does not interfere with larval functions of filter-feeding and swimming. As Fretter[15] has so aptly stated, "the success of the veliger as a larva depends on its ability to feed and grow unhampered by the development of structures esssential to the adult and, indeed, essential to the young snail at metamorphosis."

Although maximal shell size attained by aplysiid larvae is a specific characteristic (Table 1), the evolutionary forces involved in determining maximal shell size are unclear. There is no direct correlation between shell size at hatching and shell size at metamorphosis (Table 1), or between the maximal shell size of veligers and ultimate adult size.[4] On the contrary, veligers of *Aplysia parvula,* the smallest species of *Aplysia* in Hawaii, hatch at a comparatively small size of 105 μm and metamorphose at the largest size—500 μm—known for any aplysiid. The proximal cause of differences in size at metamorphosis may be related to length of the growth phase of larval life. However, evidence thus far shows that all species cultured successfully have minimal larval phases of 4 to 5 weeks under laboratory conditions during which considerable differences in shell growth occur.[4] Unfortunately, *Aplysia parvula* has not been cultured through its entire larval phase. Perhaps the ultimate cause(s) of differences in maximal size reached by larvae relates to their ability to exploit their postlarval food sources. Algal species fed upon by juvenile aplysiids differ in cell size, toughness of walls, and other physical and chemical characteristics, as well as their abundance in time and space.

Only recently have larvae of any nudibranch species with planktotrophic development been reared through metamorphosis. In species whose larvae undergo shell growth during the larval phase, the sequence of developmental stages (appearance of eyes, appearance of larval heart, mantle retraction, propodial development) is similar to that of aplysiids. In the dorid nudibranch *Doridella obscura,* which has a relatively short larval phase of nine days, all developmental stages of aplysiid larvae are present except development of the larval heart.[16] The short planktonic period and relatively small size at metamorphosis (190 to 195 μm) of *D. obscura* may obviate the need for a larval heart. This structure does not develop in aplysiids until they are near maximal size about three weeks into the larval period. *Tritonia diomedea,* a dendronotid nudibranch, also shows a developmental sequence similar to aplysiids.[17] The chronology of development of another planktotrophic dorid, *Rostanga pulchra,* contains the same developmental steps as aplysiids with the additional development of rhinophores and spiculated dorsal papillae prior to metamorphosis.[18] In aplysiids, rhinophores develop one to two weeks after onset of metamorphosis depending on the species.

Retraction of the mantle fold from the shell aperture in veligers is normally associated with shell loss in nudibranchs.[16-21] However, this same process occurs in aplysiids (Fig. 1) which retain the shell at metamorphosis. Presumably changes occur in the mantle fold while it is retracted in aplysiids because when it resumes a position at the shell aperture after metamorphosis its direction of shell deposition is very different than in the larva (Fig. 8).

METAMORPHOSIS OF APLYSIIDS
COMPARED WITH OTHER GASTROPOD MOLLUSCS

Larvae of many other invertebrate groups metamorphose preferentially on some component of their adult habitat (see reviews by Crisp[22,23] and Scheltema[24]). Not surprisingly aplysiids metamorphose on species of algae which serve as adult foods for these herbivorous animals (Table 2). What aplysiid larvae perceive in the algae (*e.g.,* chemicals, texture) and how (with what sensory structures) such factors are sensed remain unknown. This topic is covered more extensively in a companion paper by M.G. Hadfield.

Retention of both shell and operculum during metamorphosis separates aplysiids from their opisthobranch relatives the nudibranchs, which lose both at metamorphosis. This fact alone makes the process of aplysiid metamorphosis appear far less cataclysmic than in nudibranchs. In fact, metamorphosis in aplysiids more closely resembles that of prosobranchs in which the main superficial events of metamorphosis are loss of the velum followed by a modified pattern of shell deposition.[12,15] Thereafter, however, the similarity to prosobranch metamorphosis decreases. Postlarval growth in aplysiids involving detorsion of the digestive system is similar to metamorphic changes described for nudibranchs.[11,19,21] The movement of the mantle cavity to the right side and possession of a single gill are other clear indications of the aplysiids' opisthobranch affinities. D.B. Bonar provides a more complete comparative examination of morphogenesis during opisthobranch metamorphosis in another paper in this symposium.

Though the shell is not a major feature of aplysiid adult morphology—it is small in comparison to the body size, and it is held internally—the shell and operculum, which is temporarily retained, may provide significant protection during metamorphosis and early juvenile development. Even *Stylocheilus longicauda,* which is without a shell in the adult stage, retains the shell during metamorphosis and early postlarval development and thus indicates its importance to these stages.

Much still remains to be learned about aplysiid ontogeny. No ultrastructural studies have been conducted on any early stages. There is little information available on larval musculature and velar innervation or the fate of both at metamorphosis. The functions of the larval heart and the various larval excretory organs are not readily evident. However, now that reliable techniques for rearing these animals are available, answers can be sought for many questions.

ACKNOWLEDGMENTS

I am most grateful to Dr. M. G. Hadfield for his constant advice and encouragement and for his critical reading of this manuscript. I also thank F. E. Perron and S. C. Kempf for their helpful comments on the text. Drs. N. E. Strenth and J. E. Blankenship, Marine Biomedical Institute, Galveston, Texas, provided their unpublished information on *Aplysia brasiliana* and S. C. Kempf supplied a copy of his forthcoming paper. The assistance of Dr. W. F. Van Heukelem with photography is greatly appreciated. I gratefully acknowledge the patience and competence of Ms. Frances Okimoto who typed several drafts as well as the final copy of this manuscript. This project was supported by Grant No. RR-01057 from the Division of Research Resources, National Institutes of Health, to Dr. M.G. Hadfield.

REFERENCES

1. Thompson, T. E. (1967) J. mar. biol. Ass. U.K., 47, 1-22.
2. Bridges, C. B. (1975) Ophelia, 14, 161-184.
3. Kriegstein, A. R., Castellucci, V. and Kandel, E. R. (1974) Proc. Natl. Acad. Sci. USA, 71, 3654-3658.
4. Switzer-Dunlap, M. and Hadfield, M. G. (1977) J. exp. mar. Biol. Ecol., 29, 245-261.
5. Carazzi, D. (1905) Archiv. Ital. di Anat. e Embriol., 4, 231-305.
6. Saunders, A.M.C. and Poole, M. (1910) Quart. J. Microsc. Sci., 55, 497-539.
7. Kandel, E. R. (1976) Cellular Basis of Behavior, W. H. Freeman and Co., San Francisco, pp. 1-727.
8. Kriegstein, A. R. (1977) J. Exp. Zool., 199, 275-288.
9. Switzer-Dunlap, M. and Hadfield, M. G. (in press) in Reproductive Ecology of Marine Invertebrates, Stancyk, S. E. ed., University of South Carolina Press, Columbia.
10. Thompson, T. E. (1961) Proc. Malac. Soc. London, 34, 233-238.
11. Thompson, T. E. (1958) Phil. Trans. Roy. Soc., B, 242, 1-58.
12. Fretter, V. (1972) J. mar. biol. Ass. U.K., 52, 161-177.
13. Kriegstein, A. R. (1977) Proc. Natl. Acad. Sci. USA, 74, 375-378.
14. Beeman, R. D. (1970) Vie Milieu, A, 21, 189-212.
15. Fretter, V. (1969) Proc. Malac. Soc. London, 38, 375-386.
16. Perron, F. E. and Turner, R. D. (1977) J. exp. mar. Biol. Ecol., 27, 171-185.
17. Kempf, S. C. and Willows, A.O.D. (in press) J. exp. mar.Biol. Ecol.
18. Chia, F. S. and Koss, R. (1977) Mar. Biol. (in press).
19. Thompson, T. E. (1962) Phil. Trans. Roy. Soc., B, 245, 171-218.
20. Tardy, J. (1970) Annls. Sci. Nat. (Zool.), 12, 299-370.
21. Bonar, D. B. and Hadfield, M. G. (1974) J. exp. mar. Biol. Ecol., 16, 227-255.
22. Crisp, D. J. (1974) in Chemoreception in Marine Organisms, Grant, P. T. and Mackie, A. M. eds., Academic Press, London, pp. 177-265.
23. Crisp, D. J. (1977) in Adaptation to Environment: Essays on the Physiology of the Marine Animals, Newell, R. C. ed., Butterworths, London, pp. 83-124.
24. Scheltema, R. (1974) Thalassia Jugoslavica, 10, 263-296.

A REVIEW OF SUBSTRATUM SELECTION
IN FREE-LIVING AND SYMBIOTIC CIRRIPEDS

Cindy A. Lewis

Department of Zoology, San Diego State University, San Diego, California 92182

Free-living acorn cirripeds respond to a variety of stimuli during settlement: (1) arthropodin, an insoluble protein found in the epicuticle and in bases of cirripeds, (2) water currents, (3) surface texture, grooves and concavities, (4) light (direction, intensity and wavelength), and (5) gravity and hydrostatic pressure. The settlement of high densities of cyprids is apparently due to tactile-chemical sensation of arthropodin resulting in responsiveness primarily to their own species. In the absence of appropriate substrata, settlement and metamorphosis are usually delayed.

Specificity of settling site, cues and behavioral responses are reviewed for free-living cirripeds and for those in symbiotic associations. Cirripeds with host specificity may utilize highly defined cues such as host-related chemicals and host breeding cycles as well as the cues used by free-living cirripeds. Chemicals that stimulate settlement of free-living cirripeds have been extracted from sponges, molluscs and a fish, which suggests that there is preadaptation to habitat radiation. Of special interest is the possibility that the host of a parent cirriped conditions the developing embryos to settle on a host of the same or a closely related species. Two parasitic cirripeds are known to prefer molting or newly molted hosts; thus, the presence of a particular molting hormone may stimulate spawning or settlement in cirripeds whose hosts molt periodically.

INTRODUCTION

Cirripeds are marine arthropods that include the true barnacles and their allies. Most cirripeds have a planktonic feeding larval stage, the nauplius, that is adapted for dispersal and a swimming-crawling, non-feeding larval stage, the cyprid, that is specialized for settlement and subsequent attachment. It was long thought that dispersal and settlement of benthic invertebrate larvae was a random process caused by tides and currents with only a small portion of the larval stock reaching an appropriate substratum and surviving by chance.[1,2] It is now known that many larvae can delay settlement for extended periods,[3] thereby increasing the probability of their reaching a satisfactory substratum. In cirripeds the cyprid larvae are known to respond to a variety of environmental cues that include larval food organisms,[4-6] proper substratum,[7] and the presence of other individuals of the same species[8,9] that may be encountered during a period of exploratory behavior. Following exploration, larvae attach by cement production[10-13] and undergo metamorphosis.[14-15]

This paper reviews the literature on the cues to which free-living as well as symbiotic cirripeds respond at settlement. As two recently published reviews have identified many of the cues employed by cyprids of free-living cirripeds,[16,17] the emphasis here is placed on the host preferences and settlement cues of symbiotic cirripeds. Some of the information presented is unpublished and is reported with permission of the authors.

SUBSTRATUM SELECTION IN FREE-LIVING CIRRIPEDS

Cyrpids respond to physical cues such as water currents,[18] light and gravity,[19-21] and hydrostatic pressure.[22] De Wolf[23] in explaining larval discontinuities in the plankton suggested that during low current velocity cyprids sink and then are redispersed in the water column when current rates increase, allowing tidal currents to transport the cyprids. These responses tend to move the larvae toward suitable settling sites.[24]

The depth at which larvae occur in the water column is apparently controlled by a combination of geotactic and photoactic responses,[19] culminating in a particular swimming behavior or perhaps in changes in buoyancy. Nauplii of littoral species appear to be primarily photopositive so that they are attracted to the surface of the water and settle in the littoral zone. Just prior to attachment some may become photonegative and others remain photopositive. Visscher[25] showed that photonegative responses may result in settlement in the shade as well as in deep water. Sublittoral cirripeds, on the other hand, are apparently photonegative for most of their larval life.[19] Several lepadomorph cirripeds float on the surface of the ocean as part of the pleuston. Although data are lacking, cyprids of pleustonic cirripeds must regulate their depth prior to settlement by photopositive and/or geonegative responses.

Hurley[26] observed that *Balanus pacificus,* a sublittoral cirriped, settles most frequently on plates nearest the ocean floor, regardless of water depth, and concluded that the cyprids crawl or swim along the bottom prior to settlement. She studied the distribution of *B. pacificus* spat that settled on asbestos plates and concluded that at least three levels of cyprid behavioral discrimination exist: (1) regulation of depth, (2) choice of a substratum, and (3) choice of a position on a given substratum. This is an opportunistic species that occurs along the coasts of California and Baja California occupying a variety of small objects, such as pebbles, bottles and arthropod skeletons in sandy areas and also along the edges of rocky areas.[27] Giltay and Boolootian reported that it also occurs on living *Dendraster,* the Pacific sand dollar.[28,29]

Ninety-four percent of the *Balanus pacificus* spat preferred the top portion of test plates.[26] This type of distribution also is observed in *B. crenatus* (Jon Standing, personal communication). After a period of photonegative swimming, the cyprids apparently undergo a photopositive period during the crawling phase. Because *B. pacificus* settles on most solid objects found on sandy bottoms, migration upward onto these objects during exploration may insure settlement above the sand, reducing the probability of scouring or burial.

Hurley[26] found significantly greater numbers of *Balanus pacificus* spat on black or white plates in preference to clear plates; dark plates usually were chosen when over a light-colored background. Hurley concluded that the contrast of the plate to the ocean floor is important and that the cyprids of *B. pacificus* may seek sites from a distance by using visual cues rather than attaching to any available solid object.

The most specialized pelagic goose barnacle, *Lepas fascicularis,* has a paper-thin shell, secretes a gas-filled float at the base of its peduncle for buoyancy, and settles on feathers, small floats of the kelp *Macrocystis* and tar lumps.[30] Thus, its cyprids may prefer small substrata over large as a method of avoiding competition at settlement.[31] *Lepas pacifica,* which does not form a float, is found only on more substantial substrata.[30] This suggests that *L. fascicularis* is stimulated to settle by a response to many small objects, while other lepadids that do not form a float delay settlement until they encounter an object that can support their adult weight. It is not known what mechanisms are used to determine the buoyancy of an object, but visual cues may be important for discriminating substrata by size. However, there is at present no direct evidence that cyprids respond visually to objects.

Newman and Ross[31] reported that virtually all floating objects are commonly fouled by species of *Lepas,* suggesting opportunism. Thus, another interpretation of the observation that *L. pacifica* is found only on large, buoyant substrata is that larvae that settle on small objects that cannot sustain their growing weight perish. Another indication of opportunism is the observation by Willemoes-Suhm[32] that few cyprids of *L. fascicularis* were present when nauplii and newly settled spat were abundant. Thus, although some lepadids enjoy a long dispersal phase,[33] the cyprid phase may show little, if any, delay of metamorphosis owing to the acceptance of a variety of substrata.

Gregarious settlements, the clustering of many individuals of a single species, and associative gregarious settlement, the grouping of individuals of more than one species,[17] are adaptive as they increase the possibility of copulation and at the same time insure the larvae a site where their species has survived. The observation of crowded aggregations of *Lepas anatifera* and *L. australis* separated by abundant unsettled spaces suggested gregariousness to Skerman.[34] The stimulant for gregarious settlement of some free-living acorn cirripeds isolated from a variety of sources is a high molecular weight acidic protein or protein-carbohydrate complex resistant to boiling in aqueous solution.[35] Arthropod-derived stimulant has been identified as arthropodin.[9,36] Larvae respond to contact with arthropodin, an insoluble quinone-tanned protein present in the cuticle of adult cirripeds, by increasing their rate of settlement.[9,36-38] The chemical composition of arthropodin apparently varies somewhat among species because many larvae show species specificity at settlement.[39]

Perception of the settlement factor is probably chemosensory. The chemosensory mechanism appears to be contact or tactile-chemical reception rather than distance chemoreception because larvae perceive the settlement factor only when it is adsorbed onto a surface, not when it is dissolved in water. This was shown by the experiments of Crisp and Meadows: when a test panel treated with arthropodin extract and an untreated panel were immersed in the same concentration of arthropodin dissolved in sea water, larvae were able to distinguish between the two, most settling on the treated panel.[36] Apparently a layer of arthropodin adheres to the panel in a manner like that in tanned epicuticle, and larvae respond only to the actual molecular configuration of the arthropodin that is detected by their sensory organs.[9,24]

Some sublittoral cirripeds have developed specialized males that are associated with females (dwarf males) or with hermaphrodites (complemental males). The presence of complemental males in species of *Scalpellum*, which usually lives isolated in deep water, suggests that male cyprids must be able to recognize settled adults. Although experimental evidence is lacking, one suspects that gregariousness due to species specific recognition of arthropodin plays a role in this encounter. It would also seem advantageous for the larvae to perceive soluble chemicals at a distance as occurs with some commensals.[40] The thoeretical problems associated with such perception by cyprids have been discussed by Crisp and Meadows.[9] If cyprids respond to gradients of soluble arthoropodin or other molecules in the ocean, it has yet to be demonstrated.

Superficial sensory receptors are commonly presumed to be either chemoreceptors or mechanoreceptors. Chemoreceptors and proprioceptors that may play a role in substratum selection have been described on the cyprid antennular attachment organ.[41] Possible mechanoreceptors have been noted on the fourth antennular segment[42] and on the valve cuticle and caudal appendages.[43] Knight-Jones[8] suggested two possible means of sensory transduction. The adsorbed layer of arthropodin may be cleaved by enzymes released from the larval antennulary glands resulting in a soluble material. Chemoreception of this material could then occur. Nott and Foster[41] agreed with this diagnosis after considering the structure of cells that appear to be chemosensory. But Crisp[16] preferred an analogy to an antigen-antibody interaction because in cyprid settlement: (1) recognition of a very large molecule is involved, (2) the reaction is sensitive to slight changes in structure, (3) phylogenetic affinity is reflected by the amount of interaction, and (4) molecular forces depend on the phsyical contact between two layers. The complementary configuration corresponds to the antennular attachment organ and the epicuticular arthropodin. Crisp points out that cyprids have antennular organs that could detect surface configurations and "register strain on the antennule." Further, cyprids move particles around the surface, perhaps to uncover a small area where "molecular interaction" can occur. Contact-chemical reception and short-distance chemoreception remain alternative hypotheses.

Cyprids also respond to rough surface texture,[44,45] surface concavities and grooves (rugophilism).[46,47] But spat of the opportunist, *Balanus pacificus*, settle on smooth or rough surfaces, suggesting that their response to texture may not be well developed.[26] Arthropodin is more influential in inducing settlement than the presence of a pitted or rough surface.[36] Crisp and Meadows presented panels with only extract or pits to cyprids and found that most did not settle unless the two types of panels were adjacent, then primarily chose panels with extract. When four types of panels were presented to cyprids (extract ± pits; no extract ± pits), more cyprids settled than before, preferring the extract plus pits. But some cyprids settled on panels with extract alone and none settled on untreated pitted panels. Crisp and Barnes[46] also determined that cyprid reactions to light are independent of rugophilism, but that the response to surface contour is generally stronger.

The ability to delay settlement until the proper substratum is found would be particularly advantageous to cirripeds with specialized settlement sites, allowing a longer searching phase with a greater probability of finding the optimum site. Even *Semibalanus balanoides*,[4] *Balanus crenatus* and *B. glandula* (Jon Standing, personal communication), which live in the rocky intertidal zone delay naupliar release until planktonic food is available, allowing for a longer dispersal stage. But there is apparently a limit to the length of time a larva can delay its choice of settlement site and metamorphosis. Aged cyprids may have less ability to discriminate for sites owing to an increased drive to settle or they may be unable to metamorphose at all after an extended period.[8]

SUBSTRATUM SELECTION IN SYMBIOTIC CIRRIPEDS

Among the cirripeds, symbiosis may be manifested as phoresis[48] (transportation of one member by the other without metabolic ties), mutualism[49,50] (both parties benefit) commensalism[51] (one member is benefitted while the other, the host, is neither benefitted nor harmed), and parasitism[48] (one member benefits at the expense of the other, the host[52]). It is often difficult to place a particular association in one of these categories owing to a lack of basic information. Thus, where cirripeds are termed commensal, it is often a presumed rather than a demonstrated relationship.

Successful commensal relationships must involve complex adaptations that insure the persistence of those relationships. The degree of host specificity is variable among symbiotic species, ranging from a single to many host species,[53] and may even change throughout the geographic range of symbionts.[54] Because a commensal or parasite's environment has become so specialized, adaptations to the host can be extreme. The biology of commensal cirripeds is not as well known as that of free-living cirripeds. Much of the information available is reviewed here.

One suspects that beyond the ability to respond to cues such as substratum texture and to arthropodin, larvae of symbiotic cirripeds might also respond to specific biological indicators of their future habitat. This is consistent with results from experiments where extracts of the adult's host have been used to promote settlement.[55-57] The less likely it is that a cyprid will encounter its future habitat by random movement, the greater is the necessity for a mechanism of recognition.[48]

In experiments designed to determine how intraspecific gregariousness is mediated, Crisp and Meadows showed that *Semibalanus balanoides* is stimulated to settle on panels soaked in extracts from a fish and two sponges as well as from a variety of arthropods.[9] Knight-Jones had shown that phylogenetic affinities are reflected in the ability of an extract to stimulate settlement.[39] Crisp[24] and Meadows and Campbell[7] reviewed examples where responses to cues from alternative attachment sites might lead to speciation, such as in *Spirorbis* where several species each have an affinity for a "chemically different habitat."[24] Crisp suggested that there is an evolutionary significance of the non-arthropod extracts: "Since sponges and vertebrates possess substances which can present molecular arrangements sufficiently similar to those of the arthropod cuticle to stimulate cyprids of non-epizoic species, it seems quite likely that, where biochemically favorable surfaces exist on an organism that would provide an ecologically suitable habitat, an epizoic and perhaps ultimately a parasitic relationship may well have evolved."[16]

Are the characteristics of other settling factors similar to those of arthropodin? Crisp felt that non-arthropod settling inducers may be very different from the arthropodins. He showed that a crude sponge extract loses most of its inductive acitvity by boiling in water for five minutes, while arthropodin boiled for twenty hours still retains half of its activity.[16] Recently Larman and Gabbott showed that ovalbumin and boiled extracts of the bivalves *Mytilus edulis* and *Ostrea edulis* promoted *S. balanoides* settlement.[35] All of the extracts that stimulated settlement in *Semibalanus balanoides* contained acidic protein or protein-carbohydrate complexes.[35] From previous work we know that the settlement inducers have a high molecular weight, are proteinaceous and are PAS positive.[36] Unfortunately, even the arthropodin extract is not highly purified and may include unrelated fractions.[38] Cirriped settlement factors in other animal groups are even less characterized.

All four species of acorn barnacles that have complemental males are epizoic (*Conopea galeata, C. calceola, C. merrilli,*[58] *Solidobalanus masignotus*[59]). The hermaphrodites of *C. galeata* settle on exposed axial skeletons of gorgonians, on the shells of other adult hermaphrodites that are completely covered with gorgonian coenenchyme and occasionally on old egg cases of the gastropod *Neosimnia* on the gorgonian.[60] Hermaphrodites clearly prefer the exposed gorgonian skeleton and apparently recognize at least three tissues: gorgonian axial skeleton, gorgonian coenenchyme and shells of its own species. Whether or not they retain a contact-chemical sense for recognition of arthropodin and/or for gorgonin or other skeletal or soft tissue components remains unknown.

Male cyprids settle only on hermaphrodites that are large enough to accommodate them and that are sexually mature. Their settlement sites in order of preference are the inside surfaces of the rostral, lateral and scutal plates.[58] Males live for only about three months, but are consistently found on hermaphrodites that survive up to two years in the field.[58] Thus, there appears to be a continual replacement of spent males with the main limiting factor being space availability.

When hermaphrodite cyprids of *Conopea* were cultured in the absence of appropriate substrata, 85% survived for two weeks retaining 62% of the original settling capacity.[58] Thus, *Conopea* can delay settlement, a behavior that is especially adaptive to substrata which are widely separated or rare. However, most hermaphrodite cyprids that lived for 30 days or more were unable to settle and metamorphose in a normal fashion.[58]

Boscia anglica, an obligatory symbiont that relies on its host cup coral *Caryophyllia smithii* for habitat and protection, is considered "semi-parasitic" because it may use solutes that diffuse from the coral[61] or it may have some physiological control over the coral.[62,63] Settlement of *Boscia* occurs primarily along the lower half of coral polyps that are attached to a vertical surface,[61] suggesting that larvae can regulate depth prior to attachment as was assumed for *Balanus pacificus.*

Boscia anglica cyprids first attach to the living coral epithelium and then "burrow" through it by an unknown mechanism.[61,64] Duerden[65] reported that *B. stokesii* cyprids attach within the polyp cavity and "bore" through both gastrodermis and living skeleton. *B. anglica* then aligns its rostrocarinal axis specifically along the sclerosceptal grooves of the coral rim.[61] This alignment may be a rugotropic response[46] after antennal "probing" through the coenenchyme distinguishes the skeletal grooves. Moyse[66] previously noted that the cyprid's antennular setae became abraded, perhaps during the searching or initial burrowing phase. "Probing" by the antennulae may actually be mechanical burrowing rather than digestion or laceration of coenenchyme as the antennule's structure appears "little different from . . . other balanids."[61]

Boscia is apparently gregarious because Moyse showed that the actual frequency distributions of *Boscia* on *Caryophyllia* differ from a random pattern and that angles formed between two or more cirripeds inhabiting the same coral host tend to be small.[61] Only about 21% of

Boscia are attached or adjacent to other *Boscia*, but the majority lie within 90° of each other.[61] *Boscia* is thus likely attracted to the coral and to its own species.

Attempts have not been successful to extract any material from *Caryophyllia smithii* that affects *Boscia anglica* settlement or in finding a gorgonian extract that stimulates *Conopea galatea* settlement.[16] However, Bourdillon[67] prepared a water soluble extract from axial skeletons of a gorgonian that stimulated metamorphosis of *Alcyonium*, an anthozoan. As gorgonin is known to have tyrosine residues, various purified proteins were given to *Alcyonium*; weak stimulation was elicited by only one, diiodotyrosine. Meadows[16] tried a tyrosine polymer on *Semibalanus* with negative results.

Cyprids of *Conopea* hermaphrodites generally attach within a few millimeters of the soft tissues of their gorgonian host, even when a considerable area of exposed skeleton or a barnacle shell is available and *Boscia* settles at a particular distance from the coral rim.[58,61] This behavior indicates a kind of territoriality. *Conopea* cyprids apparently "pace off" from the coenenchyme before attaching. Cirripeds that settle too close to regenerating coenenchyme are smothered and killed prior to initiation of feeding. Those that settle too far from the coenenchyme will be exposed for an extended period, and the regenerating coenenchyme may not reach them before they are eaten or become fouled. *Boscia*'s settling site allows a "vantage point" for feeding and frees the cirriped from interspecific competition. Literature on behavior known as "spacing out" or territoriality, apparently a mechanism for avoiding competition, has been reviewed by Meadows and Campbell[7] and Crisp.[17] A modification of this behavior may be used for site selection on the host tissue by *Conopea* and *Boscia*.

The genera *Acasta*, *Armatobalanus*, *Membranobalanus* and *Pyrgopsella* inhabit sponges.[68,69] *Membranobalanus orcutti*, an obligate commensal, inhabits the burrowing sponges *Spheciospongia confoederata* and *Cliona celata* in southern California. After settlement and metamorphosis, the barnacle is covered completely by sponge tissue except for its aperture.

Because *Membranobalanus orcutti* resides on two species of host, questions may be raised concerning chemical recognition of the host during settlement. Jones[70] has observed the amount of recruitment onto sponges with or without adult cirripeds present and cyprid preferences for the two host sponges. The presence of adult cirripeds influences recruitment at only one of the two field sites and not at all in the laboratory; thus, gregariousness toward its own species may not be important at settlement. All cues may come from the host.

A very intriguing possibility is that the developing cirriped embryos and larvae are conditioned by their environment. Jones showed that *Membranobalanus* cyprids did distinguish between the two host sponge species in the laboratory.[70] Cyprids from parents living in one of the host species preferred that species to the alternate host species for a settlement site.[70] This suggests responsiveness to a host-specific attractant. In another experiment Jones removed adults from their host sponges and allowed them to produce several broods of embryos for up to three months in the absence of the sponge. Then a brood of released nauplii was removed from the adult and allowed to develop in isolation. These larvae still preferred the sponge species occupied by their parents. When developing embryos were conditioned by the presence of the alternate host sponge, attractiveness to that species increased, but preference for the parental host did not change. This suggests that there is a genetic predisposition to a certain host. Perhaps the cirriped's behavior is due to a combination of inheritance and conditioning by a host factor.

Acrothoracicans are small commensal cirripeds having normal-sized females and dwarf males. During metamorphosis the female bores into calcareous shells of bivalves, gastropods, echinoids, chitons, cirripeds, live or dead corals, or limestone in shallow water. Later a male attaches on the outside of her mantle cavity. Only a small opening to the burrow of the female indicates its presence.[68,71] Little information is available regarding their larval settlement.[72]

212

Cryptophialus melampygos invades a variety of littoral molluscs. Female cyprids that leave the maternal mantle cavity have the ability to crawl, but not to swim.[73] Thus, dispersal is very limited and may help to explain the high density of spat. The female cyprid invades the outer surface of the shell, especially where abrasion has removed the periostracum,[73] allowing the thick calcareous prismatic layer to be exposed. Female cyprids adhere to the host shell by the antennular attachment organ, but whether a tactile or chemical response is elicited by the host is not known. Male cyprids apparently are attracted to the mantle cavity or to the pit bored by the attached femlae. Once cemented, the male cyprid metamorphoses into a dwarf male.

Female cyprids of *Trypetesa lampas* and *T. nassarioides*, burrowers commensal in the shells of hermit crabs, discriminate between species of hermit crabs and between size and age of shells.[74] They choose only inhabited shells when given the choice of shells with or without crabs. Turquier[74] concluded that *Trypetesa* embryos are sensitized to a specific biochemical environment because when the parent was removed from its host shell, fewer of its progeny settled than progeny from parents left in their host shells. The assumption is that some factors in the host environment affected later settlement success. Turquier found no evidence to suggest that chemoreception occurs at a distance as in some other commensals,[40] but inferred that the choice of host was made on contact. He concluded that a settlement-stimulating factor resides within the periostracum of the mollusc shell.[74]

Cryptophialus burrows into the eroded outer part of mollusc shells, while *Trypetesa* burrows into the central part of gastropod shells from the interior. Cyprids of both genera contact periostracum during their exploratory period and prismatic shell layers during burrowing. An organic compound present in the periostracum of gastropod shells has been shown to play a role in the attraction of certain sea anemones that are commensal on them.[75,76] Anemones respond to periostracum stripped from shells, but not to shells that are boiled in alkali to remove their organic content.[75] The factor is insoluble in organic solvents and is not digested by proteases.[77] Thus, the factor that stimulates anemones to settle on the shell is "highly stable and insoluble, perhaps a quinone-tanned protein or a mucopolysaccharide"[75] and may be the same factor important to acrothoracican cirripeds. Another possibility is that an active material and/or binding sites are provided for the cyprid by exposing the prismatic layer during burrowing.

Whale and "turtle" barnacles generally are assumed to be phoretic or commensal, depending on the emphasis one places on the presumed dependence of the cirriped on host-contributed water currents. "Turtle" barnacles include a variety of cirripeds that settle on sea snakes and crustaceans as well as on turtles. Cirripeds may synchronize their breeding season with that of their hosts, but little information is available to document this notion.

A variety of cirripeds, mostly balanomorphs, reside on whales. They include: *Coronula*, primarily on the humpback whale[78]; *Cryptolepas* on the California grey whale[78]; *Tubicinella* on the southern right whale[79]; and *Xenobalanus* on porpoises, dolphins and blackfish.[80] Two species of *Conchoderma* occur on whale barnacles: *C. auritum* and *C. virgatum*. *Conchoderma auritum* is found on *Coronula* while *C. virgatum* is opportunistic.[81]

The evidence of settling cues available to whale barnacles is indirect. California grey whales caught during their migration south toward the breeding and calving grounds have only large cirripeds attached, while northbound whales carry two discrete size and, presumably, age groups.[78] These whales spend the winter in aggregations in shallow lagoons in Baja California at which time their young are born. Because the greatest opportunity for finding a new host probably occurs at this time, *Cryptolepas* may synchronize its breeding cycle with that of its host.[82] Cirripeds would then be available to "infect" all whales, young and old, every winter. The synchronization may be cued to a release of whale pheromones, changes in sea water temperature or salinity, or some other factor. Once settled, the cirripeds orient themselves for maximal water flow through the cirri.[83-85].

The Order Rhizocephala constitutes a group of entirely parasitic cirripeds that have evolved a complicated life cycle. Delage[86] artificially infected a host crab with cyprids and found that they attach to the base of a setal hair on a walking leg or in a cavity on the carapace. Other species attach only in the branchial chamber of the host.[87] After attachment, the cyprid metamorphoses into a sac-like body, the kentrogon, that develops a hollow stylet to penetrate the host exoskeleton. The site of female cyprid fixation on the host and the manner of metamorphosis may differ for each rhizocephalan species. Veillet[88] suggested that the evolution of attachment and metamorphosis of the parasite's cyprid parallels the evolution of its host.

Settlement of the cyprid larvae of *Lernaeodiscus porcellanae* on their host, the porcelain crab, *Petrolisthes cabrilloi,* has been documented by observations in the laboratory on the infection of unparasitized crabs by the larvae.[89,90] Cyprid attachment takes place 30 minutes or so after their introduction to a crab. Although cyprids may explore the crab's outer carapace, no attempt is made to attach and metamorphose there. Instead, larvae swim back to the water column. Female cyprids attach primarily on the branchiae within the host branchial chamber after being caught in the respiratory current.

Lernaeodiscus porcellanae is found in about 11% of the host crabs in Ritchie's study.[89] Because *Petrolisthes cabrilloi* commences grooming behavior upon encountering cyprids, the parasite's success must take advantage of the decreased effectiveness of the crab's grooming behavior. Thus, the loss or damage of one or both appendages (1.5% of field populations) may be incipient to parasitism. Veillet noted that the cyprid of one sacculinid attaches only in areas of the branchial chamber of its host that are not accessible to the cleaning appendages.[87] Experiments where Ritchie removed or damaged one or both appendages showed greater cyprid attachment and subsequent invasion ($\bar{x}$ of 25% to 75%) than in untreated control crabs ($\bar{x}$ of 10%).[89]

Recognition of the host by the female rhizocephalan cyprid is required if parasitism is to be initiated. What cues are utilized and what are the responses by female cyprids? Female cyprids apparently recognize their prospective host primarily by tactile stimuli and do not use distance reception. This conclusion is based on the following observations and experiments of Ritchie.[89]

Larval behavior did not change when female cyprids were introduced to *Petrolisthes cabrilloi.* The larvae neither swam toward nor became "excited" by the presence of the crabs. Larvae were frequently oberved landing on crabs and "walking" by their antennules, but the behavior was the same as if they were on the bottom of a vessel. Larvae were also observed lying on the crab as though resting, and then swimming away without any sign of host recognition. It appears that a cyprid never intentionally enters the branchial chamber, nor does a cyprid that lands on a crab walk systematically toward or into the gill area. Most cyprids walking on the crab detach or are removed by grooming behavior and return to the water column. More often larvae enter the crab's inhalent gill current and are swept into the branchial chamber. If the cyprid does not come out the exhalent side, it is considered to have settled and is later observed for metamorphosis.

Ritchie placed female larvae in vessels with dead host crabs with or without carapaces (removal of a carapace compensates for the lack of a gill current). Female larvae did not prefer live to dead hosts. Considerably more cyprids attached to dead crabs without carapaces than with them. Parasites were found on dead crabs with carapaces only on the peripheral gills. One interpretation is that female cyprids do not actively search for prospective hosts, but depend on random encounters that are facilitated by the crab's gill currents. More cyprids attached to controls with ablated cleaning appendages than to dead crabs with their carapaces intact,[89] also suggesting that gill currents actively transport cyprids. Thus, host recognition and discrimination probably occur at the time of contact.

The strongest indication that female larvae did not employ distance reception to locate their prospective host is the observation that in the presence of other decapods, the larvae do not react any differently than in the presence of their host species. Larvae were just as likely to be trapped and swept into the branchial chambers of non-hosts as into hosts.[89] Ritchie offered 18 decapod species not normally parasitized by *Lernaeodiscus porcellanae* to the female cyprids.[89] All of the porcelain crab species were colonized. Thus, specificity of host recognition is not refined, and probably encompasses the family Porcellanidae. However, more kentogrons attached to species of *Petrolisthes* than *Pachycheles*.[89]

When a female cyprid contacts any object, it must determine whether or not the object is a host. The decision to settle on species that are normally not hosts would presumably be based on phylogenetic affinity as is thought in balanomorphs.[39] Thus, the greater the recognition of a non-host species as the host, the closer its affinity to the host species. If such recognition exists, the mechanism of discrimination would likely be a chemical cue. The chemical cue may be non-specific and present in varying quantities in species closely related to the host. Thus, the more distant the phylogenetic affinity, the smaller the amount of the cue, resulting in a decreased likelihood of host recognition. The fact that none of the non-porcellanid crabs were recognized supports this contention.

A non-specific chemical cue used by female larvae of *Lernaeodiscus porcellanae* female larvae to determine whether to metamorphose could possibly be ecdysones, molting hormones, since female cyprids prefer crabs that are late in their molt cycle.[89] In addition, three of 37 *Petrolisthes gracilis* (non-host crabs) underwent ecdysis during an exposure to larvae and three times the number of kentogrons attached to molting than non-molting crabs.[89] Female *Heterosaccus californicus* cyprids also discriminate in favor of newly molted crabs.[91] Alternatively, newly tanned arthropodin that is apparently present at ecdysis[9] could induce settlement.

After the female cyprid invades the host, it cannot reproduce until a male cyprid settles in the female mantle cavity and injects its cells into the female. Ritchie's settling experiments in the laboratory showed that virgin externae elicited an immediate response from male larvae. Within seconds after introduction, the externae were covered with walking males. A distinctive feature of this behavior is a clustering of the males about the aperture of the virgin female where they literally "fight" to gain entrance into the mantle cavity. All virgin externae placed in the field and recovered after 24 hours had at least three male cyprids attached.[89] Only virgin externae elicit this response, suggesting the possible production of a male attractant that would probably be released during the transitory ecdysis from the unattractive juvenile to the attractive virgin stage when the internal environment of the mantle cavity (*i.e.*, the male-cell receptacles) is open to the external environment. Thus, ecdysone could be an attractant.

Reinhard[92] reported on a female *Peltogaster* whose mantle cavity was packed with hundreds of male larvae. Apparently the female's body wall had been perforated and partially healed without closing. The males entered by this route in such numbers that the visceral mass atrophied. This mass gathering of males supports the suggestion that males may respond to a concentration gradient of some substance released by the female.

It is not known whether female receptivity to males is due to a developmental mechanism, hormonal activity of the female externa, or site selectivity and recognition by the male cyprid. Why should male and female rhizocephalan cyprids have different mechanisms to recognize settlement sites? There is a higher probability of females randomly encountering host crabs than of males encountering virgin females (only 11% of the host crabs are infected by *Lernaeodiscus*). Thus, chemoreception, if used, may be an adaptation to insure a greater frequency of males encountering females.

CONCLUSIONS

Symbiotic cirripeds probably respond to most of the cues that are used by free-living cirripeds (Table 1), but they may respond in a unique manner, as with territoriality. In addition,

TABLE 1

Comparison of Cues Probably Used by Cirripeds at Settlement

	Free-living	References	Symbionts	References
Depth regulation by the swimming cyprid				
Phototaxis, geotaxis, pressure	+	19-22,25,26,93-97	+	61
Response to water currents	+	18	?	
Exploration by the crawling cyprid				
Gregariousness				
Response to tactile-chemical stimulation				
by arthropodin/other stimulants	+	8,9,36-39	+	24,89
Species specificity	+	8,34,39	+	58,61
Territoriality	+	36,98	+	58,61
Associative gregariousness	−		+	48,51,58,61,70-74
Host "conditioning"	−		+	70,74
Surface concavities and grooves	+	46,47	+	61
Surface texture	+	44,45	?	
Water currents	+	83,99	+	83-85
Delay of attachment and metamorphosis	+	4,8	+	58,89
Reproductive synchronization with host	−		+	78,82

they may be able to detect certain chemical and/or physical characteristics of their prospective host and to synchronize their reproductive cycle with that of their host. Most of the information currently available for settlement of symbiotic cirripeds is based on circumstantial evidence. We know little regarding actual mechanisms used.

ACKNOWLEDGMENTS

I gratefully acknowledge the following who spent considerable time and effort making suggestions for this manuscript: Arnold Ross, William Newman, Jon Standing, Andy Olson, Jack Tomlinson and Gayle Kidder. Special thanks are due to Linda Jones, Ed Gomez and Jack O'Brien for allowing the author to include unpublished observations from their Ph.D. dissertations.

Larry Ritchie died tragically without completing his Ph.D. dissertation. His work was supervised by William Newman, Scripps Institution of Oceanography, who graciously allowed the author to read the first draft of Mr. Ritchie's dissertation and has given permission to relate parts of it here.

Contribution No. 31 from the Center for Marine Studies, San Diego State University, San Diego, California 92182.

REFERENCES

1. Nelson, T. C. (1928) Biol. Bull., 40, 180-192.
2. Colman, J. S. (1933) J. mar. biol. Ass. U.K., 18, 435-476.
3. Wilson, D. P. (1932) Phil. Trans. R. Soc. B, 221, 231-334.
4. Barnes, H. (1957) Annee biol., 33, 67-85.
5. Crisp, D. J. (1956) Nature, Lond., 178, 263.
6. Crisp, D. J. and Spencer, C. P. (1958) Proc. R. Soc. B, 148, 278-299.

7. Meadows, P. S. and Campbell, J. I. (1972) Adv. Mar. Biol., 10, 271-382.

8. Knight-Jones, E. W. (1953) J. exp. Biol., 30, 584-598.

9. Crisp, D. J. and Meadows, P. S. (1962) Proc. R. Soc. B, 156, 500-520.

10. Walker, G. (1971) Mar. Biol., 9, 205-212.

11. Walker, G. (1973) J. exp. mar. Biol. Ecol., 12, 305-314.

12. Holland, D. L. and Walker, G. (1975) J. Cons. Int. Explor. Mer, 36, 162-165.

13. Fyhn, U.E.H. and Costlow, J. D. (1976) Biol. Bull., 150, 47-56.

14. Costlow, J. D. (1968) in Metamorphosis in Crustaceans, Etkin, W. and Gilbert, L. I. eds., Appleton Century Croft, New York, pp. 3-41.

15. Walley, L. J. (1969) Phil. Trans. Roy. Soc. B, 256, 237-280.

16. Crisp, D. J. (1974) in Chemoreception in Marine Organisms, Grant, P. T. and Mackie, A. M. eds., Academic Press, New York, pp. 177-265.

17. Crisp, D. J. (1976) in Adaptation to Environment: Studies on the Physiology of Marine Animals, Newell, R. C. ed., Butterworths Pub., Inc., London, pp. 83-124.

18. Crisp, D. J. (1955) J. exp. Biol., 32, 569-590.

19. Thorson, G. (1964) Meddr. Kommn Havunders Kbh. Ser. Plankton, 4, 1-523.

20. Scheltema, R. S. (1975) in The Ecology of Fouling Communities, Proc. U.S.-USSR Workshop within the Program "Biological Productivity and Biochemistry of the World's Oceans," Costlow, J. D. ed., U.S. Government Printing Office, Washington, D.C., pp. 27-47.

21. Forbes, L., Seward, M.J.B. and Crisp, D. J. (1971) in Fourth European Marine Biology Symposium, Crisp, D. J. ed., Cambridge Univ. Press, Cambridge, pp. 539-558.

22. Knight-Jones, E. W. and Morgan, E. (1966) Oceanogr. mar. Biol. A. Rev., 4, 267-299.

23. De Wolf, P. (1973) Netherlands J. Sea Res., 6, 1-129.

24. Crisp, D. J. (1965) in Botanica Gothenburgensia III. Proc. 5th Marine Biology Symposium, Goteborg, pp. 51-65.

25. Visscher, J. P. (1928) Biol. Bull., 54, 327-335.

26. Hurley, A. C. (1973) J. Anim. Ecol., 42, 599-609.

27. Hartline, A. C. (1972) Ecology of the Subtidal Acorn Barnacle *Balanus pacificus* Pilsbry, Unpublished Ph.D. dissertation, Scripps Institution of Oceangraphy.

28. Giltay, L. (1934) Mus. Hist. Nat. Belgique Bull., 10, 1-7.

29. Boolootian, R. A. (1964) Bull. South. Calif. Acad. Sci., 63, 185-191.

30. Cheng, L. and Lewin R. (1974) Fishery Bull., 74, 212-217.

31. Newman, W. A. and Ross, A. (1971) Antarctic Cirripedia, vol. 14 in Antarctic Research Series, American Geophysical Union of the National Academy of Sciences-National Research Council, pp. 1-257.

32. von Willemoes-Suhm, R. (1876) Phil. Trans. R. Soc. Lond., 166, 131-154.

33. Moyse, J. (1963) J. Cons. Int. Explor. Mer, 28, 175-187.

34. Skerman, T. M. (1958) New Zealand J. Sci., 1, 383-390.

35. Larman, V. N. and Gabbott, P. A. (1975) J. mar. biol. Ass. U.K., 55, 183-190.

36. Crisp, D. J. and Meadows, P. S. (1963) Proc. R. Soc. B., 158, 364-387.

37. Crisp, D. J. (1961) J. exp. Biol., 38, 429-446.

38. Gabbott, P. A. and Larman, V. N. (1971) in Fourth European Marine Biology Symposium, Crisp, D. J. ed., Cambridge Univ. Press, Cambridge, pp. 143-153.

39. Knight-Jones, E. W. (1955) Nature Lond. 175, 266.

40. Davenport, D. (1950) Biol. Bull., 98, 81-93.

41. Nott, J. A. and Foster, B. A. (1969) Phil. Trans. R. Soc. B., 256, 115-134.

42. Gibson, P. H. and Nott, J. A. (1971) in Fourth European Marine Biology Symposium, Crisp, D. J. ed., Cambridge Univ. Press, Cambridge, pp. 227-236.

43. Walker G. and Lee, V. E. (1976) J. Zool. Lond., 178, 161-172.

44. Barnes, H. (1956) Arch. Soc. Zool. Bot. Fenn. Vanamo, 10, 164-168.

45. Pomerat, C. M. and Weiss, C. M. (1946) Biol. Bull., 91, 57-65.

46. Crisp, D. J. and Barnes, H. (1954) J. Anim. Ecol., 23, 142-162.

47. Gregg, J. H. (1948) Biol. Bull., 94, 161-168.

48. Gotto, R. V. (1969) Marine Animals Partnerships and Other Associations, The English Univ. Press, London, pp. 1-96.

49. Newman, W. A. (1961) Veliger, 4, 99-107.

50. Jones, E. C. (1968) Crustaceana, 14, 312-314.

51. Zann, L. P. (1975) in Biology of Sea Snakes, Dunson, W. A.. ed., Univ. Park Press, Baltimore, pp. 268-286.

52. Odum, E. P. (1971) Fundamentals of Ecology, 3rd ed., W. B. Saunders, Philadelphia, pp. 1-574.

53. Ross, A. and Newman, W. A. (1973) Trans. San Diego Soc. Nat. Hist., 17, 137-173.
54. Davenport, D. (1955) Q. Rev. Biol., 30, 29-46.
55. Crisp, D. J. and Williams, G. B. (1960) Nature, Lond., 188, 1206-1207.
56. Williams, G. B. (1964) J. mar. biol. Ass. U.K., 44, 397-414.
57. Hayward, P. J. and Harvey, P. H. (1974) J. mar. biol. Ass. U.K., 54, 665-676.
58. Gomez, E. D. (1973) The Biology of the Commensal Barnacle *Balanus galeatus* (L.) with Special Reference to the Complemental Male-Hermaphrodite Relationship, Unpublished Ph.D. dissertation, Scripps Institution of Oceanography.
59. Henry, D. P. and McLaughlin, P. (1967) Crustaceana, 12, 43-58.
60. Patton, W. K. (1963) Am. Zool., 3, 522.
61. Moyse, J. (1971) in Fourth European Mar. Biol. Symp., Crisp, D. J. ed., Cambridge Univ. Press, Cambridge, pp. 125-141.
62. Ross, A. and Newman, W. A. (1969) Pacif. Sci., 23, 252-256.
63. Patton, W. K. (1976) in Biology and Geology of Coral Reefs, vol. 3, Biol. 2, Jones, O. A. and Endean, R. eds., Academic Press, New York, pp. 1-36.
64. Utinomi, H. (1943) Annot. Zool. Japan, 22, 15-22.
65. Duerden, J. E. (1904) Carnegie Inst. Wash. Publ. Mem. No. 20.
66. Moyse, J. (1961) Proc. zool. Soc. Lond., 137, 371-392.
67. Bourdillon, A. (1954) C. R. hebd. Seanc. Acad. Sci., Paris, 239, 1434-1436.
68. Newman, W. A., Zullo, V. A. and Withers, T. H. (1969) in Treatise on Invertebrate Paleontology Part R. Arthropoda, Moore, R. C. ed., Geol. Soc. Am., Univ. Kansas, pp. R206-R295.
69. Newman, W. A. and Ross, A. (1976) San Diego Soc. Nat. Hist., Memoir 9, pp. 1-108.
70. Jones, L. The Life History Patterns and Larval Host Selection Behavior of the Commensal Barnacle *Membranobalanus orcutti* (Pilsbry), Ph.D. dissertation, Scripps Institution of Oceanography.
71. Tomlinson, J. T. and Newman, W. A. (1960) Proc. U.S. Nat. Mus., 112, 517-526.
72. Tomlinson, J. T. (1969) U.S. Nat. Mus. Bull. 296, 1-162.
73. Batham, E. J. and Tomlinson, J. T. (1965) Trans. Roy. Soc. N.Z., 7, 141-154.
74. Turquier, Y. (1922) Arch. Zool. exp. gén., 113, 499-551.
75. Ross, D. M. and Sutton, L. (1961) Proc. Roy. Soc. B, 158, 266-281.
76. Ross, D. M. and Sutton, L. (1967) Can. J. Zool., 45, 895-910.
77. Ross, D. M. (1965) Am. Zool., 5, 573-580.
78. Rice, D. W. and Wolman, A. A. (1971) The Life History and Ecology of the Gray Whale (*Eschrichtius robustus*), Am. Soc. Mammolgoists, pp. 1-142.
79. Pilsbry, H. A. (1910) Am. Nat. 44, 304-306.
80. Pope, E. (1958) Proc. R. Zool. Soc. New South Wales, 1956-57, 159-161.
81. Roskell, J. (1969) Crustaceana, 16, 103-104.
82. Ross, A. and Emerson, W. K. (1974) Wonders of Barnacles, Dodd, Mead and Co., New York, pp. 1-78.
83. Crisp, D. J. and Stubbings, H. G. (1957) J. Anim. Ecol., 26, 179-196.
84. Briggs, K. T. and Morejohn, G. V. (1972) J. Zool. Lond., 167, 287-292.
85. Kasuya, T. and Rice, D. W. (1970) Scient. Rep. Whales Res. Inst., 22, 39-43.
86. Delage, Y. (1884) Arch. Zool. exp. gén., 2, 417-736.
87. Veillet, A. (1963) c. r. Acad. Sci., Paris, 256, 1609-1610.
88. Veillet, A. (1964) Zool. Meded. Rijk. Nat. Hist. Leiden, 39, 573-576.
89. Ritchie, L. (1977) The Coevolution of a Parasite/Host Relationship: Life History of the Rhizocephalan Parasite *Lernaeodiscus porcellenae* and its Relationship to the Crab *Petrolisthes cabrilloi*. With notes on the Evolution of Parasitism, Unifinished Ph.D. dissertation, Scripps Institution of Oceanography.
90. Ritchie, L. and Newman, W. A. (in preparation).
91. O'Brien, J. (in preparation).
92. Reinhard, E. G. (1942) J. Morph., 70, 389-402.
93. Visscher, J. P. and Luce, R. H. (1928) Biol. Bull., 54, 336-350.
94. Pomerat, C. M. and Reiner, E. R. (1942) Biol. Bull., 82, 14-25.
95. Gregg, J. H. (1945) Biol. Bull., 88, 44-49.
96. Weiss, C. M. (1947) Biol. Bull., 93, 240-249.
97. Crisp, D. J. and Rite, D. A. (1973) Mar. Biol., 23, 327-335.
98. Knight-Jones, E. W. and Stevenson, J. P. (1950) J. mar. biol. Ass. U.K., 29, 281-297.
99. Smith, F.G.W. (1946) Biol. Bull., 90, 51-70.

ECHINODERM METAMORPHOSIS: FATE OF LARVAL STRUCTURES

Fu-Shiang Chia and Robert D. Burke

Department of Zoology, University of Alberta, Edmonton, Canada T6G 2E9

From the literature we have compiled a list of 121 echinoderm species for which the development through metamorphosis has been observed. We have also tabulated the fates of larval structures for the five echinoderm classes. During metamorphosis the preoral lobe of the lecithotrophic larva of the asteroid *Leptasterias hexiactis* is first histolysized and then the degenerated tissue is incorporated into the digestive tract of the juvenile. The resorption of larval structures in planktotrophic larvae is briefly reviewed. We report our observations on the resorption of the larval epidermis and digestive tract of the echinopluteus of the sand dollar *Dendraster excentricus*. In this case, the larval epidermis and larval gut undergo histolysis and degenerate, the epidermal cells become necrotic and are phagocytized by the dedifferentiated cells of the larval gut. Subsequently, the gut redifferentiates into the adult digestive tract.

INTRODUCTION

The larval forms within each echinoderm class are: doliolaria in crinoids, auricularia, doliolaria and pentactula in holothurians, bipinnaria and brachiolaria in asteroids, ophiopluteus in ophiuroids, and echinopluteus in echinoids. With the exception of the doliolaria and pentactula, which are radially symmetric, the rest of the echinoderm larvae are bilaterally symmetric and most of them are pelagic and planktotrophic.

Echinoderm metamorphosis involves, in most cases, the transformation of a bilaterally symmetric and pelagic larva into a radially symmetric and benthic adult. This radical event has attracted the attention of many zoologists throughout the past century.

In a literature survey, we have found that 121 echinoderm species have been observed through metamorphosis. Of the 121 species only 30 have been studied since 1938, the year that Mortensen published the last of his series of papers on echinoderm development. Mortensen's main objective, like that of many of his contemporary embryologists, was to seek the relationships between the adults and the larvae in order to understand the phylogeny within the phylum as well as the phylogenetic relationships between echinoderms and other phyla. Therefore, it is not surprising that very little attention was paid in that period to the mechanisms and the cellular processes of metamorphosis. It is also true that not much information on this topic has been added since that time.

Many earlier workers were impressed by the process of histolysis of larval structures during metamorphosis but its significance was not clearly recognized. For example, Bury[1] thought that histolysis, which commonly occurs during the metamorphosis in echinoderm larvae, was, in fact, "pathological." This paper reviews the general fate of larval structures and pays particular attention to cellular processes and possible mechanisms of metamorphosis.

GENERAL SURVEY

We have attempted to compile a list of echinoderm species which have been studied through metamorphosis (Table 1), and we have also summarized the fate of larval structures during and after metamorphosis (Table 2). This effort, we hope, will be informative to those who are interested in the overall view of echinoderm metamorphosis. It is noted from Table 1 that all the crinoids studied so far are limited to the suborder Comatulida. We know virtually nothing about the metamorphosis of isocrinids, the stalked sea lilies.

219

TABLE 1

List of Echinoderm Species Which Have Been Studied Through Metamorphosis

Crinoids	Echinoids	Ophiuroids
Antedon rosacea[6]	*Arbacia lixula*[24]	*Amphipholi abditis*[66]
Comanthus japonicus[7]	*Arbacia punctulata*[25]	*Amphiura chiajei*[67]
Compsometra serrata[5]	*Astriclypeus mani*[26]	*Axiognathus (Amphipholis) squamata*[68]
Heterometra savignyi[8]	*Clypeaster humilis*[9]	*Gorgonocephalus caryi*[69]
Isometra vivipara[5]	*Dendraster excentricus*[27]	*Ophiocoma lineolata*[9]
Lamprometra klunzingeri[9]	*Diadema setosum*[9]	*Ophiocoma nigra*[70]
Tropiometra carinata[5]	*Echinarachnius parma*[28]	*Ophiocoma pica*[9]
	Echinocardium cordatum[29]	*Ophiocoma scolopendrina*[9]
	Echinocyamus pusillus[30]	*Ophioderma brevispina*[71]
Asteroids	*Echinodiscus auritus*[8]	*Ophioderma longicaudum*[72]
	Echinometra lacunter[26]	*Ophiolepus cincta*[8]
Acanthaster brevispina[39]	*Echinometra mathaei*[9]	*Ophiolepus elegans*[73]
Acanthaster planci[40]	*Echinus acutus*[3]	*Ophiomaza cacaotica*[9]
Asterias amurensis[41]	*Echinus esculentus*[32]	*Ophionereis squamulosa*[26]
Asterias rubens[42]	*Eucidaris metularia*[9]	*Ophiopholis aculeata*[74]
Asterias pallida[43]	*Fibularia craniolaris*[9]	*Ophiothrix fragilis*[75]
Asterina coronata japonica[44]	*Heliocidaris erythrogramma*[26]	*Ophiothrix osterdi*[76]
Asterina gibbosa[45]	*Heterocentrotus mammillatus*[9]	*Ophiothrix savignyi*[8]
Asterope carinfera[9]	*Laganum depressum*[8]	*Ophiothrix triloba*[9]
Astropecten aurantiacus[46]	*Lytechinus pictus*[33]	*Ophiura brevispina*[77]
Astropecten latespinosus[44]	*Lytechinus variagatus*[33]	
Astrpoecten polychanthus[9]	*Mellita quinquesperforata*[34]	
Astropecten scoparius[26]	*Mellita sexiesperforata*[35]	**Holothuroids**
Astropecten valitaris[9]	*Mellita testudinata*[36]	
Certonarda semiregularis[47]	*Nudichinus gravieri*[9]	*Actinopyga serratidens*[9]
Crossaster paposus[48]	*Parechinues angulosus*[37]	*Caudina chilensis*[10]
Ctenodiscus australis[49]	*Paracentrotus lividus*[24]	*Cucumaria echinata*[12]
Culcita novaequineae[50]	*Peronella lessieuri*[35]	*Cucumaria elongata*[13]
Echinaster echinophorus[51]	*Prionocidaris baculosa*[8]	*Cucumaria frondosa*[14]
Echinaster purpureus[8]	*Psammechinus microtuberculatus*[1]	*Cucumaria normani*[15]
Echinaster sepositus[52]	*Strongylocentrotus purpuratus*[33]	*Cucumaria planci*[16]
Fromia ghadagana[8]	*Temnotrema scillae*[9]	*Cucumaria saxicola*[15]
Gomophia egyptiaca[53]	*Tripneustus esculentus*[38]	*Holothuria arenicola*[9]
Henricia leviuscula[54]	*Tripneustus gratilla*[9]	*Holothuria difficilis*[8]
Henricia sanguineleuta[55]		*Holothuria floridana*[17]
Leptasterias aequalis[56]		*Holothuria impatiens*[8]
Letpasterias hexactis[57]		*Holothuria spinifera*[9]
Letpasterias polaris[58]		*Labidoplax digitata*[18]
Leptasterias ochotensis[59]		*Leptosynapta clarki*[19]
Linkia laevigata[50]		*Leptosynapta inhaerens*[20]
Linkia multiforma[8]		*Psolus chitonoides*[21]
Luidia ciliaris[60]		*Psolus phantapus*[14]
Luidia sarignyi[8]		*Stichopus variegatus*[9]
Luidea sarsi[60]		*Thyone briareus*[22]
Mediaster aequalis[61]		*Thyonepsolus nutriens*[23]
Parania pulvillus[62]		
Pentaceraster mammillatus[8]		
Pteraster tesselatus[63]		
Pycnopodia helianthoides[64]		
Solaster endeca[65]		

The fate of larval structures during metamorphosis, as shown in Table 2, varies from class to class. It must be cautioned that this table represents our interpretation of the literature; some statements may be oversimplified and will not hold true for all species. This table indicates that the digestive system in the planktotrophic larvae is retained more or less entirely in the holothurians. This may be due to the fact that the feeding methods between the adults and the larvae are similar; the larvae are planktotrophic and the adults are either suspension feeders or deposit feeders. On the other hand, both the esophagus and intestine are resorbed in starfish larvae. This is perhaps because most of the starfish adults are carnivores while the larvae are suspension feeders. In the ophiuroids, both adults and larvae are suspension feeders but the larval intestine and anus are resorbed. This is because the adult ophiuroids have neither an intestine nor an anus.

The locomotory ciliary bands, highly developed in the echinopluteus, ophiopluteus, bipinnaria and brachiolaria, and less well developed in the doliolaria, are, with few exceptions, resorbed during metamorphosis. The adult epidermis in all larval forms, except the echinopluteus, is formed directly from larval epidermis. In the echinopluteus most of the larval epidermis is resorbed while the adult epidermis is formed from the vestibule wall. A vestibule exists only in the larvae of echinoids, holothurians and crinoids; in all three classes it contributes directly to the adult epidermis.

Habitat requirements of metamorphosis differ among the five classes. The brachiolaria and crinoid doliolaria possess adhesive organs; they are usually attached during metamorphosis. On the other hand, the ophiopluteus and bipinnaria are known to undergo metamorphosis in the plankton. The echinopluteus settles first and metamorphosis follows. In holothurians, metamorphosis of auricularia to doliolaria to pentactula takes place in plankton and from pentactula to juvenile after settlement.

The metamorphosis of echinopluteus is rapid; it takes less than an hour to transform from a bilaterally symmetric pluteus to a radially symmetric urchin. The development of the adult rudiment, however, takes more than a month. The metamorphic changes of all other echinoderm larvae are slow. The development of the adult rudiment in echinoids and asteroids is on the left side of the larval body but in the other three classes it is on the ventral side.

The metamorphic processes and the fate of larval structures are quite different among echinoids and ophiuroids although the morphology of the larvae is similar (Table 2). This similarity may provide further support to the belief that the similarities of the two larval forms is due to convergent evolution and does not indicate closer phylogenetic relationships.

METAMORPHOSIS OF LECITHOTROPHIC LARVAE

In each of the extant echinoderm classes some species are known to undergo lecithotrophic development. This group includes all crinoid species whose development has been studied. A few of the lecithotrophic larvae are pelagic and many are demersal or brood-protected, but most of them have developed from large, yolky eggs. All of them possess a preoral lobe or hood. The size of the preoral lobe in proportion to the rest of the larva varies; in most cases it measures from one-third to one-half of the larva. The adult rudiment may develop in the posterior part of the larval body. The preoral lobe may function as an adhesive organ for attachment (as in some crinoids), or it may function as an adhesive organ for attachment (as in some crinoids and asteroids), or it may function as a floating device increasing the buoyancy of the larva (as in *Cucumaria elongata*).[13] In all cases the preoral lobe serves as a nutrient store, not unlike a yolk sac, providing energy resources during larval development and metamorphosis. In *Mediaster aequalis* metamorphosis can be arrested for over a year if the proper substrate is not available; this prolonged larval life is presumably supported by the nutrients in the preoral lobe.[61]

The most obvious changes during metamorphosis of lecithotrophic larvae are the resorption of the preoral lobe and the differentiation of adult structures; both are slow and gradual. The resorption of the preoral lobe in *Leptasterias hexactis* was studied histologically in some detail

221

TABLE 2

Fate of Larval Structures during Metamorphosis in Echinoderms

Larval Structure	Class: Echinoids Larval Types: Echinopluteus	Ophiuroids Ophiopluteus	Asteroids Bipinnaria	Asteroids Brachiolaria	Holothuroids Auricularia Doliolaria	Crinoids Doliolaria
Digestive Tract: Mouth	Resorbed	Adult mouth	Resorbed	Resorbed	Adult mouth	
Esophagus	Resorbed	Adult esophagus	Resorbed	Resorbed	Adult esophagus	
Stomach	Adult stomach	Adult stomach	Adult stomach	Adult stomach	Adult stomach	
Intestine	Adult intestine	Resorbed	Resorbed	Resorbed	Adult intestine	
Anus	Adult anus	Resorbed	Resorbed	Resorbed	Adult anus	
Skeletal rods	Discarded	Discarded or resorbed				
Hydrophore	Resorbed or Madreporite	Madreporite	Resorbed	Resorbed	Resorbed or Madreporite	Resorbed
Epidermis (Ciliary bands, arms, apical tuft, adhesive pits, etc.)	Resorbed	Mostly resorbed; remainder forms all adult epidermis	Mostly resorbed or discarded; remainder forms all adult epidermis	Mostly resorbed; remainder forms all adult epidermis	Ciliary bands resorbed. Adult epidermis formed directly from larval epidermis	Mostly resorbed; remainder forms all adult epidermis
Vestibule	Adult epidermis				Adult oral epidermis	Adult oral epidermis, mouth, esophagus
Preoral coelom or part of axocoel	Resorbed or pharyng. coel.	Resorbed	Resorbed	Resorbed	Resorbed	Resorbed
Preoral lobe	Resorbed	Resorbed	Resorbed	Resorbed	Resorbed	Resorbed
Other notes	Metamorphosis on substrate, a fast process; adult rudiment on left side of larva	Metamorphosis in plankton, a slow process; adult rudiment on ventral side of larva	Metamorphosis in plankton, or on substrate, a slow process; adult rudiment on left side of larva	Metamorphosis on substrate, a slow process; adult rudiment on left side of larva	Metamorphosis of Auricularia-Doliolaria-pentacula planktonic, juvenile benthic, a slow process; adult rudiment on ventral side of larva	Metamorphosis on substrate, a slow process; adult rudiment on ventral side of larva

by Chia,[78] who reported that it undergoes autolysis. This process begins at the inner ends of the ectoderm cells where the cytoplasm appears first to be liquified and then to shrink to a collection of fibers (Fig. 1). The mesoderm cells of the preoral coelomic epithelium as well as the mesenchyme cells are condensed into a syncytial mass. All the disintegrated and broken cells of the preoral lobe are phagocytized by amoebocytes which originate from either the endoderm or the mesenchyme. It was postulated that the amoebocytes transport the degenerated cells of the preoral lobe to the larval gut which has become a syncytial mass. This assumption is based on the fact that that the larval gut increases substantially in volume, coincident with the resoprtion of the preoral lobe. At the end of metamorphosis, the preoral lobe appears as a small knob between two arms, and it is eventually incorporated into the adult epidermis. The retraction of the preoral lobe is attributed to the contraction of the muscle fibers in the coelomic epithelium and the elastic fibers of the connective tissue. At the same time the adult digestive tract undergoes redifferentiation.

The syncytial nature of the larval digestive system during later development has been noted by Gemmill[48] in *Crossaster papossus.*

METAMORPHOSIS OF PLANKTOTROPHIC LARVAE

The cellular processes involved in the resorption of larval structures during metamorphosis of planktotrophic larvae are not clear. As mentioned earlier, histolysis has been observed during metamorphosis of all four classes of echinoderms with planktotrophic larvae.

In asteroids, metamorphosis takes place either at the bipinnaria or brachiolaria stage. The bipinnaria of the families Luidiidae and Astropectinidae metamorphose while still planktonic. More commonly, a brachiolaria develops, settles, and the attached larva metamorphoses. The transformation in most cases involves the resorption of the larval body into the adult rudiment, which is formed on the left and posterior side of the larva. Gemmill[42] described the resorption

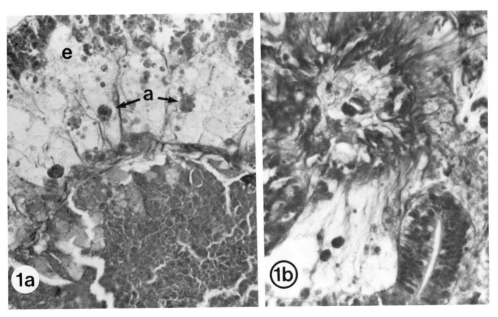

Fig. 1. Resorption of preoral lobe in *Leptasterias hexactis* during metamorphosis. 1a. Early stages of resorption showing the breaking down of the ectoderm cells (e) while the mesothelium (m) of the preoral coelom is a syncytium. Note also the multi-nucleated amoebocytes (a). 1b. Latter phase of resorption. Note that the autolysized cells have disappeared (incorporated into the digestive tract) and only fibrous materials are left. From Chia (1968)[78]

of the larval epidermis as a gradual process of indrawing and histolysis. Hyman[79] stated that there is considerable reorganization of the gut including ingestion of disintegrated tissues.

Planktotrophic holothurian larvae metamorphose gradually, retaining the majority of larval tissues intact. The bilaterally symmetric auricularia is first transformed to a radially symmetric doliolaria then to a pentactula; this change is accomplished by a reorganization of auricularian tissues. Hyman[79] refers to degenerative changes in the ciliary band resulting in the ciliary rings and the preoral lobe being gradually reduced in size. The metamorphosis is definitively accomplished by the direct reorganizatin of larval structures of the pentactula into a juvenile. In holothurians there appears to be no extensive histolytic nor resorptive phases of metamorphosis.

Metamorphosis of ophiuroids, as outlined by Hyman,[79] is a slow process which occurs while the ophiopluteus is still planktonic. The larval body is gradually resorbed into the adult rudiment which develops on the ventral surface of the larva. The larval gut goes through a phase of reorganization during which the anus and intestine are lost. The stomach for a time is folded and is filled with dissociated cells prior to the formation of the adult stomach. The adult rudiment, in many species, separates from the posterolateral arms and sinks to the bottom. The posterolateral arms with their ciliary bands remain in the plankton for some time.

Metamorphosis of the echinopluteus occurs after settlement and is, initially, a rapid process. The adult rudiment develops within an epidermal invagination, the vestibule, on the left side of the larva. At metamorphosis the adult rudiment is everted, and the larval body is resorbed into the aboral surface of the juvenile. Bury,[1] MacBride[32] and Von Ubisch[24] describe extensive histolysis of the larval epidermis and gut during the resorption of the larval body and the subsequent differentiation of the adult form. Cameron[80] has recently studied the metamorphosis of *Lytechinus pictus* with electron microscopy and has noted autolysis of the larval epidermis.

We have undertaken a histological and ultrastructural investigation of the metamorphosis of the Pacific sand dollar, *Dendraster excentricus,* with the intention of providing more detailed information on the atrophy and reorganization of larval structures.

D. excentricus larvae, when reared in the laboratory, become competent to metamorphose in five to seven weeks. Prior to metamorphosis the left side of the larva is distended due to the bulk of the rudiment (Fig. 2). Competent larvae can be induced to metamorphose by providing them with a few grains of sand which have been conditioned by the presence of an adult.[81] Stimulated larvae show a sustained reversal of ciliary beating and the adult spines and tube feet can be seen to move within the vestibule.

The first phase of metamorphosis in *D. excentricus* is the eversion of the adult rudiment. The left postoral and posterodorsal arms spread apart and over the dorsal surface to the right side of the larva, and the vestibule opens (Fig. 3).

Prior to metamorphosis the rudiment is concave, the spines and tube feet are oriented radially toward the incipient adult mouth. After eversion, the oral surface of the rudiment is convex with the spines and tube feet directed away from the mouth. The extrusion of the rudiment can be compared to the reversal of concavity observable at the fingertip of a rubber glove. This process in *D. excentricus* is completed in 3 to 5 minutes (Fig. 3).

With the extrusion of the adult rudiment the larval body begins to collapse. The epidermis is pierced at the tips of the larval arms by the skeletal rods and the epidermis is drawn to the base of the arms, forming a mass of tissue on the aboral surface of the juvenile (Figs. 3, 4). The larval skeleton is soon discarded. The epidermis which originally lined the vestibule now covers the aboral surface of the juvenile and becomes the adult epidermis. The collapse of the larval body in *D. excentricus* takes from 15 to 30 minutes.

After these initial phases of metamorphosis, the juvenile undergoes a period of extensive reorganization and differentiation, the result of which is an independently feeding young sand dollar. This phase of development in *D. excentricus* lasts for about seven days.

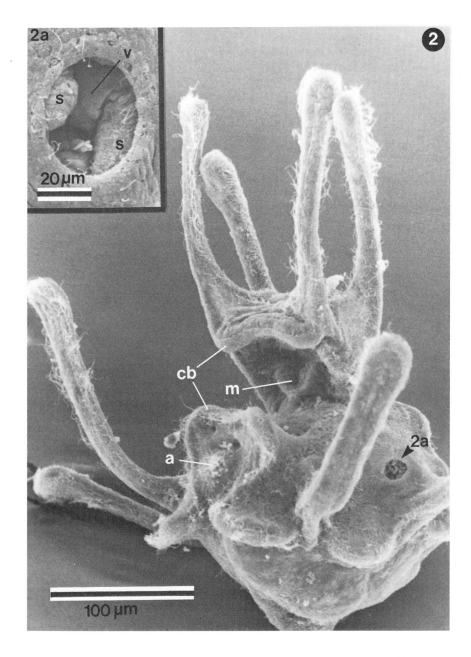

Fig. 2. Scanning electron micrograph of an eight-arm pluteus larva of *Dendraster excentricus*. The insert (2a) shows more detail of the opening of the vestibule (V) and the adult rudiment within. a, anus; cb, ciliary bands; m, mouth; s, adult spines.

In examining the cellular process of metamorphosis in *D. excentricus*, we have followed the fates of two larval structures: the epidermis and the digestive tract.

The epidermis prior to metamorphosis is simple squamous epithelium underlined by a thin basal lamina. The ciliary bands are comprised of spindle-shaped ciliated cells and nervous tissues. The larval gut is comprised of an esophagus, stomach and intestine. The esophagus is made of simple cuboidal and columnar epithelium and is surrounded by bands of smooth muscles

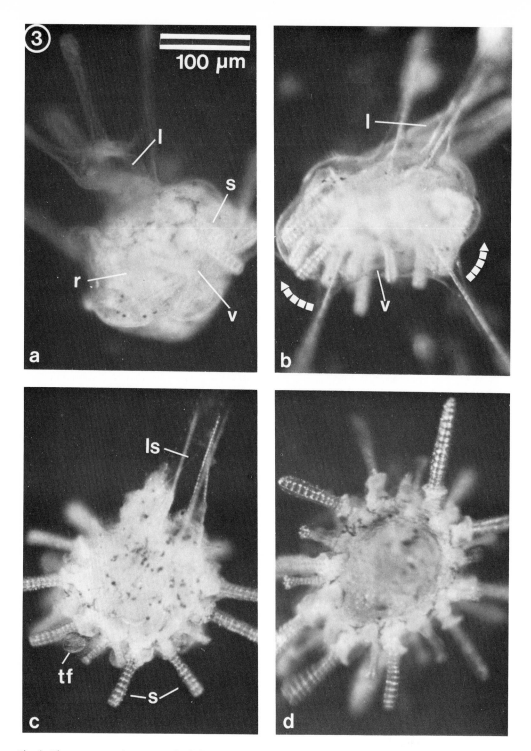

Fig. 3. The sequence of metamorphosis in *Dendraster excentricus:* a. A competent larva, showing the large size of the adult rudiment within the vestibule. b. The eversion of the adult rudiment, as seen from the right side of the larva. The dotted lines indicate the direction of retraction of the vestibular epithelium. c. The collapse of the larval body, showing that the epithelium of the larval arms has been retracted from the larval skeleton. d. Aboral view of a juvenile 48 hours after metamorphosis. l, larval body; ls, larval sekelton; r, adult rudiment; s, adult spines; tf, tube feet; v, vestibule.

which are responsible for the peristaltic contractions of swallowing. The larval stomach consists of a single layer of columnar cells that are characterized by large lipid droplets in their basal regions (Fig. 5). The lipid droplets accumulate during the growth of the larva. The intestinal epithelium is similar to that of the stomach with the notable absence of lipid droplets.

During metamorphosis, while the larval body is collapsing into the aboral surface of the juvenile, the cells of the larval epidermis and the larval digestive tract undergo extensive degenerative changes.

The epidermal cells break contact with each other, round up, and begin to disintegrate. Some cells undergo cytolysis, releasing their cytoplasmic contents (Fig. 6). Others can be seen to have retracted their cilia into their cytoplasm (Fig. 7). The nuclei of some epidermal cells become pyknotic and fragmented.

The cells of the digestive tract similarly begin a series of degenerative changes (Fig. 4). The cells dissociate and lose their polarity; there are no longer defined apical and basal regions. The cells lose their specialization; the elaborated apical surfaces disappear, and their cilia retract into the cytoplasm (Fig. 8). In contrast to the epidermal cells, the gut cells do not undergo cytolysis. The nuclei remain as they were prior to metamorphosis.

Forty-eight hours after metamorphosis, the gut cells reorganize into a tube which convolutes throughout the young sand dollar (Fig. 9). It is still limited by the squamous epithelium of the somatocoels. The gut cells do not, however, form a tissue proper; there are no identifiable cell-junctions and the intercellular spaces are large and vary considerably (Fig. 10).

A number of vacuoles, containing either larval epidermal cells or fragments of epidermal cells, are found within the cytoplasm of most of the endodermal cells. Frequently two or more epidermal cells are enclosed within the same vacuole and each gut cell may contain two or more

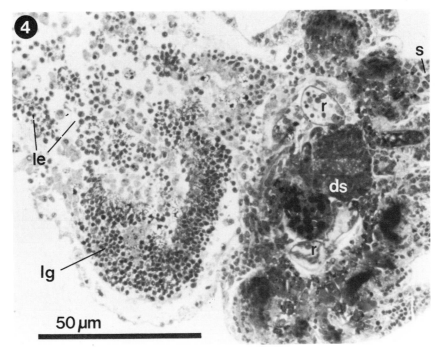

Fig. 4. A section parallel to the oral-aboral axis of a metamorphosing sand dollar; the adult rudiment has been everted and the larval body has begun to collapse. ds, dental sac; le, larval epidermis; lg, larval gut; r, radial canal; s, adult spine.

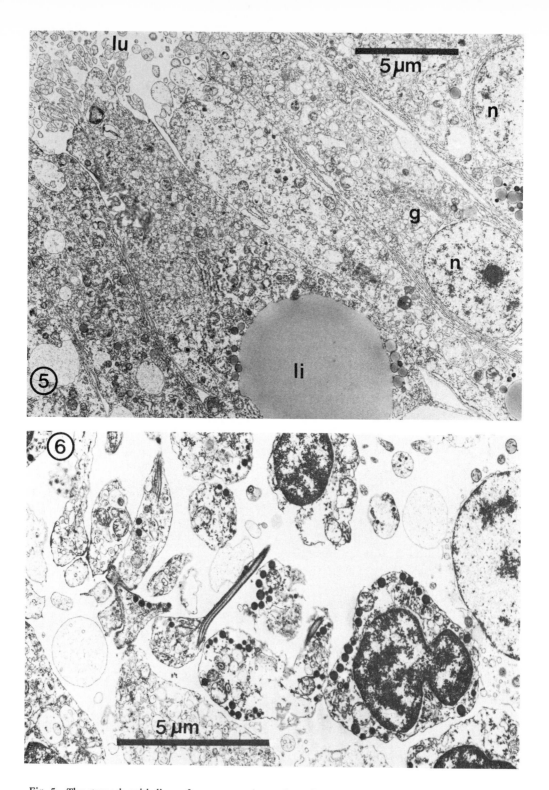

Fig. 5. The stomach epithelium of a competent larva of *Dendraster excentricus*, showing the polarization of the cytoplasm and the lipid droplets (li). g, Golgi apparatus; lu, lumen of the stomach; n, nucleus.

Fig. 6. Larval epidermis during metamorphosis, showing the dissociation and degeneration of the epidermal cells.

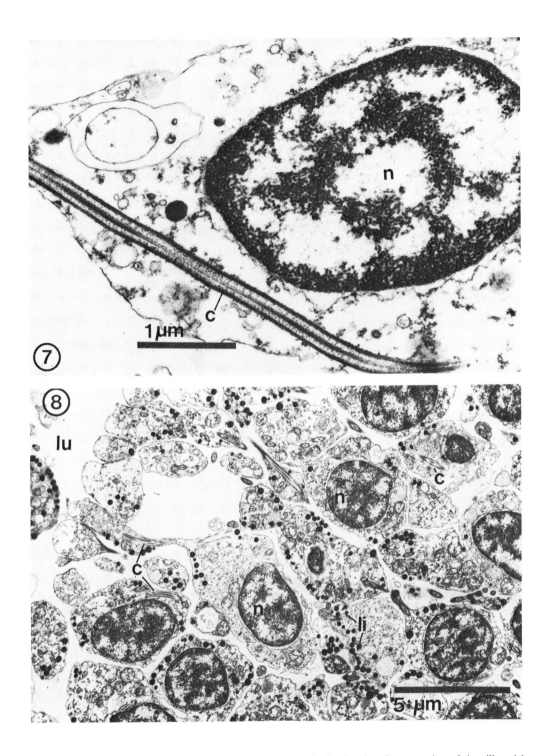

Fig. 7. Ciliated cell from larval epidermis during metamorphosis, showing the resorption of the cilium (c) and the pyknosis of the nucleus (n).

Fig. 8. The digestive tract of a metamorphosing sand dollar, showing the dissociation of the epithelium and the degenerate state of the cells. c, cilium; l, lipid droplets; lu, lumen of the digestive tract; n, nucleus.

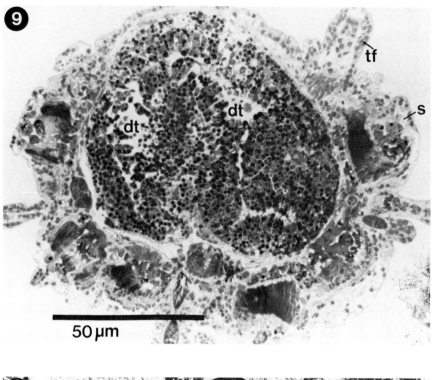

50 μm

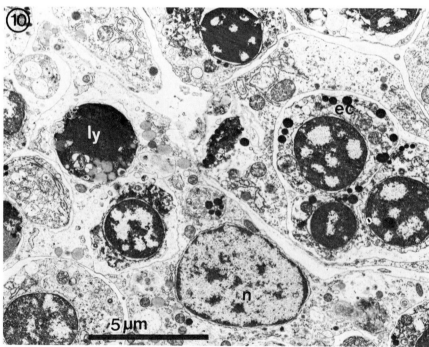

5 μm

Fig. 9.   A juvenile sand dollar, 48 hours after metamorphosis; the section is at right angles to the oral-aboral axis: dt, digestive tract; s, spine; tf, tube foot.

Fig. 10. Gut cells, 48 hours after metamorphosis, showing the phagocytosis of the larval epidermal cells (ec): ly, lysosome; n, nucleus of gut cell.

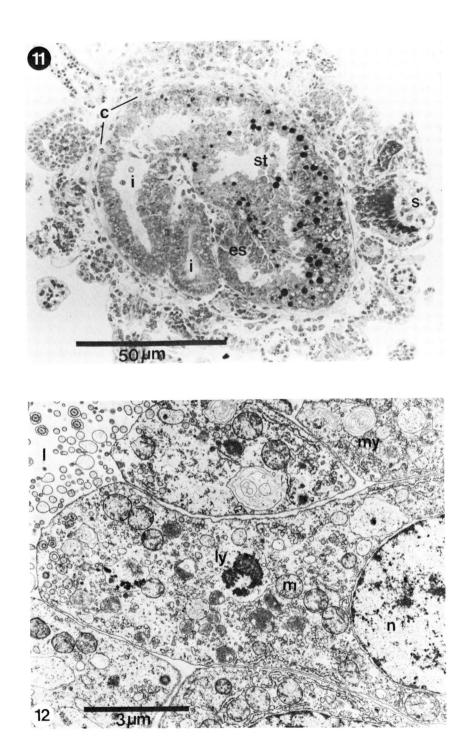

Fig. 11. A section taken at right angles to the oral-aboral axis of a young sand dollar, 7 days after metamorphosis: es, esophagus; c, coelomic epithelium; i, intestine; s, spine; st, stomach. Note the stomach still contains lipid droplets (li).

Fig. 12. The apical portion of a stomach cell of a sand dollar 7 days after metamorphosis: lu, lumen; ly, lysosomal vesicles; m, mitochondria; my, myelin body; n, nucleus.

vacuoles (Fig. 10). The contents of the vacuoles appear to undergo a sequence of changes leading ultimately to their condensation into an electron opaque lysosome. (Fig. 11).

Seven days after metamorphosis, juvenile *D. excentricus* have a fully formed and apparently functional gut (Fig. 11). The cells have reapposed one to another, forming a simple columnar and ciliated epithelium (Fig. 12). The gut cells have regained their polarity; cisternae of rough endoplasmic reticulum are found along the peripheries and the basal regions, and small clumps of polysomes are dispersed throughout the cytoplasm. Remnant lysosomal vacuoles are also present.

In short, both the epidermis and the digestive tract undergo histolysis at metamorphosis. The epidermis then apparently begins a sequence of necrotic degeneration. The gut cells, after dissociation, dedifferentiate from their larval condition and begin phagocytizing the remanants of the epidermal cells. After incorporating the epidermal cells, the gut cells begin reforming the adult digestive tract and redifferentiating to the adult condition.

CONCLUDING REMARKS

It is clear from this review that although echinoderm metamorphosis has been amply described in the literature, little is known about the mechanisms and cellular processes of the resorption of larval structures. We can only report details on the studies of two species. From such scanty material it is difficult to draw any generalized conclusions. However, what appears to be consistent is that larval structures, which do not transform directly to a specific adult structure, are disintegrated at metamorphosis, and subsequently incorporated into the redifferentiating digestive tract of the juvenile. The larval structures resorbed during metamorphosis are thus utilized, possibly supplying resources for post-metamorphic development. This general statement needs to be verified by additional studies.

ACKNOLWEDGMENTS

Preparation of this paper and the original research reported here were supported by an operating grant to F. S. Chia and a pre-doctoral fellowship to R. D. Burke, both from the National Research Council of Canada.

REFERENCES

1. Bury, H. (1895) Quart. J. Microsc. Sci., 38, 45-135.
2. Seeliger, O. (1892) Zool. Jahrb. Abth. f. Morph., 6, 481-483.
3. Thomson, W. (1863) Nat. Hist. Rev., 11, 395-415.
4. Bury, H. (1889) Quart. J. Microsc. Sci., 29, 409-449.
5. Mortensen, T. (1920) Vidensk. Medd. dansk. naturh. foren., 71, 133-160.
6. Seeliger, O. (1892) Zool. Jahrb. Anat. Ontog., 6, 257-301.
7. Dan, J. and Dan, K. (1941) Jap. J. Zool., 9, 565-574.
8. Mortensen, T. H. (1938) Memoirs de l'Academie Royale des Sciences et des lettres de Danemark, Copenhague, Section des Sciences, 9me serie, t. VII no. 3, 1-59.
9. Mortensen, T. H. (1927) Memoirs de l'Academie Royale des Sciences et des lettres de Danemark, Copenhague, Section des Sciences, 9me serie, t. VII no. 1, 1-65.
10. Inaba, D. (1930) Sci. Repts. Tohoku (ser. 4), 5, 215-248.
11. Clark, H. L. (1910) J. Exp. Zool., 9, 497-516.
12. Oshima, H. (1921) Quart. J. Micr. Sci., 65, 173-246.
13. Chia, F. S. and Buchanan, J. (1969) J. mar. biol. Ass. U.K., 49, 151-159.
14. Runnstrom, S. and Runnstrom, J. (1918) Bergens. Mus. Aarbok. Naturv. Vidensk., 2, 1-99.
15. Newth, G. H. (1969) Proc. Zool. Soc. London, 2, 631-641.
16. Selenka, E. (1876) A. Wiss. Zool., 27, 155-178.
17. Edwards, C. L. (1909) J. Morph., 20, 211-230.
18. Semon, R. (1888) Jena. Zeit. f. Naturwissen., 22, 175-309.
19. Everingham, J. W. (1961) M.Sc. Thesis, Zool. Univ. of Washington, 1-70.
20. Thomson, W. (1862) Quart. J. Micr. Sci., 2, 131-146.

21. Jones, S. A. (1960) M.Sc. Thesis, Zool. Univ. of Washington, 1-56.
22. Ohshima, H. (1925) Science, 61, 420-422.
23. Wootton, D. M. (1949) Ph.D. Thesis, Stanford Univ., 1-92.
25. Garman, H. and Colton, B. (1883) Studies Biol. Lab. Johns Hopkins Univ., 2, 247-255.
26. Mortensen, T. (1921) Copenhagen, G.E.C. Gas., 1-266.
27. Moore, A. P. (1927) Proc. Soc. Exp. Biol. Med. New York, 25,37-38.
28. Fewkes, J. W. (1886) Bulletin of the Museum of Comparative Zoology, Harvard Univ., 12, 105-152.
29. MacBride, E. W. (1914) Quart. J. Microsc. Sci. (new series), No. 236, 59(4), 471-486.
30. Theil, H. (1892) Nova Acta Regiae Societatis Scientarium Upsaliensis, Seriei Tertiac Vol. XV(6), Fasc. I, 1-47.
31. Shearer, C., DeMorgan, W. and Fuchs, H. M. (1912) Quart. J. Microsc. Sci. (new series), No. 230, 58(2), 337-352.
32. MacBride, E. W. (1903) Phil. Trans. Roy. Soc. London ser. B, 195, 285-330.
33. Hinegardner, R. T. (1969) Biol. Bull., 137, 465-467.
34. Caldwell, J. W. (1972) M.Sc. Thesis, Zoology Dept. Univ. of Florida, 1-63.
35. Mortensen, T. (1914) Annot. Zool. Japan, 8, 543-552.
36. Grave, C. (1902) Johns Hopkins Univ. Circ. No. 157, 57-79.
37. Cram, D. L. (1971) Trans. Roy. Soc. S. Africa, 39, 321-337.
38. Lewis, J. B. (1958) Can. J. Zool., 36, 607-621.
39. Lucas, J. S. and Jones, M. M. (1976) Nature, 263, 409-412.
40. Henderson, J. A. and Lucas, J. S. (1971) Nature, 232, 655-657.
41. Dan, K. (1968) in Invertebrate Embryology, Kume, M. and Dan, K. eds., Nolit Publishing House, Belgrade, pp. 280-332.
42. Gemmill, J. F. (1914) Phil. Trans. Roy. Soc. London ser. B., 205, 213-294.
43. Goto, S. (1898) J. of Coll. of Sci. Imper. Univ. of Tokyo, Japan, 10, 239-278.
44. Komatsu, M. (1975) Proc. Jap. Soc. Syst., Zool., 11, 42-48.
45. MacBride, E. W. (1896) Quart. J. Microsc. Sci., 38, 339-411.
46. Hörstadius, S. (1926) L. Ark. Zool., 186(7), 1-5.
47. Hayashi, R. and Komatsu, M. (1971) Proc. Jap. Soc. Syst. Zool., 7, 74-81.
48. Gemmill, J. M. (1920) Quart. J. Microsc. Sci., 64, 155-190.
49. Lieberkind, I. (1926) Vidensk. Meddr. dansk. Naturh. Foren. ser. 8, 82, 183-196.
50. Yamaguchi, M. (1973) in Biological and Geological Coral Reefs, Vol. 2, Biol. 1, Academic Press, New York, pp. 369-387.
51. Atwood, D. G. (1973) Biol. Bull., 144, 1-11.
52. Nachtsheim, H. (1914) Zool. anz., 44, 600-606.
53. Yamaguchi, M. (1974) International Symposium. Tropical Reef Biol. at Guam (Abstract).
54. Chia, F. S. (1966) Amer. Zool., 6, 331-332 (Abstract).
55. Masterman, A. T. (1902) Trans. Roy. Soc. Edinb., 40, 373-418.
56. Gordon, I. (1929) Phil. Trans. Roy. Soc. London ser. B, 217, 289-334.
57. Osterud, H. L. (1918) Puget Sound Mar. Stn. Publ., 2, 1-15.
58. Emerson, C. J. (1977) Scanning Electron Microscopy, 2, 631-638.
59. Kubo, K. (1951) J. Fac. Sci. Hokkaido Univ. ser. 6 Zool., 10, 97-105.
60. Tattersal, W. M. and Sheppard, E. M. (1934) in J. Johnston Mem. Vol. Liverpool, 35-61.
61. Birkeland, C., Chia, F. S., and Strathmann, R. R. (1971) Biol. Bull., 141, 99-108.
62. Gemmill, J. F. (1915) Quart. J. Microsc. Sci., 61, 27-50.
63. Chia, F. S. (1966) Proc. Calif. Acad. Sci., 34(4) 505-510.
64. Greer, D. L. (1962) Pacif. Sci., 16(3), 280-285.
65. Gemmill, J. F. (1912) Trans. Zool. Soc. London, 20, 1-72.
66. Hendler, G. (1973) Ph.D. Thesis Univ. of Connecticut, 1-255.
67. Fenaux, L. (1963) Vie Milieu, 14(1), 41-46.
68. Fell, H. B. (1946) Trans. Proc. Roy. Soc. N.Z., 75, 419-464.
69. Patent, D. H. (1970) Mar. Biol., 6, 262-267.
70. Mortensen, T. H. (1913) J. mar. biol. Ass. U.K., 10(1), 1-18.
71. Grave, C. (1916) J. Morph., 27, 413-453.
72. Mortensen, T. H. (1898) Ergebmisseder Plankton-Exped. der Humboldt-Stiftung, 2(1), 1-120.
73. Stancyk, S. E. (1973) Mar. Biol., 21, 7-12.
74. Olson, H. (1942) Berg. Mus. Aarbok. Natur. Rekke, 6, 1-107.
75. MacBride, E. W. (1907) Quart. J. Microsc. Sci., 51, 557-606.
76. Mladenov, P. V. (1976) M.Sc. Thesis McGill Univ., 1-136.

77. Grave, C. (1900) Mem. Biol. Lab. Johns Hopkins Univ., 4, 83-100.
78. Chia, F. S. (1968) Acta Zool., 49, 321-364.
79. Hyman, L. H. (1955) The Invertebrates: Echinodermata, The Coleomate Bilateria, McGraw-Hill, New York, Vol. 4, 1-763.
80. Cameron, R. A. (1975) Ph.D. Dissertation Univ. of Calif., Santa Cruz. 1-81.
81. Highsmith, R. (1977) Amer. Zool. 17(4), 935 (Abstract).

LARVAL SETTLEMENT IN ECHINODERMS

Richard R. Strathmann

Friday Harbor Laboratories and Department of Zoology, University of Washington,
Friday Harbor, Washington 98250

Echinoderms vary greatly in specificity of factors inducing settling. The most careful laboratory observations are on echinoids and asteroids. Delay of settling in the absence of a suitable substratum has not been demonstrated beyond doubt for ophiuroids, holothuroids, or crinoids but probably occurs in these classes also. Taxa which depend on podia for attachment at settling appear to vary in swimming ability after podial development and this may affect the capacity to explore the substratum at settling. Diverse mechanisms of attachment to the substratum are employed. Non-homologous epithelia are involved in sensing the substratum prior to settling in different taxa. Lecithotrophic asteroid larvae can delay settling for long periods and the effect of larval planktotrophy on the capacity for delayed settling needs more careful examination in echinoderms. Studies on distribution, mortality, and feeding of juveniles suggests that substratum preferences do guide settling larvae to favorable habitats, though not infallibly.

INTRODUCTION

This review is organized around the following topics: (1) delay of settling and attractive factors, (2) anatomy, development, and behavior at settling, and (3) distribution of juveniles in the field and causes of juvenile mortality. Within each section the classes Echinoidea, Asteroidea, Ophiuroidea, Holothuroidea, and Crinoidea are treated in that order, and within classes planktotrophic (feeding) larvae are discussed usually before lecithotropic (non-feeding) larvae.

Because observations on larval settling are scattered in many papers not dealing primarily with this subject, it is difficult to achieve a complete review of the topic. I have attempted to cover the major studies on settling by echinoderms and shall make several statements on what is known or not known about choice of substrata in the hope of inducing additions or corrections by other biologists. No attempt has been made to correct species names used by earlier authors.

DELAY OF SETTLING AND ATTRACTIVE FACTORS

Echinoidea

Mortensen[1] first observed delay of settling in the absence of a suitable substratum. He found that larvae of the sand dollar *Mellita sexiesperforata* with a sand substratum completed metamorphosis whereas those in a jar without sand did not.

Caldwell[2] found that *Mellita quinquesperforata* can delay settling up to two weeks in the absence of suitable substratum, but that the percentage which can settle and complete metamorphosis declines. Larvae were tested against attractive sediments at intervals. A high percentage of larvae metamorphosed between 7 to 15 days of age. At 15 days 98% could still metamorphose; at 21 days 54% metamorphosed. Selectivity declined over time, a few eventually settling on the plastic containers. Caldwell analyzed factors affecting attractiveness of the sediment. He used sediments from areas with adult populations. In sieved samples the most abundant sediment fraction induced settling most effectively. Attractiveness of the sediment was reduced by incinerating, oven drying, and heating the wet sediments (in order of greatest to least effect of treatment). Washing the sediments also reduced their attractiveness. Incubating the incinerated sediments with unfiltered sea water restored their attractiveness to a greater extent than incubating incinerated substratum with filtered sea water. Microorganisms therefore appear to be involved in the attractiveness of the substratum. Incubation in the dark for two weeks did not reduce the attractiveness of the sediment, so that the presumed changes in the microflora under these conditions had little effect. The presence of juvenile sand dollars did not greatly increase the attractiveness of incinerated sediments, but the test does not rule out the possibility of gregarious settling.

The sand dollar *Dendraster excentricus* delays metamorphosis in the absence of a suitable substratum or inducing substance. Highsmith[3] obtains a higher percentage of metamorphosis with sediments from beds of the adults than with sediments adjacent to beds of adults. He also can induce settling with a factor from adults in the absence of sediments. Because the factor passes through dialysis tubing to induce metamorphosis of larvae outside the tubing, the material must be a substance soluble in sea water. Selectivity declines during the period of delayed metamorphosis.

Cameron and Hinegardner[4] have conducted the most detailed study of factors inducing settling in regular echinoids. Metamorphosis of *Lytechinus pictus* and *Arbacia punctulata* was induced by a bacterial film. Attractive factors resided both in the bacterial film itself and in the water but appear to be of bacterial origin. The active factor passed a dialysis membrane, indicating a molecular weight of less than 5000. The active factor was not passed through air. The roughness of the substratum made little difference. Hinegardner's[5] echinoid larvae did not metamorphose if kept in clean containers, but deteriorated after about two months.

Larvae of *Parechinus angulosus* settled on a fragment of *Patella* shell when 56 days old.[6] When the shell was removed, no more larvae metamorphosed until the shell was returned on the 67th day. The shell had been attached to an adult echinoid.

Larvae of *Prionocidaris baculosa* in a bowl with algae metamorphosed before larvae in a bowl without algae.[7]

Settling larvae of *Tripneustes esculentus* exhibited no preference between a bare dish, ground algae, sand, and rock.[8] In each of the four dishes about 90% of the larvae metamorphosed. This result does not prove a lack of selectivity in the species under all conditions, nor does Lewis[8] make this inference. Special care must be taken in proving a lack of sensitivity by negative laboratory results, because selectivity may have declined before the first test, or a bare bowl may have been fouled with microorganisms.

The ability to delay metamorphosis is thus clearly established for a variety of echinoids. The attractive factors vary according to source but are not fully characterized in any case. In at least some echinoids the ability to complete metamorphosis declines after a period of delayed settling. All of the species listed above have planktotrophic larvae.

Asteroidea

Of species with planktotrophic brachiolariae, *Acanthaster planci* has received the most study. Henderson and Lucas[9] reported a wide variety of substrata which induce metamorphosis. The presence of algae (*Cladophoropsis*), serpulid tubes, fine sand, encrusting coralline algae, live coral (*Pocillopora damicornis*), *A. planci* spines and podia and recently eaten *P. damicornis* induced metamorphosis. Brachiolariae kept in clean dishes did not metamorphose and some regressed. Time for development from fertilization to young seastar ranged from 28 to 47 days, and indicated capacity to delay metamorphosis for at least two weeks. Lucas[10] also asserted a lack of specificity in settling requirements, stating that settling larvae of *A. planci* require bacteria and encrusting algae, calcareous or non-calcareous. Lucas added that the larvae require little water movement in the settlement area, prefer down-facing substrata, and prefer substrata which tend to enclose the metamorphosing larva.

Yamaguchi[11] studied settling of *A. planci* at another location. In his cultures larvae settled mostly on dead coral encrusted with coralline algae and other algae. Bleached coralline algae or beach rock with filamentous algae were ineffective. Yamaguchi's results thus suggest a greater degree of specificity than the study of Henderson and Lucas.

Yamaguchi[11] stated that settling behavior in *Culcita novaeguineae* and *Linckia laevigata* is similar to that in *Acanthaster planci*. To my knowledge, earlier settlement in the presence of a preferred substratum has not been demonstrated for other feeding brachiolariae. Only tropical species have been studied in this regard. Mead[12] described *Asterias* spp. settling on eel grass, branched algae, and other objects but did not infer selectivity. On the contrary, he stated that when advanced larvae are put in a dish they attach to any object they happen to strike.

Substratum preferences and delayed settling have been reported for several asteroid species with lecithotrophic brachiolariae. Larvae of *Mediaster aequalis* settled on tubes of the polychaete *Phyllochaetopterus prolifica* (fouled with small organisms) and in lesser numbers on sand, whereas none metamorphosed in empty dishes or with the prey sea pen *Ptilosarcus gurneyi*.[13] Subsequent tests on *P. prolifica* tubes or bare glass showed the larvae have a remarkable capacity to delay metamorphosis. Some larvae survived in bare glass bowls and metamorphosed with worm tubes as much as 14 months after fertilization and more than a year after the first individuals in the culture were induced to metamorphose. Late metamorphosing individuals were more transparent. No tests of differences in viability were made.

Gomophia egyptiaca settled on diatom-covered glass as early as 9 to 10 days after fertilization but could delay metamorphosis four more weeks in glass bowls.[15] Individuals with late metamorphosis formed smaller juveniles. More larvae of *G. egyptiaca* were induced to settle by coralline algae *(Porolithon)* or a branching red alga *(Polysiphonia)* than by another species of branching red alga *(Gelidium)* or a sponge preyed upon by the seastar *(Tethya* sp.), but the latter two substrata were more effective than a clean bowl.

Ophidiaster granifer settled on algal substrata after 9 days.[15] Larvae 45 days old kept in clean beakers appeared normal. One-third of the larvae could still metamorphose after a 10-week delay, but most of the juveniles were abnormal.

Thus lecithotrophic asteroid larvae can delay settling for considerable periods, although one cannot yet conclude that their capacity typically equals or exceeds the capabilites of planktotrophic brachiolariae. It may be significant that longer periods of delay were observed with *Mediaster aequalis* in cold water than in the tropical *Gomophia egyptiaca* and *Ophidiaster granifer* in warm water. Development is slower in *M. aequalis,* and metabolic rates and consumption of nutrient reserves are probably reduced also.

Chia and Spaulding[16] mentioned that tubes of *Phyllochaetopterus* induce settling of the lecithotrophic brachiolariae of *Pteraster tesselatus.* Mortensen[7] described varying times to settling in his *Fromia ghardaqana* culture and this has been cited as an example of delayed settling in the absence of a suitable substratum, but no comparison among substrata has been reported for this seastar.

Ophiuroidea, Holothuroidea, Crinoidea

To my knowledge, delay of settling in the absence of a suitable substratum has not been directly demonstrated for animals in these classes. Chia and Spaulding[16] mentioned unpublished observations on induction of settling of *Cucumaria miniata* and *Psolus chitinoides* by tubes of *Phyllochaetopterus.* Both of these holothuroid species are dendrochirotes with lecithotrophic development. No data were provided. Mortensen[17] added bits of shell or coral to induce settling by larvae of the crinoids *Tropiometra audouini* and *Lamprometra klunzingeri,* but he did not mention prolonged swimming in the absence of these substrata. Mortensen[7] found great variation in time from fertilization to attachment in his culture of the crinoid *Heterometra savigny,* but he did not mention that attachment was earlier when more attractive substrata were provided.

Bury[18] and Seeliger[19] both stated that larvae of the crinoid *Antedon rosacea* are not fastidious about their place of attachment. Seeliger found them attaching to glass, stones, and various parts of plants. Seeliger did mention that these larvae appear to avoid places of even delicate water movement. Larvae of *Comanthus japonicus* attach at the most lighted side of a bowl.[20]

Other observations which indirectly suggest the capacity to delay settling and select substrata are reviewed below.

ANATOMY, DEVELOPMENT, AND BEHAVIOR AT SETTLING

Attachment organs and the transition from larva to juvenile vary among members of different classes or orders of echinoderms. These differences suggest different capacities for delayed settling and substratum selection, but these suspected differences are not yet confirmed.

Echinoidea

In the echinoids the oral side of the juvenile develops at the lower left side of the larva or very rarely at the right side (Fig. 1). Before settling and completion of metamorphosis the five primary podia and a varying number of pairs of podia have developed within the larva and will be functional when the left side of the larva opens. The five primary podia extend the farthest and appear to be the most effective organs of attachment.

Echinoplutei can feed even after the primary podia extend from the larval body,[21] so starvation probably does not limit the period of delay of settling.

At least some echinoplutei appear to explore the substratum before the podia emerge from the larval body. Caldwell,[2] in his study of *Mellita quinquesperforata*, distinguishes (1) a testing stage, in which the pluteus sinks to the bottom and moves about, (2) a pre-emergent stage, in which the larva comes to rest in one place and the spines of the echinus rudiment begin to move inside the larval body, and (3) an emergent stage, in which the larval body wall ruptures and spines and podia emerge. In this account exploration and selection of a settling site occur before the podia emerge. Cameron and Hinegardner[4] examined the stages of metamorphosis in *Lytechinus pictus* and *Arbacia punctulata*, beginning with protrusion of the podia and followed by the bending of the larva arms relative to the anterior-posterior axis and then attachment by the primary podia and protrusion of spines of the echinus rudiment. These steps are reversible. The larva can return to its original shape and swim away. This in itself suggests that the primary podia may be used in testing the substratum, but Cameron and Hinegardner provide additional evidence. The larvae of these two species would not metamorphose in about 80% of the trials in which larvae were exposed to a metamorphosis-inducing substance but the podia were not touching a solid surface. A brief touch on the larval surface or podia did not induce metamorphosis, but if the podia were allowed to hold onto an object several minutes the animals completed metamorphosis. Thus both a chemical stimulus and contact by the podia were necessary

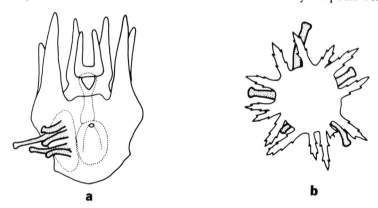

a   **b**

Fig. 1. a. Echinopluteus (unusual in having echinus rudiment on its right side).   b. Newly metamorphosed juvenile. The primary podia are stippled.

for completion of metamorphosis in most cases. A discrete area of the larval surface was ruled out as a sensory organ for the chemical inducer because all portions of the larval epithelium had been removed from settling larvae in one experiment or another. If all echinoplutei can explore the bottom and select attachment sites before the primary podia extend, as in *Mellita*, but attachment of the primary podia is necessary before the individual irreversibly commits itself to metamorphosis, as in *Lytechinus* and *Arbacia,* then both the larval epithelium and the primary podia are involved in selecting a favorable site for metamorphosis. Alternatively, there may be differences among echinoplutei in sensory structures employed at settling and ability to select substrata before emergence of primary podia.

Few species of echinoids have pelagic lecithotrophic larvae. In those reared through metamorphosis, the podia emerge while the larva is still capable of lcomotion and apparently attach the larva to the bottom as in planktotrophic forms. Williams and Anderson[23] mention that larvae of *Heliocidaris erythrogramma* swim near the surface of culture vessels after primary podia emerge but settle within one to three days after this.

Asteroidea

The Valvatida, Spinulosida, and Forcipulatida have a brachiolaria larval stage with larval attachment organs (Fig. 2). Both lecithotrophic and planktotrophic larvae have brachiolar arms, used in temporarily attaching to the bottom during exploration, and an adhesive disk, used in a more permanent attachment during resorption of the larval body at metamorphosis. The adhesive disk is at the anterior end of the larva, with a pair of brachiolar arms lateral and posterior to it and an anterior arm anterior to it. Papillae are at the tips of the brachiolar arms and in a row on each side of the adhesive disk. The anterior arm is modified to varying degrees from the anterior median lobe of the bipinnaria stage. In the Forcipulatida the anterior brachiolar arm resembles the other two brachiolar arms more closely than in the Spinulosida, in which the anterior median lobe is less modified. In planktotrophic larvae the extension of the coelom into the brachiolar arms is clearly visible.

Gemmill's[23] description of settling by the planktotrophic larvae of *Asterias rubens* is the most detailed and is consistent with most other reports. Gemmill's suggested mechanisms for the larval movements require experimental confirmation and are presented as hypotheses. Gemmill attributed extension of the brachiolar arms to variation in coelomic pressure. The brachia can be extended and moved independently. They can adhere to the bottom firmly but can release voluntarily. On final fixation the brachia are spread and the adhesive disk bulges outwards. Gemill claimed that contractions of the body wall increase coelomic pressure to effect this bulging and that it is probably on slackening of these contractions that the adhesive disk becomes cup-like with a thickened rim. The hold of the sucker can be released by a subsequent contraction. While the adhesive disk attachment is being loosened and renewed, shifting the spot of attachment, brachia adhere to the substratum so that the larva crawls about by means of brachia and sucker. A secretion helps the brachia to adhere to the substratum. At a later stage a secretion replaces suction in attachment of the adhesive disk. Gemmill noted that the adhesive material is visible. The adhesive disk is composed largely of secretory cells.[21] Once the adhesive disk is attached by its secretion, the larval body is resorbed, with a remnant of the anterior portion of the larva remaining as an attachment stalk. In *Asterias* the adhesive disk and part of the stalk are left behind as the larva crawls away,[23] but in larvae of *Solaster* the sucker is detached and resorbed.[24] Exploration with brachiolar arms and attachment by the adhesive disk during metamorphosis have been observed in planktotrophic larvae of the spinulosid *Acanthaster planci*[9,11] as well as in other forcipulates (such as *Pycnopodia helianthoides*[25]). Among lecithotrophic forms this pattern of exploration and attachment is mentioned for the

239

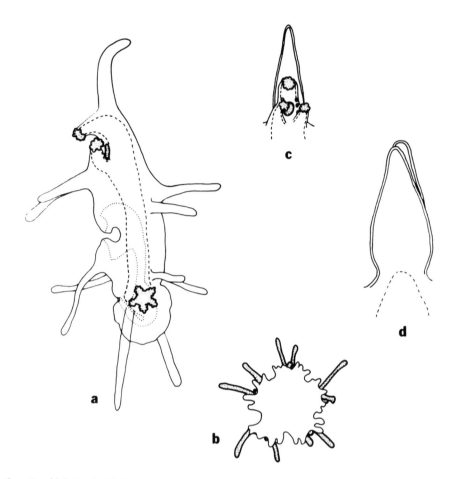

Fig. 2. a. Brachiolaria. b. Newly metamorphosed juvenile. c. Anterior end of a brachiolaria. d. Anterior end of *Luidia foliolata* at stage of settling. Attachment organs, papillae, and podia are stippled. In b and c the ciliary band is indicated by a double line. Anterior extension and the coelom into the brachiolar arms is indicated by a double line. Anterior extension of the coelom into the brachiolar arms is indicated by a dashed line. a, b, and c are from *Pisaster ochraceus*.

spinulosids *Solaster endeca*,[26] *Crossaster papposus*,[24] and the valvatid *Certonardoa semiregularis*.[27] In describing development of *Asterina gibbosa* MacBride[28] contrasted the larvae attached "by a cupping action of the muscles of the preoral lobe" with "truly sessile larvae" attached by the adhesive disk. MacBride said that the temorarily attached larvae are easily detached if taken unaware but difficult to detach once disturbed, and he noted that the "truly sessile" larvae cannot reattach when removed. Gemmill[26] found that *Solaster endeca* larvae could reattach if detached during the first day or the two following settling, but that older ones were unable to orient for attachment if detached.

A few authors give a different account of settling by larval forms of this type. Goto[29] said brachiolaria larvae of *Asterias pallida* do not attach, despite the presence of well-developed brachiolar arms in his figures of the larva. Komatsu[30] described weak attachment by the brachiolar arms of the lecithotrophic larva of *Asterina coronata japonica* but said the adhesive disk does not aid attachment in this species. Goto was concentrating on morphological changes and may have observed metamorphosis in a detached larva. Komatsu's observations cannot be so easily explained and suggest some variation among species with lecithotrophic larvae.

The observations on settling behavior suggest that the anterior epithelium of the larva, perhaps on the brachiolar arms, is the site which tests the suitability of the substratum in the valvatids, spinulosids, and forcipulates. The podia cannot be concerned, as they are in echinoids, with sensing the substratum. In the planktotrophic forms feeding can continue until the larval body is resorbed, so that starvation should not limit the period during which settling can be delayed.

In the Luidiidae and Astropectinidae there are no brachiolar arms, adhesive disk, or other larval organs of attachment (Fig. 2d). *Luidia sarsi* and *Luidia ciliaris* attach by the juvenile podia. The larval body is then resorbed by *L. ciliaris* but cast off almost intact by *L. sarsi*.[31] Wilson[32] has observed partial resorption of the larval body by some *L. sarsi*. In these seastars the podia become functional whereas the larvae retain their locomotory capacity, and these species are therefore likely to possess the capacity to explore the substratum while delaying metamorphosis. In *Astropecten* spp. the timing of resorption of the larval body relative to development of the podia suggests a limited capacity to explore the substratum before metamorphosis. Hörstadius's[33] account of *Astropecten arancius* indicates that resorption of the larval body is far along when the ring canal of the water vascular system is completed, although resorption of the larval body is not complete when the juveniles begin to creep about by the podia. It is not clear whether the larva can still carry the juvenile about at the time the podia become functional. Oguro, Komatsu, and Kano[34] found that the larval body of *Astropecten scoparius* was almost entirely resorbed before the podia became functional. Unless the larvae find a site and remain there before metamorphosis is far advanced, the ability to move from a less preferred site is greatly reduced in the *Astropecten*. *Astropecten* larvae have no obvious means of attachment other than podia, although Komatsu[35] described the lecithotrophic larva of *Astropecten latespinosus* as attaching by its anterior end, despite the lack of brachiolar arms or adhesive disk. It may be that larvae of *Astropecten* have a greater capacity to search out favorable sites than these observations suggest, but the stage at which this is accomplished is not clear from available laboratory observations.

Ophiuroidea

In ophiuroids attachment is by the juvenile podia. In most species the primary podia and two pairs of podia are formed at metamorphosis. In ophioplutei resorption of the larval body begins before the podia are functional (Fig. 3). If selection of a favorable site is not possible before the podia are functional, then the ability to feed during the search for a favorable substratum must be greatly reduced. Because there are no observations on delay of settling in ophiuroids, it is not possible to say whether the capacity for delay of settling is diminished in ophiuroids relative to asteroids or echinoids with planktotrophic larvae. The swimming ability of ophioplutei may be reduced to varying degrees as the larval body is resorbed. In some ophiuroids (*Ophiothrix* species,[1] *Ophiopholis aculeata*[36]) the posterolateral arms are not resorbed but are cast off after the juvenile attaches by the podia. In these forms the ability to swim is not greatly diminished. (According to MacBride[37] the tissue of posterolateral arms of *Ophiothrix fragilis* is resorbed after settling.) In many other ophioplutei the arms are resorbed entirely and in a different sequence, often with the posterolateral arms and the right anterolateral arms resorbed last.[1] Ophioplutei with resorption of arms far progressed and even completely metamorphosed juveniles are frequently found in the plankton. It is not known if these are unusual individuals. It is possible that most of the juveniles settle before the larval arms are resorbed and the ciliary band eliminated. The timing of resorption relative to attachment needs study.

Pelagic lecithotrophic larvae of doliolaria type are found in the families Ophionereididae and Ophiodermatidae and the subfamily Ophiolepidinae of the Ophiuridae. Grave[38] and Stancyk[39] described attachment by the podia in *Ophiura brevispina* (no longer *Ophiura*) and (*Ophiolepis elegans*. In both species the transverse ciliary bands and the podia are simultaneously functional.

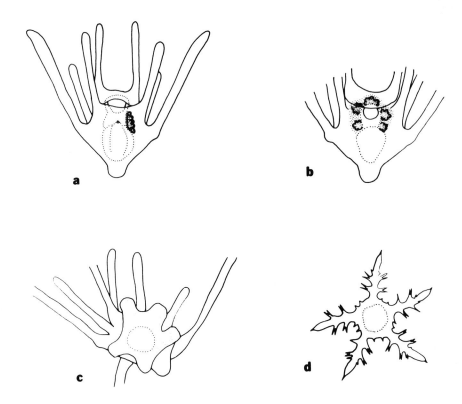

Fig. 3. a. Ophiopluteus. b. and c. Stages in metamorphosis prior to attachment. d. Metamorphosed juvenile (terminal podia and paired podia not shown). Denser strippling is hydrocoel and podia. Light stippling in b and c is thickened tissue of forming juvenile. All from an unidentified planktonic form.

Stancyk described the larvae at one stage as either swimming or walking about in what appeared to be a testing period before the final settlement. It is likely that these forms can delay settling for at least a day while searching for a favorable substratum.

Holothuroidea

In holothuroids with either lecithotrophic or planktotrophic development, larvae metamorphose to a pentactula with five primary tentacles before settling (Fig. 4). Some larval ciliation is retained during this transformation.

Although species of several orders have been reared through metamorphosis, accounts of settling are not detailed. Runnström and Runnström[40] described newly settled larvae of the dendrochirote *Cucumaria frondosa* as creeping with the five primary tentacles and the two ambulacral podia. In settling *Cucumaria elongata,* however, the primary tentacles are used in locomotion, and ambulacral podia do not develop until well after settling.[41] Thus, even within the Dendrochirotida, ambulacral podia are not always involved in attachment. So far as is known all dendrochirote larvae are lecithotrophic.[1] It should be noted that the five primary tentacles are probably not homologous to the five primary podia in echinoids. The five primary tentacles remain in an oral position whereas the five primary podia of echinoids and other echinoderms remain at the terminal (aboral or distal) ends of the ambulacra. *Cucumaria* species and also *Cucumaria echinata*[42] develop functional primary tentacles while the larva is still capable of swimming, so it is likely that the larvae probably can search for a favorable substratum for at least one or perhaps several days.

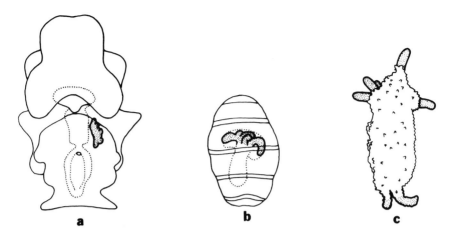

Fig. 4. Larval stages of *Parastichopus californicus*. a. Auricularia stage. b. Doliolaria stage before emergence of tentacles. c. Settled pentactula. Hydrocoel, five primary tentacles, and two podia are stippled.

Nyholm[43] and Runnström[44] gave some information on setttling in lecithotrophic larvae of two species of synaptids. According to Nyholm the pentactula of *Labidoplax buskii* can swim three or four days. The five primary tentacles are used in burrowing into the sediments. Runnstrom said that settled pentactulae of *Leptosynapta inhaerens* use the five primary tentacles in creeping. The synaptids are apodous with no development of ambulacral podia.

More definitive accounts of substratum selection by holothuroids, particularly in species with planktotrophic development, are needed.

Crinoidea

The crinoids have a strictly larval attachment organ, the adhesive pit, at the anterior end. Even forms which are not stalked as adults have a sessile stalked juvenile stage at settling. Seeliger[19] said the cells of the adhesive pit of *Antedon rosacea* are secretory. Holland and Kubota[45] reviewed development of *Comanthus japonica* and noted the secretion released from the adhesive pit at settling. I was unable to consult Kubota's[46] account of attachment by *Comanthus japonica*. Podia do not emerge until after settling and therefore are not involved in sensing the substratum or attachment.

DISTRIBUTION OF JUVENILES IN THE FIELD
AND CAUSES OF JUVENILE MORTALITY

Echinoidea

Several recent studies bear on settling and recruitment of juveniles to populations of *Dendraster excentricus*. This sand dollar usually occurs in dense aggregations. Birkeland and Mesmer[47] transplanted adult *D. excentricus* to a shore without sand dollars and subsequently observed enhanced recruitment of juveniles. Highsmith[3] found that larvae of *D. excentricus* are induced to settle by sediments from beds of the adults but not by sediments nearby. Highsmith also observed that a species of tanaid crustacean was much more abundant outside beds of the sand dollars than within the beds. In the laboratory, Highsmith observed that the tanaids ate newly settled sand dollar larvae but that sand dollar adults greatly reduced the number of tanaids in the sediments. Highsmith concluded that tanaid densities are reduced by physical disturbance of the sediments. Because adult sand dollars are a major and reliable disturbing factor, a substance associated with the adults is a reliable cue for settling larvae of *D. excentricus*.

Timko[48] found, however, that adult *D. excentricus* are capable of eating larvae which are near metamorphosis and that recruitment may be low in very dense aggregations of adults. Timko based the latter conclusion on the dominance of a few year classes and spacing of year classes in dense aggregations. Taken together, these observations suggest that the risks of settling within dense aggregations are less than the risks of settling outside beds of adults, but that recruitment may be poor at both very low and very high densities. It should be noted, however, that Birkeland, Mesmer, and Highsmith observed *D. excentricus* in or near Pugent Sound, whereas Timko observed populations in southern California.

Juveniles of regular echinoids have been found in close association with adults, although substances from adults have not been implicated in induction of settling and metamorphosis. Tegner and Dayton,[49] studying populations in California, found juveniles of *Strongylocentrotus purpuratus* in varied habitats but juvenile *Strongylocentrotus franciscanus* almost exclusively sheltering under the spines of adult *S. franciscanus*. C. Jones[50] has observed the same type of sheltering of juveniles under adult spines in *S. franciscanus* in the Straits of Juan de Fuca, but in this region *S. purpuratus* juveniles also can be found sheltering under adults.[51] Tegner and Dayton suggested that the spines of adult *S. franciscanus* are an important refuge from predators for the juveniles, and that survival of juveniles is lower in the absence of adults. Either larvae settle in response to a factor associated with adults, or juveniles become associated with adults by the time they are visible in the field. This association is not a form of brooding because the larvae have undoubtedly been dispersed widely and few could be associated with their parents. There should be little selection for protection of juveniles by the adult.

Asteroidea

Adult *Acanthaster planci* prey on corals, but Yamaguchi[52] found that the same corals can severely injure juveniles of this asteroid. The juveniles feed on algae for several months after metamorphosis and transfer to coral prey after considerable growth. The preference for algal substrata by settling *A. planci* is not surprising. The utlization of safe and abundant prey species permits a broad array of suitable settling sites. Yamaguchi[11] and Lucas[10] doubt that settling sites are limiting for this species. Birkeland noted a variety of asteroid juveniles on tubes of the polychaete *Phyllochaetopterus*. Juveniles of *Mediaster aequalis, Luidia foliolata, Crossaster papposus, Henricia leviuscula, Solaster stimpsoni, S. dawsoni,* and *Pteraster tesselatus* were commonly found in this habitat.[13] These juveniles appear to be occupying a different microhabitat and feeding on a quite different array of prey organisms than the adults. (All but *L. foliolata* have pelagic lecithotrophic development.)

There are no forcipulate asteroids in the above list. There are few observations of feeding by newly metamorphosed forcipulate asteroids. Cannibalism among juveniles of *Asterias* species has been observed and may be partly responsible for the rapid decline following heavy sets.[53] Juveniles of *Asterias* species are described eating small barnacles and bivalves.[12,52] After a little growth, a wide variety of animal prey would be available to a juvenile forcipulate asteroid, but immediately after metamorphosis many prey items may be too large. In forcipulates with planktotrophic larvae the juveniles formed at metamorphosis are about 0.5 mm in diameter. Newly metamorphosed bivalves are often between 0.2 and 0.3 mm long and would be an appropriate size of prey. Most barnacle cyprids fall between 0.5 and 1.2 mm in length and may be too large to be effectively attacked. Some invertebrates which are too small to be suitable prey items for adult seastars may be an abundant and reliable food for newly metamorphosed juveniles. One possibility is newly settled larvae of *Spirorbis* species, which are well under 0.5 mm long.

Ophiuroidea

Thorson,[54] using unpublished data from Muus, described the appearance of juveniles of an *Amphiura* species on mud earlier and in greater numbers than on sand. Subsequently mortality proved the sand to be a much less favorable substratum for these ophiuroids. This observation suggests that delayed settling may occur in this ophiuroid and may be effective in guiding many individuals to a more favorable habitat.

Ophiothrix fragilis has a planktotrophic larva but Warner[55] found small juveniles on the disks, on the arms, and in the genital bursae of adults. Larvae may settle directly onto large adults, which occur in dense beds. Warner added that although the small juveniles of this species may be semi-parasitic on adults, the medium-sized individuals move away from the dense aggregations perhaps to avoid competition with adults. As in the *Strongylocentrotus* species discussed above, it is extremely unlikely that parents are harboring their own offspring.

The life cycle of the basket star *Gorgonocephalus* is unusual in that young stages are found within polyps of the octocorallian *Gersemia*. Patent[56] said that juveniles do not appear to damage the polyps. At a later stage juveniles attach to the disks of *Gorgonocephalus*. *Gorgonocephalus* is free spawning with lecithotrophic development. How the larvae enter the polyps is not known. It is doubtful that the juveniles found clinging to adults are associated with their parents.

Another study of substratum selection by asteroid larvae was published too recently to be included in this review. Barker[57] finds that *Stichaster australis* larvae preferentially settle, metamorphose, and feed on an encrusting coralline alga, but *Coscinasterias calamaria* will settle on almost any hard substratum with a surface film of microoragnisms. Both of these species are forciculates, family Asteriidae.

ACKNOWLEDGMENTS

My interest in functional morphology of echinoderm larvae began in Robert L. Fernald's course in invertebrate embryology. Megumi Strathmann corrected portions of this paper. Work on the review was supported by NSF grant DES-74-21498.

REFERENCES

1. Mortensen, T. (1921) Studies of the Development and Larval Forms of Echinoderms, G.E.C. Gad, Copenhagen, pp. 1-261.
2. Caldwell, J. W. (1972) Development, Metamorphosis, and Substrate Selection of the Larvae of the Sand Dollar, *Mellita quinquesperforata* (Leske, 1778). Master's Thesis, University of Florida, pp. 1-63.
3. Highsmith, R. (1977) Amer. Zool. 17(4), 935 (Abstract) and personal communication
4. Cameron, R. A. and Hinegardner, R. T. (1974) Biol. Bull., 146, 335-342.
5. Hinegardner, R. T. (1969) Biol. Bull., 137, 465-475.
6. Cram, D. L. (1971) Trans. Roy. Soc. S. Africa, 39, 321-337.
7. Mortensen, T. (1938) Kgl. Dan. Vidensk Selsk. Biol. Skr., 7(3), 1-59.
8. Lewis, J. B. (1958) Canad. J. Zool., 36, 607-621.
9. Henderson, J. A. and Lucas, J. S. (1971) Nature, Lond., 232, 655-657.
10. Lucas, J. S. (1975) in Crown-of-thorns Starfish Seminar Proceedings, Australian Government Publishing Service, Canberra, pp. 109-121.
11. Yamaguchi, M. (1973) in Biology and Geology of Coral Reefs, vol. 2, Biology 1, Academic Press, New York, pp. 369-387.
12. Mead, A. D. (1899) Bull. U. S. Fish. Comm., 19, 203-224.
13. Birkeland, C., *et al.* (1971) Biol. Bull., 141, 99-108.
14. Yamaguchi, M. (1974) Micronesica, 10, 57-64.
15. Yamaguchi, M. (1974) International Symp. Tropical Reef Biol. at Guam, June 1974 (Abstract).
16. Chia, F.-S. and Spaulding, J. G. (1972) Biol. Bull., 142, 206-218.
17. Mortensen, T. (1937) Kgl. Dan. Vidensk Selsk. Biol. Srk., 7(1) 1-65.
18. Bury, H. (1888) Phil. Trans. Roy. Soc. B, 179, 257-301.

19. Seeliger, O. (1892) Zool. Wahr. Abt. Anat. Ontog., 6, 161-440.
20. Dan, J. C. and Dan, K. (1941) Jap J. Zool., 9, 565-574.
21. Strathmann, R. R. (1971) J. exp. mar. Biol. Ecol., 6, 109-160.
22. Williams, D.H.C. and Anderson, D. T. (1975) Aust. J. Zool., 23, 317-404.
23. Gemmill, J. F. (1914) Phil. Trans. Roy. Soc. Lond., 205, 213-294.
24. Gemmill, J. F. (1920) Quart. J. Micros. Sci., 64, 155-189.
25. Greer, D. L. (1962) Pacific Sci., 16, 280-285.
26. Gemmill, J. F. (1910) Trans. Zool. Soc. Lond., 27, 1-71.
27. Hayashi, R. and Komatsu, M. (1971) Jap. Soc. Syst. Zool., 7, 74-80.
28. MacBride, E. W. (1896) Quart. J. Micros. Sci., 38, 339-411.
29. Goto, S. (1898) J. Coll. Sci. Imp. Univ. Tokyo, Japan, 10, 239-278.
30. Komatsu, M. (1975) Jap. Soc. Syst. Zool., 11, 42-48.
31. Tattersal, W. M. and Sheppard, E. M. (1934) in James Johnstone Memorial Volume, Tinling and Co., Liverpool, pp. 35-61.
32. Wilson, D. P. (1967) J. mar. biol. Ass. U.K., 47, 751.
33. Hörstadius, S. (1939) Publ. Staz. Zool., Napoli, 17, 222-321.
34. Oguro, C., et al. (1976) Biol. Bull., 151, 560-573.
35. Komatsu, M. (1975) Biol. Bull., 148, 49-59.
36. Olsen, H. (1941) Bergens Mus. Aarbok, Natur, Rekke (6), 1-107.
37. MacBride, E. W. (1907) Quart. J. Micros. Sci., 51, 557-606.
38. Grave, C. (1900) Mem. Biol. Lab. Johns Hopkins Univ. 4, 83-100.
39. Stancyk, S. E. (1973) Mar. Biol. Berl., 21, 7-12.
40. Runnström, J. and Runnström, S. (1919) Bergen Mus. Aarbok, Natur. Rekke (5), 1-99.
41. Chia, F.-S. and Buchanan, J. B. (1969) J. mar. biol. Ass. U.K., 49, 151-159.
42. Ohshima, H. (1921) Quart. J. Micros. Sci., 65, 173-246.
43. Nyholm, K. G. (1951) Zool. Bidrag Uppsala, 29, 239-254.
44. Runnström, S. (1927) Bergens Mus. Aarbok, Natur. Rekke (1) 1-80.
45. Holland, N. D. and Kubota, H. (1976) Trans. Amer. Micros. Soc., 94, 58-70.
46. Kubota, H. (1969) Jap. J. Develop. Biol., 23, 92-93.
47. Birkeland, C. and Mesmer, K., personal communication.
48. Timko, P. (1978) in Reproductive Ecology of Marine Invertebrates, Stancyk, S. E. ed., Univ. South Carolina Press (in press) and personal communication.
49. Tegner, M. J. and Dayton, P. K. (1977) Science, 196, 324-326.
50. Jones, C., personal communication.
51. Paine, R. T., personal communication.
52. Yamaguchi, M. (1974) Pacific Sci., 28, 123-138.
53. Hancock, D. A. (1958) J. mar. biol. Ass. U.K., 37, 565-589.
54. Thorson, G. (1964) Ophelia, 1, 167-208.
55. Warner, G. F. (1971) J. mar. biol. Ass. U.K., 51, 267-282.
56. Patent, D. H. (1970) Ophelia, 8, 145-160.
57. Barker, M. F. (1977) J. Mar. Biol. Ecol., 30, 95-108.

GROWTH AND METAMORPHOSIS OF PLANKTONIC LARVAE OF *PTYCHODERA FLAVA* (HEMICHORDATA: ENTEROPNEUSTA)

Michael G. Hadfield

Kewalo Marine Laboratory, Pacific Biomedical Research Center, University of Hawaii, 41 Ahui Street, Honolulu, Hawaii 96813

Enteropneust larvae collected from inshore Hawaiian plankton have been used to calculate larval growth rates and a dispersal time of 3 to 9 months for *Ptychodera flava*. Evidence is presented that post-tornariae taken in plankton hauls have been stimulated to metamorphose by the plankton net and are thus not typically planktonic stages. Mechanisms are suggested for retention of larvae for long periods around oceanic islands.

INTRODUCTION

The tornaria larva of enteropneust hemichordates has long held a special fascination for marine zoologists because of its elegant form and its large size. However, most tornariae are so long-lived that laboratory rearing has been considered impossible. Strathmann and Bonar,[1] for example, maintained *Ptychodera flava* in culture for 7½ months without seeing completion of the larval phase. Thus, most of our knowledge of tornariae has been gained from studies of planktonic specimens, many preserved on expeditions, and all presenting problems of identification, age, and stage. Only poor estimates of the age of planktonic tornariae have been possible and estimates of minimal and maximal dispersal time are weakly based.

In the present study, advantage is taken of the very restricted breeding season of *Ptychodera flava* in Hawaii,[2] the isolation of the Hawaiian Islands from other habitats for this species, and the relative abundance of tornariae in local plankton, in order to study larval size as a function of age, dispersal times, and the biology of metamorphosis. The studies of tornariae by Stiasny,[3,4] Stiasny-Wijnhoff and Stiasny,[5] and Bjornberg,[6] together with summaries of reproduction and larval development provided by Van der Horst,[7] Hyman[8] and Hadfield,[2] provide background for the data and discussions presented here.

MATERIALS AND METHODS

Surface plankton hauls were taken with a net 80 cm in diameter (area = .5m^2) and 2 m long. Minimum mesh size was 0.163 mm. Most tows were taken between 8 and 10 a.m. about 100 m outside of the surf zone at Ala Moana Park, Honolulu, Hawaii. Duration of each tow was about 20 minutes. Water depths in the area do not exceed 12 to 13 m. Tows were generally made between the mouths of two harbors, Ala Wai and Kewalo, with care taken not to tow across the harbor mouths. It was empirically determined that the richest plankton hauls (a subjective evaluation of the greatest density and diversity) were obtained early in the morning, with calm surface conditions, on falling tides. Hauls with enteropneust larvae were usually those rich in larvae of gastropod molluscs and Sipuncula.

On completion of each tow, plankton was immediately diluted in 4 liters of sea water in plastic buckets and carried back to the laboratory for sorting. Most zooplankters were in a healthy, vigorous state when sorted with the aid of dissecting microscopes in the laboratory.

Larvae recognized as tornariae or post-tornariae were removed from the plankton, placed in small dishes of fresh sea water and measured. Usually at least three measurements were recorded: overall length, greatest width, and length of the pre-oral segment. Post-tornariae are very elastic, sometimes contracting to 50% of their extended length. All larvae were measured several times in order to obtain an average length. The larvae collected as described above were vigorous and survived for many days in small dishes of sea water. Post-tornariae buried themselves and completed metamorphosis if exposed to sand from a suitable adult habitat.

Records were kept of all plankton tows as well as the number of enteropneust larvae in each tow. Measurements of most larvae were made. Many of the larvae were preserved in 70% ethanol or 4% formalin in sea water for subsequent examination.

Four post-tornariae collected in July 1977 were measured, placed in bowls of fine sand mixed with organic detritus and covered with water, and observed periodically for nearly two months. Water and sand were changed on alternate days for the first week and thereafter at weekly or biweekly intervals. Growth was monitored by gently washing the juvenile worms from the sand and measuring them, in water, with an ocular micrometer in a dissecting microscope.

RESULTS

1. Occurrence of Larvae

Between 26 July 1973 and 21 May 1975, 68 plankton hauls were taken. An additional 12 hauls were made between 13 July and 27 October 1977. On these 80 tows, 23 (29%) contained one or more larvae of *Ptychodera flava* (Table 1). Most larvae were collected between March and July, with few being taken before or after that time and none from October through January. More than half of the 46 larvae seen were taken in April and May.

Post-tornariae probably belonging to species distinct from *Ptychodera flava* were collected occasionally. These larvae were smaller than those of *P. flava* and had distinct and characteristic pigmentation. Although I have never encountered any enteropneust species in Hawaii that was not assignable to *P. flava*, Edmondson[9] refers to two additional unknown species. He suggests that of these two a species of *Spengelia*, a genus known to develop via a pelagic larva,[2] is probably more common.

TABLE 1

Collecting Data and Sizes of Larvae

Month	Tows with larvae / Total Tows Taken	Number of Larvae	Size Range (mm)	Mean Size (mm)
Jan.	1/5	1 (T)[a]	1.25	—
Feb.	2/7	2	2.08	2.08
Mar.	4/5	6 (2T)	T-1.25	—
			1.33-2.00	1.65
Apr.	7/11	13	1.75-2.50	2.06
May	4/7	13	1.83-2.50	2.33
June	2/3	6	2.67[b]	—
July	1/5	4	1.83-2.75	2.32
Aug.	2/12	2	1.66-2.58	2.12[c]
Sept.	1/9	1	—	—
Oct.	0/7	0	—	—
Nov.	0/4	0	—	—
Dec.	0/5	0	—	—

[a]T, tornariae; all others listed are post-tornariae.
[b]Only one animal was measured.
[c]Mean with smallest animal removed is 2.27.

2. Larval Stages

Two stages were seen in these planktonic larvae: tentaculate tornariae (Fig. 1), and post-tornariae characterized by a large, distinct proboscis separated from the posterior part of the larva by a deep groove (Fig. 2). The inflated trunk of post-tornariae is encircled by a telotrochal band of powerful cilia, by means of which the larva swims. Tentaculate tornariae ("Krohn stage," see Hadfield[2]) were collected only in January, February and March. All had approximately the same appearance and dimensions (Fig. 1). Post-tornariae occurred as early as February and accounted for all recorded occurrences of enteropneust larvae from April throughout the remainder of the year.

3. Sizes and Growth of Larvae

All tornariae collected in this study measured about 1.25 mm high and 1.1 mm wide. Sizes of post-tornariae are given in Table 1 and Figure 3. The small post-tornariae, less than 1.75 mm long, were caught in February and March, and through July the average and maximum sizes of trapped larvae increased progressively to a maximum of 2.75 mm.

In Figure 3 the sizes of larvae are plotted against the month of collection. Because the spawning season of *Ptychodera flava* in Hawaii extends from mid-November through late December (Hadfield[2] and unpublished observations), a point in mid-January represents larvae averaging about one month of age. However, because larvae are produced over a period of about 1½ months, calculations such as average monthly size, at least in the early months, are not very meaningful because the 30-day period of March, for instance, could encompass from 22% to 33% of the life spans of larvae collected in that month.

Growth rates can be calculated for early development of planktonic enteropneust larvae if one assumes that the post-tornaria collected earliest in the year developed from an egg that was

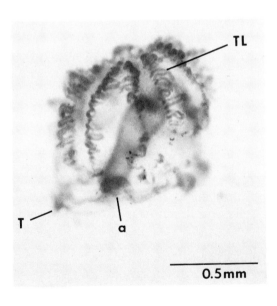

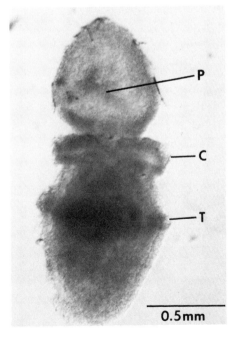

Fig. 1. Tentaculate Krohn-stage tornaria of *Ptychodera flava*. a, anus; T, telotroch; TL, tentacles.

Fig. 2. Post-tornaria of *Ptychodera flava*, 12-14 hr after collection. C, collar; P, proboscis; T, telotroch.

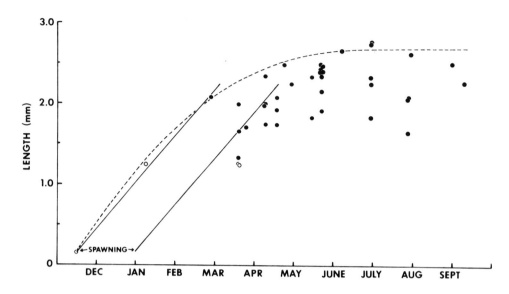

Fig. 3. Size distribution of larvae of *Ptychodera flava* plotted by date of collection. Dotted line estimates maximum growth rate; solid lines bracket early growth period over the range of the spawning season. Open circles represent tornariae; solid circles represent post-tornariae.

fertilized at the beginning of the spawning season. In Figure 3, parallel solid lines bracket probable "birth" dates and growth periods of post-tornariae collected in February, March and April. Twenty-four hours after hatching, larvae of *Ptychodera flava* measure about 160 μm long.[2] A larva of 2 mm length thus would have increased by about 1.84 mm; this growth would probably have occurred in 3 to 3.5 months (Fig. 3). Using an average of 1.84 mm increment in 3.25 months, a growth rate of 0.57 mm per month is extrapolated. Because no larvae surpass lengths of 2.75mm after a minimum of six months of planktonic life, it can be assumed that the rate of growth declines rapidly after the first three months. The dashed line in Figure 3 thus estimates a growth curve for the larvae of *Ptychodera flava* collected in this study. It should be noted that many larvae never achieve the maximum size, even very late in the collecting season. The extrapolated growth rate must therefore represent a maximum rate rather than an average.

4. Metamorphosis

Nearly all larvae seen in this study were clearly post-tornariae, and should probably be regarded as stages in metamorphosis ("Agassiz stage" is considered "regressive" in the scheme of Stiasny-Wijnhoff and Stiasny; see Hadfield[2]). However, it has been usual[6] to regard only the morphogenetic changes occurring after such stages have become benthic as metamorphosis, and this convention is followed here (see Discussion, however, for an alternative hypothesis).

Measurements of the four larvae raised in sand are presented in Table 2. Animal No. 4 had progressed the least into metamorphosis at the time of collection and its very rapid length increase—nearly 1 mm in 12 hours—is obviously due to change in shape and not to growth. The posterior segment of this animal was an anteroposteriorly flattened sphere at the time of collection; 12 hours later this region was an anteroposteriorly elongated cylinder.

Two of the larvae, Nos. 1 and 3, buried themselves in sand immediately; early "growth" was greatest in these two. Larva No. 2 was damaged during collection and had two spherical flaps of tissue hanging from the posterior segment 24 hours after collection. This larva did not initiate burrowing until after the third day when both flaps had fallen off and it looked completely normal. Its recovery is indicated by its belated, but nonetheless large, size increase after two weeks.

TABLE 2

Growth of Four Metamorphosed *Ptychodera flava* collected in July 1977
(All measurements are in mm)

Animal No. Region	Collection	+12 hr	+2d	+5d	+14d	+28d
1. Proboscis	1.25	–	1.67	1.67	1.67	–
Trunk[a]	0.92	1.17	1.67	–	2.50	–
Total L	2.17	–	3.34	–	4.17	–
2. Proboscis	1.0	–	1.00	1.25	1.67	–
Trunk	1.08	–	1.50	2.50	3.33	–
Total L	2.08	–	2.50	3.75	5.00	6.00
3. Proboscis	0.92	1.0	1.25	1.25	1.67	–
Trunk	1.66	3.0	3.42	4.45	4.16	–
Total L	2.58	4.0	4.67	4.70	5.83	5.83
4. Proboscis	0.83	1.0	1.0	1.68	1.67	–
Trunk	0.83	1.67	3.67	3.83	4.16	–
Total L	1.66	2.67	4.67	5.50	5.83	5.83

[a]Trunk includes collar.

The major immediate morphological events of metamorphosis in these larvae were elongation of the posterior segment and development of the collar-trunk groove. Development of the groove was apparent 24 hours after initiation of burrowing; by 48 hours it had the definitive form. Elongation of the trunk occurred nearly equally both anteriorly and posteriorly to the site of the telotroch; that site is clearly marked, long after metamorphosis, by a line of reddish pigment.

One pair of gill slits, probably formed before settling, was visible 24 hours after larvae began burrowing. Two gill pairs were present after 48 hours; three pairs were present after two weeks; and five pairs could be seen after one month.

At the end of one month, the juvenile enteropneusts had developed sufficient diagnostic characters to be clearly assigned to the species *Ptychodera flava*.

DISCUSSION

Growth Rates

The rate of growth of planktonic organisms is usually difficult to measure because age at time of collection is impossible to determine and size/age relationships are known for few planktonic forms, larvae or adult. Because the spawning period of Hawaiian *Ptychodera flava* is restricted to less than two months of the year,[2] and because the Hawaiian Islands are so distantly isolated from other shallow-water land masses with populations of this species, it is probably safe to conclude that most, if not all, larvae collected in nearshore Hawaiian waters are offspring of the local population(s). Indeed, the Hawaiian *P. flava* have been designated a distinct subspecies, *P. flava laysanica*.[10] Ages of larvae of *P. flava* collected in Hawaiian waters thus can be roughly determined and the size/age calculation can be extrapolated to produce a growth rate. For *P. flava* in Hawaii, larvae grow about 0.57 mm in length per month for approximately the first 3½ months, after which growth rate declines to much less than half this rate (Fig. 3). These growth rates are maximal and many (most?) larvae apparently never reach the 2 mm length which some achieve in 3½ months.

Estimates of larval growth rates usually are obtained from laboratory studies of larvae in culture.[11] In another study,[2] it was shown that laboratory-reared larvae of P. flava reached a height of 1 mm in four months. Their growth rate appears to have been much slower than occurs in the plankton. They were, in addition, considerably less developed; the four-month old laboratory-grown larvae were in the "Metschnikoff stage," whereas planktonic tornariae representing a maximum age of two months were in the more advanced "Krohn stage," called tentaculate tornariae (stages are outlined in Hadfield[2]).

Apparently it is widely believed that the longer tornariae, and perhaps other larvae, exist in the plankton, the larger they become.[6,11] From the data in this study, larvae appear to achieve a nutritional level sufficient to maintain only a particular mass. There is a "plateau size" around 2.5 mm long for the P. flava larvae seen here which is achieved at 4 to 5 months of age and maintained for perhaps 4 to 5 months (Fig. 3). There seems no a priori reason to suspect that size plateaus may not also occur in larvae of other species.

Dispersal

Dispersal time is the period from spawning to the minimal or maximal possible ages for metamorphosis to occur in pelagic larvae of a given species. Dispersal time for P. flava can be calculated by making the following assumptions: (1) the post-tornariae collected earliest in the year were "born" earliest in the reproductive season; and (2) the latest post-tornariae in local inshore plankton were "born" at the end of the reproductive season. The youngest metamorphosable larva is 3½ months old, the oldest 9 months old, and these are then the limits for minimal and maximal dispersal times. These periods are considerably greater than the 45 to 50 days estimated by Rao[12] for the larval period of P. flava. His methods for making the estimate are unclear but appear to have been based on extrapolations from laboratory rearing of only the very earliest, nonfeeding stages.

The results presented here suggest that long-lived larvae may be maintained in the vicinity of the Hawaiian Islands for at least 8 to 9 months. Indeed, all the enteropneust larvae seen were netted within a few hundred meters of the reef. (We have empirically determined that the density of larvae of benthic shallow water invertebrates declines precipitously only a short distance from shore.) Mechanisms for maintaining long-lived larval populations near shore, although unclear, must be greatly important in maintaining adult populations of species represented on the shores of oceanic islands. Patzert[13] described cyclonic and anticyclonic eddies of two or more months duration on the SW side of the Hawaiian chain. These eddies, tidal current oscillations,[14] and localized near-shore currents[15] may provide the mechanisms for long-term maintenance of local planktonic populations.

Metamorphosis

Major works on tornariae[5,6,7] have treated succeeding larval stages—Müller, Heider, Metschnikoff, Krohn, Spengel, and Agassiz—as though all occurred normally in the plankton, each stage having some predetermined duration of its own. Such set durations are probably true only for the progressive series, Muller through Krohn stages. These are the true tornariae and they possess the ciliary loops, saddles, lobes and tentacles which are the feeding apparatus of these larvae.[1] The so-called "regressive stages," Spengel and Agassiz, are those wherein the anterior hemisphere of the larva is transforming into the proboscis of the juvenile worm. In so doing, it first loses all traces of ciliary bands; the area of the larval "food groove" forming the oral field contributes to the proboscis-collar groove of the juvenile. Such larvae, devoid of anterior ciliary apparatus, have lost none of their swimming ability since the telotroch persists. They have, however, lost their ability to feed. In most cases where Agassiz stage tornariae have been kept alive, they completed metamorphosis within 24 hours following capture (this study; Bjornberg[6]; Rao[16,17]).

The observations and comments presented here suggest that all Spengel and Agassiz stage larvae are, in fact, in the process of metamorphosing. Because it is most unlikely that these or any other larvae metamorphose in mid-water, the unavoidable conclusion is that they have already undergone metamorphic induction when observed. The only obvious induction mechanism is the plankton net. I suggest that all post-Krohn stage larvae seen in plankton samples were Krohn tornariae at capture. Passage into the net and tenure in the cod-end would ably simulate the turbulent passage through shallow water that must precede settling in *Ptychodera flava* and most other known enteropneust species. This hypothesis would also explain Bjornberg's[6] observations that Spengel and Agassiz stages were taken far from shore.

If the conclusion presented above is correct, then late Krohn larvae represent the stage of metamorphic competence and the tentaculate tornariae collected in the present study must have been precompetent when netted; that is, they were either too young or too small, or both, to respond to the metamorphic stimulus of the net.

The size differences between the tornariae and the post-tornariae reported here may be deceptive. Several authors[6,7,16,17] have reported that Krohn stages are larger than ensuing Spengel and Agassiz stages; the larvae shrink or condense during the early phase of metamorphosis. If this is true in larvae of *P. flava*, then the Krohn stages represented by the post-tornariae collected in this study were much larger, perhaps double the sizes indicated in Figure 3. Krohn tornariae 4 mm high are known from Hawaiian waters.[2]

Larval Life History

The significant events of larval life of *Ptychodera flava* are now envisioned as follows: (1) Fertilization and early embryology, within the egg membrane, occur in proximity to parent populations and last about 48 hours. (2) Soon after hatching, larvae swell to a maximum dimension of 160 μm, they begin swimming and feeding, and they grow, during the next 3 to 4 months, to a height exceeding 1.50 mm. This minimal dispersal phase is sufficient for very long distance transport. (3) During the dispersal phase, larvae feed and grow, and, while growing, continue to elaborate the ciliary feeding mechanism of the anterior hemisphere; the larger the larva, the more elaborate the ciliary tracts become. (4) The apex of larval development, the Krohn stage, is presumably achieved at the end of the minimal dispersal phase, about 3 to 4 months in *P. flava*. Larvae achieving this stage become competent to metamorphose. If carried by currents and waves into shallow water the turbulence and/or surface contacts initiate metamorphosis and the larvae responds by rapidly losing its anterior ciliary apparatus (minimally within two hours in our studies), transforming the anterior hemisphere into the muscular proboscis of the young worm, yet retaining the telotroch for locomotion. (5) When such partially metamorphosed larvae (Agassiz stage) finally land in substrata soft enough to penetrate, they burrow immediately. Burrowing is initiated and accomplished solely by the proboscis. This action would presumably be significantly hindered if the anterior hemisphere of the larva still bore the looped and tentaculated ciliary feeding apparatus. The second stage of metamorphosis, consisting of loss of the telotroch and elongation of the posterior part of the larva, is completed beneath the surface of the sand; these latter transformations occur in the following 24+ hours.

Division of metamorphosis into two distinct and temporally separated phases makes good sense in functional analysis. The first step prepares the larva for burrowing while leaving it motile; the second rids the larva of remaining, superfluous larval characters after it has entered the adult habitat.

ACKNOWLEDGMENTS

The patient, plankton-sorting assistance of K. Milisen, M. Switzer-Dunlap, and W. Van Heukelem is gratefully acknowledged. Dr. E. A. Kay's constructive criticism of the manuscript served to identify (and I hope I have corrected) many ambiguities. This work was partially supported by NIH research grant 1 RO1 RR01057-01 AR.

REFERENCES

1. Strathmann, R. and Bonar, D. (1976) Mar. Biol., 34, 317-324.
2. Hadfield, M. G. (1975) in Reproduction of Marine Invertebrates, Vol. 2, Giese, A. C. and Pearse, J. D. eds., Academic Press, New York, pp. 185-240.
3. Stiasny, G. (1914) Z. Wiss. Zool., 110, 36-74.
4. Stiasny, G. (1914) Mitt. Zool. Sta. Neapel, 22, 255-290.
5. Stiasny-Wijnhoff, G. and Stiasny, G. (1927) Ergeb. Fortschr. Zool., 7, 38-104.
6. Bjornberg, T.K.S. (1959) Bol. Inst. Oceanogr. Sao Paulo, Brazil, 10, 1-104.
7. van der Horst, C. J. (1932-1939) in H. G. Bronn's Klassen und Ordnungen des Tier-Reichs, 4(Abt. 4, Buch 2), 1-737.
8. Hyman, L. H. (1959) The Invertebrates: Smaller Coelomate Groups, Vol. 5, McGraw-Hill, New York, 783 pp.
9. Edmondson, C. H. (1946) Reef and Shore Fauna of Hawaii. B. P. Bishop Museum, Honolulu, 381 pp.
10. Spengel, J. W. (1903) Zool. Jahrb. Abt. Anat., 284, 271-326.
11. Thorson, G. (1961) in Oceanography, Publ. 67, Sears, M. ed., AAAS, Washington, D.C., pp. 455-474.
12. Rao, K. P. (1951) The Anatomy of *Ptychodera flava* and studies on other Enteropneusta from Madras. Doctoral Dissertation, Madras Univ. Not seen, work cited in Rao.[16]
13. Patzert, W. C. (1969) Hawaiian Inst. Geophys. (Univ. Hawaii) Rpt. HIG-69-8, 51 pp.
14. Laevastu, T., Avery, D. E., and Cox, D. C. (1964) Hawaiian Inst. Geophys. (Univ. Hawaii) Rpt. HIG-64-1, 101 pp.
15. Wyrtki, K., Graef, V., and Patzert, W. (1969) Hawaiian Inst. Geophys. (Univ. Hawaii) Rpt. HIG-69-15, 27 pp.
16. Rao, K. P. (1953) J. Moprh., 93, 1-18.
17. Rao, K. P. (1955) Hydrobiologia, 7, 269-278.

ASCIDIAN METAMORPHOSIS: REVIEW AND ANALYSIS

Richard A. Cloney

Department of Zoology, University of Washington, Seattle, Washington 98195.

There are many structural variations in ascidian larvae especially in the development of the viscera and organs of attachment, but all species are lecithotrophic. The swimming period is generally limited to a few hours in compound ascidians and a few days in solitary species. Organs or organ rudiments of significance in the post-larval phase of the life cycle do not grow or differentiate in the larva.

Metamorphosis usually begins with settlement and is followed by a series of coordinated morphogenetic movements that rearrange organs, tissues and cells. The axial complex of the tail, visceral ganglion and sensory organs of the cerebral vesicle are destroyed and engulfed by phagocytes.

Metamorphosis can be characterized by a set of relatively rapid developmental events that transform the larva into a sessile organism. Some species may be able to begin feeding immediately; others require an additional period of development.

The principal events of metamorphosis are discussed and compared in species representing several taxa. Five types of everting papillae are compared. Six forms of tail resorption are identified. In some forms of tail resorption the caudal epidermis forces the tail into the trunk. In other forms there is evidence that tail resorption is caused by contraction of the notochordal cells.

INTRODUCTION

Metamorphosis is a brief phase in the development of ascidians characterized by settlement, rapid changes in shape, reorientation of organs, emigration of cells, and disintegration of larval nervous, sensory, muscular and skeletal tissues. Recent investigations have delineated some of the basic mechanisms involved in metamorphosis at the tissue and cellular levels of organization, but generalizations must be made cautiously because comparative analyses are beginning to uncover striking variations in these mechanisms.

This paper includes a brief review of the microscopic and submicroscopic anatomy of ascidian larvae, a definition of ascidian metamorphosis and a discussion of some discrete morphogenetic phenomena and problems associated with metamorphosis in representative species of the major taxa.

MODES OF DEVELOPMENT

Larvae of species in ten of thirteen families (classification of Monniot and Monniot[1]) have been described since Kowalevsky's[2] incisive work on *Ciona intestinalis* and *Phallusia mammilata* in which he inferred that the ascidians are chordates. The spectrum of larval morphology has been illustrated in reviews by Berrill[3] and Millar,[4] including obvious differences in dimensions, state of differentiation of visceral organs and the configurations of ampullae and papillae. In some compound ascidians, blastozooids within the trunk significantly alter the larval morphology (*Hypsistozoa fasmeriana, Diplosoma macdonaldi, Polysyncraton amethysteum*). All species except a few anurans with direct development[5] have non-feeding, lecithotrophic larvae. In general, small eggs and oviparity are characteristic of solitary ascidians, but a few are ovoviviparous; they retain embryos in their atrial chambers until hatching (*Corella willmeriana, Boltenia echinata, Molgula sluiteri*). Nearly all compound ascidians have relatively large eggs and are ovoviviparous, but *Diazona violacea*[3] is oviparous and *Hypsistozoa fasmeriana* is truly viviparous.[6]

Larvae of solitary ascidians range from about 0.6 to 1.4 mm in length; those of compound ascidians range from about 1.0 to 11.0 mm in length.[4] Volume measurements give a better indication of differences in size; the larva of *H. fasmeriana* (6.5 mm long) is about 3500 times greater in volume than that of *Molgula occidentalis* (0.6 mm long).

STRUCTURE OF SOLITARY ASCIDIAN LARVAE

The least complex larvae, based on histological characteristics, are found in the phlebobranch families Cionidae, Ascidiidae, Corellidae, and the stolidobranch families Styelidae, Pyuridae and Molgulidae. In *Ciona intestinalis*,[2] *Ascidia callosa, Corella willmeriana, Styela partita*,[7] *Boltenia villosa*[8] and *Molgula manhattensis*[9] for example, the epithelial cells of the rudiments of the digestive tract are yolky and relatively undifferentiated. The endostyle, stigmata and blood vessels of the pharynx are not yet formed. The heart can be identified in *Ciona*[2] and *Molgula*[9] as a vesicle, but it is not functional. The branchial and atrial siphons and atrial invaginations (paired in phlebobranchs and single in stolidobranchs) are present, but in most cases can be identified only in sections. The smooth muscle fibers of the siphons and the body wall have not differentiated. There may be several types of cells in the hemocoel, but the definitive blood cells have not yet undergone cytomorphosis.

The central nervous system of the trunk consists of a visceral ganglion associated with a cerebral (sensory) vesicle and a rudiment of the post-larval cerebral ganglion and the subneural gland[10,11]—the neurohypophysis. Details of the otocyst, ocellus and other putative sensory organs have been described in *Ciona* by Eakin and Kuda.[12] Although molgulid and some styelid larvae have no pigmented ocelli, they are not necessarily insensitive to light. *Molgula occidentalis*, for example, responds to lowered light intensities or shadows by swimming vigorously, but it does not display the phototactic responses characteristic of species with ocelli.

The axial complex of the tail consists of a central turgid notochord, paired lateral muscle bands, a dorsal tubular nerve cord and a ventral endodermal strand. The notochord in most fully developed larvae is a slender, tubular, rod-shaped structure; in *B. villosa*[8,13,14] it is about 0.7 mm long and 18 μm in diameter. About 40 squamous epithelial cells joined together in a single layer by zonulae occludentes form the wall of the tube. The tube is closed at both ends; the lumen extending along its axis contains a matrix of low viscosity. The surface of the notochord is covered with a sheath composed of a basal lamina and circumferentially arranged banded filaments about 110 Å in diameter. Thirty-six striated muscle cells are arranged in three rows on each side of the notochord. The dorsal nerve cord consists of a simple, ciliated tubular epithelium with two lateral grooves. In *B. villosa* and *A. callosa* each groove contains about seven naked motor fibers that grow out into the tail from the visceral ganglion at the tail bud stage. These fibers adjoin and innervate the dorsal rows of muscle cells; there are no segmentally arranged, peripheral motor fibers. There is one sensory nerve dorsal to the nerve cord and another one ventral to the endodermal strand; each extends the length of the tail in a channel formed by epidermal cells.

The epidermis is a simple epithelium composed of squamous cells in the tail and cuboidal and columnar cells in the trunk. One ventral and two dorsal coniform groups of epidermal cells form adhesive papillae at the anterior end of the trunk, except in molgulids. Ampullae are usually not formed, but may be represented by thickenings in the wall of the trunk (*Molgula citrina*[9]).

Larvae are covered with two transparent cuticular layers of tunic. The space between these two layers and between the epidermis and the inner cuticle contains matrix material, filaments and a few amoeboid cells. The outer cuticular layer forms the dorsal, ventral and caudal fins.[8]

STRUCTURE OF COMPOUND ASCIDIAN LARVAE

Compound ascidian larvae have been described in detail by Salensky,[15] Ivanova-Kazas,[16] Trason,[17,18] Brewin,[6] Levine,[19] Grave,[20,21] Scott,[22] Sebastian,[23] Abbott[24] Grave and Woodbridge[25] and others.

The state of differentiation of the digestive and circulatory systems is only slightly advanced beyond that of solitary ascidians in polystyelids (*Metandrocarpa*,[24] *Distomus* and *Stolonica*[3]); but it is well advanced in aplousobranchs (*Distaplia*,[15,26] *Aplidium*,[21] *Diplosoma*[27]), botryl-lids (*Botryllus*,[20] *Botrylloides*[28]) and perophorids (*Ecteinascidia*, *Perophora*[29]).

Distaplia occidentalis (Polycitoridae) is an example of a species with well-developed visceral organs (Fig. 1). As in the adult, four rows of stigmata are present on each side of the branchial basket. The siphons, atrium, endostyle, esophagus, stomach, pyloric gland, intestine, body wall musculature, and heart can function without further differentiation. The heart pumps in the larva and at least eight different types of blood cells are identifiable. The oozooid begins feeding as soon as the siphons open, usually within 30 minutes after settlement.

The tail, as in nearly all aplousobranchs and perophorids, is twisted 90° to the left during development; consequently, the fins are oriented in the frontal plane, relative to the trunk. The notochord is tubular and composed of squamous epithelial cells, a filamentous sheath and matrix, similar to that in solitary ascidians. The notochord supports two bands of about 750 striated muscle cells. These cells are joined together by gap junctions and transverse myomuscular junctions. Only the dorsal rows of muscle cells adjacent to the nerve cord are innervated.[30] Detailed accounts of the fine structure of the caudal muscle cells and neuromuscular junctions of other species have been published.[27,31,32]

The caudal epidermis of *D. occidentalis* is a simple squamous epithelium about 1.5 mm in length and 0.5 to 2.0 μm in thickness, bordered by a basal lamina and apically by the tunic. The caudal epidermis makes an abrupt bend at the base of the tail (the zone of transition) (Fig. 1). and is continuous with the epidermis of a large chamber in the posterior part of the trunk that receives the axial complex during metamorphosis. The epidermis follows the contours of the viscera and sensory vesicle and forms three bulbous ampullae and three adhesive papillae in a triangular field. The entire tadpole is covered by two closely opposed cuticular layers of tunic.

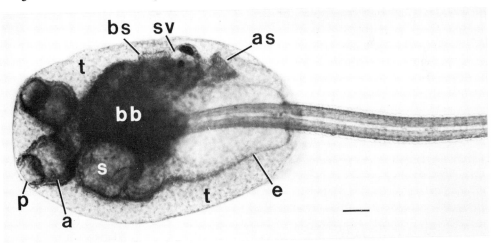

Fig. 1. *Distaplia occidentalis*. View of the left side of a living larva. The paired dorsal and single papillae (p) are continuous at their bases with bulbous ampullae (a). The entire larva is covered with a complex tunic (t). The fins of the tail are in the frontal plane and are not visible. The sac-like posterior chamber bordered by epidermis (e) receives the axial complex when the tail is withdrawn. The branchial basket (bb), branchial (bs) and atrial siphons (as), stomach (s) and sensory vesicle (sv) are visible. Bar = 0.1 mm.

The outer layer forms the fins. The space between the cuticular layers and the epidermis contains matrix material, filaments, and unidentified symbiotic spiriform bacteria; large vacuolated cells cushion the viscera and provide the trunk with a smooth ovoid configuration. Two or three isolated probuds in the anterior ventral part of the trunk grow into functional blastozooids several weeks after metamorphosis.[16,33]

The histology of the nervous system has been described in *D. magnilarva*[34] and other species.[17,18,21,22] Ultrastructural details of the sensory organs associated with the sensory vesicle have been described in *D. occidentalis*[12] and *Aplidium (Amaroucium) constellatum*.[35]

DEFINITION OF ASCIDIAN METAMORPHOSIS

Metamorphosis is a term used to categorize a variety of disparate developmental phenomena among different taxa. It does not suggest specific mechanisms, but refers to a set of more or less radical changes in structure and function that occur in a brief period of time relative to the life cycle of a species. These changes generally are irreversible rather than cyclical and usually occur in a definite sequence. Movements of tissues, changes of cellular shape, separation of cells and phagocytosis of larval structures are often important aspects of metamorphosis. Histogenesis is included in the concept, for example in studies of bryozoans,[36] insects[37] and amphibians.[38]

In ascidians, differentiation of the viscera and parts of the nervous system may occur in different species, either before or sometime after the period of rapid change, immediately following settlement; in other words, histogenesis and cytomorphosis may be premetamorphic or delayed developmental phenomena. I consider histogenetic phenomena to be parametamorphic, not strictly part of metamorphosis. According to this view, metamorphosis plus histogenesis must occur before the onset of feeding in some ascidians. Because, so far as is known, histogenesis of the viscera does not occur during the free-swimming stage of any species, release from inhibition, or stimulation of organs or tissues in an arrested state of development is an important aspect of metamorphosis.

The following list of the principal events of ascidian metamorphosis may also serve as an ostensive definition of metamorphosis for this group of organisms.
1. Secretion of adhesives by papillae or other parts of the epidermis.
2. Papillary eversion and retraction.
3. Resorption of the tail.
4. Loss of the outer cuticular layer of tunic.
5. Emigration of blood cells or pigmented cells.
6. Rotation of organs through an arc of about 90 degrees; expansion of branchial basket; elongation of oozooid.
7. Expansion, elongation or reciprocation of ampullae; reorientation of test vesicles; expansion of the tunic.
8. Retraction of the cerebral vesicle.
9. Phagocytosis of visceral ganglion, sensory organs and cells of axial complex.
10. Release of organ rudiments from an arrested state of development.

It should not be assumed that identical mechanisms are involved in each of these categories or that all of these events can be demonstrated in a single species.

Table 1 provides a schedule of the principal events that occur during metamorphosis in the didemnid *Diplosoma macdonaldi* (Fig. 11).

In the following pages the principal events will be discussed, but without any attempt to give equal attention to each topic. Our knowledge in many areas, such as the development of larval organs, is in an arrested state.

TABLE 1

Specific Events Occurring during Metamorphosis in *Diplosoma macdonaldi*

Elapsed Time	Cellular Events	Manifestations in Organism
Start	Adhesive papillae rapidly evert; adhesive adhesive secreted	Larva attaches to substratum
5-10 sec.	Inhibition of muscular contraction in tail	Quiescence
1.5-2 min.	Caudal epidermis begins to contract	Tail begins to shorten
2-3 min.	Notochordal cells contract; gaps form between cells	Axial complex loses turgor and begins to coil under compression
4-5 min.	White pigment cells begin to emigrate to surface of tunic	Color lightens
4-6 min.	Epidermal cells covering sensory vesicle begin to contract	Ocellus and otocyst are retracted
4-6 min.	Epidermal cells of papillary stalks contract	Papillae retract pulling animal closer to substratum
6-9 min.	Continued contraction of caudal epidermis cells	Tail reduced to short knob; caudal epidermis begins invagination
7-12 mins.	Epidermal cells anterior to caudal epidermis begin to contract	Coiled axial complex becomes tightly compressed
15 min.-many hours	Cyclic contractions and relaxations of epidermal cells of ampullae	Ampullae reciprocate expanding contact of tunic with substratum
15 min. to several hours	Unknown	Rotation of viscera of oozooid, about 90°
60 min. to several days	Blood cells begin phagocytosis	Destruction of axial complex, visceral ganglion and larval sensory organs

STRUCTURE AND FUNCTION
OF THE ADHESIVE PAPILLAE

Adhesive papillae enable larvae to attach, at the onset of metamorphosis, to a variety of substrata. Although attachment usually signals the onset of metamorphosis, in culture dishes or on slides metamorphosis may proceed without attachment. Attachment may later be effected by papillae if they have not completely retracted, or by sticky parts of the tunic, usually associated with ampullae.

At least eight types of adhesive papillae can be identified[39] in addition to the sensory papillae and "holdfast" mechanism of *Botryllus,* described by Grave.[20] In this discussion emphasis will be given to papillae that evert.

The Simple Coniform Papilla

In the families Cionidae, Ascidiidae, Corellidae, Styelidae and Pyuridae, the papillae consist of coniform groups of glandular cells and non-secretory cells with long apical processes. The papillae of *Boltenia villosa*[8] secrete an adhesive shortly before attachment without visibly altering shape, but within two to three minutes, while other metamorphic changes are occurring, the papillae retract, and pull the organism closer to the substratum. This sequence is widespread, if not universal among solitary ascidians with papillae. When the papillae of *Halocynthia roretzi* (Pyuridae) retract, however, they form deep inpocketings in the thick epidermis of the anterior part of the trunk.[40]

The Scyphate Papilla with an Axial Protrusion

All three papillae of *Distaplia occidentalis* completely evert within 30 seconds at 15°C, a process that brings about the secretion and exposure of adhesives[39] (Fig. 2). These papillae are suitable for study because they can be easily caused to evert. Freshly prepared working solutions of 2 to 4% dimethylsulfoxide (DMSO) and $10^{-5} - 10^{-4}$M acetylcholine (ACH) in sea water are effective inducers of papillary eversion. A drop of either solution added to a whole mount preparation stimulated papillary eversion, as well as all the other events of metamorphosis, without producing abnormalities. About 70% of the larvae respond in the late morning and early afternoon when they are most actively swimming in laboratory aquaria. Pinching the tail

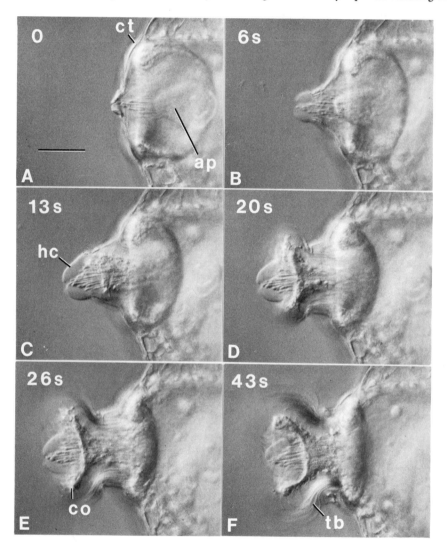

Fig. 2. Eversion of the papilla of *Distaplia occidentalis*. The total elapsed time in seconds is marked on the upper left of each frame. The axial protrusion (ap) moves forward about 60 μm during eversion exposing the hyaline cap (hc) of adhesive. During eversion a collar (co) is formed by the collocytes. Tubules (tb) from the papillary sulcus are extruded. In this papilla maximum eversion occurred in 26 seconds. Retraction is beginning in the last frame. Differential interference microscopy. Bar = 50 μm.

or parts of the trunk also immediately stimulates eversion; more than 95% of the larvae of *D. occidentalis* and *Diplosoma macdonaldi* respond. This method is useful when it is appropriate to fix specimens without permitting them to attach to a surface, or for timing papillary eversion under different conditions.

Each scyphate papilla of *Distaplia occidentalis* is a semi-transparent, radially symmetrical structure with a prominent coniform axial protrusion[39] (Fig. 2A). The protrusion extends from the base of the papilla, bordering the hemocoel, to the level of the papillary rim. It bears a hyaline cap of extracellular adhesive material and long cellular processes that extend into a fenestration of the cuticular layer of tunic. Except for nerve fibers the entire papilla is composed of epidermal cells joined by apical zonulae occludentes. The marginal (outer) layer of the cup recurves at the rim to form the parietal (inner) epithelium. This layer is continuous with the axial protrusion. The papillary sulcus, an extension of the tunic compartment within the papilla, contains tubular arrays of extracellular filaments (Fig. 3).

Six distinct types of cells can be identified: marginal cells, myoepithelial cells, anchor cells, type 1 collocytes, type 2 collocytes and axial columnar cells. The marginal cells are unspecialized epidermal cells of the marginal layer. Type 1 collocytes contain large membrane-bound granules of adhesive; they are distributed on the surface of the parietal layer and the lower two-thirds of the axial protrusion. Type 2 collocytes with small secretory granules are limited to the upper third of the axial protrusion beneath the hyaline cap. The axial columnar cells form the central part of the axial protrusion and have long apical processes extending into the hyaline cap. The mechanism of eversion depends on the arrangement of the other two cell types. About 260 myoepithelial cells, 90 μm in length, extend from the rim of the cup beneath the collocytes of the parietal layer to the margins of the axial protrusion. Together they form a basket-like layer of contractile tissue around the entire papilla. These cells contain arrays of both thick and thin filaments, but no sarcomeres are evident. Many myoepithelial cells are associated with small, naked motor fibers from the papillary nerve. Eleven pairs of anchor cells distributed symmetrically around the rim of the cup adjoin the cell bodies of the myoepithelial cells. Each of these cells has a thick apical process with a single axial cilium and many microvilli with bulbous termini. The microvilli diverge from the apices of the anchor cells, extend through the tunic, and attach to the inner cuticular layer. These cells are visible in living larvae and remain fixed in position throughout papillary eversion and retraction.

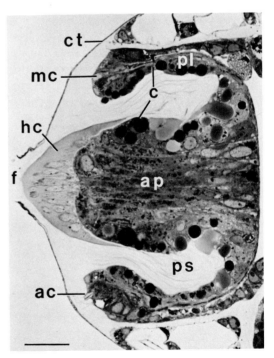

Fig. 3. Uneverted papilla of *Distaplia occidentalis*. The papilla is surrounded by tunic and protrudes into a fenestration (f) of the cuticle (ct). The wall of the cup is formed by the parietal layer (pl) and the marginal layer. Myoepithelial cells (mc) surround the cup and extend from the rim to the base within the parietal layer. Anchor cells (ac) in the rim hold the papillae to the cuticle Collocytes (c) border the papillary sulcus (ps). The space beneath the axial protrusion (ap) is the hemocoel. A hyaline cap (hc) of adhesive material covers the axial protrusion. Bar = 25 μm.

During eversion the tip of the axial protrusion, with its hyaline cap of adhesive, extends through the cuticular fenestration until it has advanced about 60 μm relative to fixed reference points in the trunk (Fig. 2). During extension, the rim of the cup at the level of the anchor cells remains stationary and does not change in diameter. During the first few seconds of eversion the type 1 collocytes almost simultaneously secrete their granules. Because these granules are refractile, their exteriorization appears as a wave of decreased optical density that spreads over the layer of collocytes. This is clearly visible with both bright field and differential interference microscopy.

As a papilla elongates, the collocytes flow forward over the surface of the axial protrusion. This movement produces an elevated collar-like fringe of epithelium around the central hyaline cap. The filamentous tubules from the papillary sulcus splay out around the sides of the papilla as it transforms from a scyphate to an hyperboloidal configuration. As the papilla reaches its full extension the axial protrusion begins to shorten, a process that draws the attached larva closer to the substratum (Figs. 2F, 12A).

During the period of elongation the myoepithelial cells shorten to about 30 μm or one-third of their length in the uneverted papilla. I infer that the anchor cells act as tendon-like structures so that the movement generated by contraction of the myoepithelial cells produces a net forward movement rather than a compression of the papilla.

Type 1 collocytes are forced outward by the contraction along the interface of the axial protrusion to form the collar; the papillary sulcus is packed with tubules so that the collocytes cannot protrude into this space. When the adhesive is secreted it comes in contact with the tubules and apparently coats their surfaces as they spread out. It can be shown that both the hyaline cap and the lateral tubules are adhesive by touching these parts with fine needles.

Papillary eversion is not inhibited by cytochalasin B (CCB) at concentrations as high as 15 μg/ml. In this respect the myoepithelial cells behave like the striated muscle of the tail and the heart and like the smooth muscle of the trunk. The application of CCB, however, reversibly inhibits the retraction of the papillae in concentrations as low as 0.25 μg/ml, the same concentrations that block tail resorption in *D. occidentalis*. These results show that the retraction of the papillae involves a separate mechanism from that of eversion; it may be the same kind of mechanism found in the caudal epidermal cells (see below).

The mechanism of eversion in the dideminid *Diplosoma macdonaldi* is similar to that described for *Distaplia occidentalis*. The papillae of this species have both myoepithelial cells and anchor cells, but the parietal wall of the papilla contains no collocytes; the collocytes are confined to the surface of the axial protrusion. Lane[41] has shown by histochemical techniques that the secretory granules of collocytes in *Diplosoma listerianum* consist of protein, rich in sulfhydryl and disulfide groups, together with a carbohydrate.

The Scyphate Papilla with an Axial Vesicle

The papillae of *Eudistoma ritteri* (Polycitoridae) superficially resemble those of *Distaplia* but are fundamentally different. They are cup-shaped, but, instead of a solid axial protrusion, each contains a hollow vesicle composed of distal columnar and lateral squamous cells (Figs. 4a, 5). The vesicle has a narrow neck that is continuous with and supported by the parietal layer of the cup. The rim of the cup is held in position by anchor cells that attach to the cuticular layer of tunic. The lumen of the vesicle is continuous with the hemocoel of the trunk, and its outer surface lies in a papillary sulcus (Fig. 5).

At the onset of metamorphosis the three papillae simultaneously shoot out to form long bulbous processes (Figs. 4a, 4b). The entire surface of each papilla is adhesive, but the tips bear distinct columnar secretory cells (collocytes). Immediately following eversion all three papillae retract and soon become flush with the surface of the tunic.

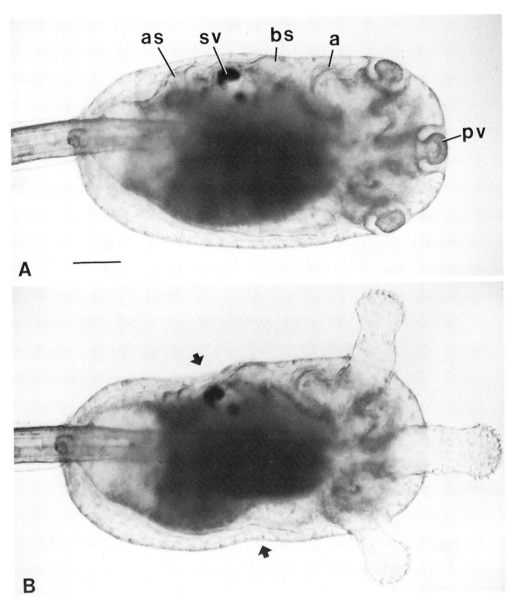

Fig. 4. Eversion of the papillae of *Eudistoma ritteri*. Prior to eversion (A) the papillary vesicle (pv) is held within a double-walled cup. During eversion each papilla shoots outward within a few seconds (B); at the same time contractions occur in the body mass (arrows). Hydrostatic pressure may cause eversion, but this hypothesis has not been tested. See Fig. 1 for other abbreviations. Bar = 0.1 mm.

The mechanism of eversion has not been investigated experimentally; observations of living specimens suggest the hypothesis that hydrostatic pressure, generated by smooth muscle in the trunk, forces the papillae outward. This should be tested by opening the body cavity of the trunk. If hydrostatic forces are involved this intrusion would be expected to dissipate the forces and inhibit eversion. Retraction is probably effected by contraction of the epidermal cells of the papillary wall.

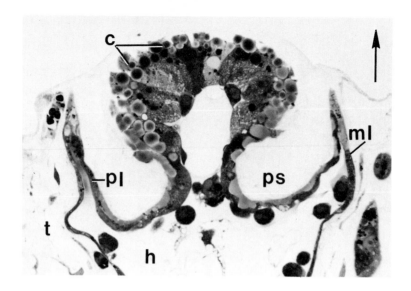

Fig. 5. *Eudistoma ritteri.* Sagittal section of an uneverted papilla. The cup-shaped papilla has a central axial vesicle composed of columnar collocytes (c) with secretory granules. The vesicle is surrounded by a papillary sulcus (ps) and is supported by the parietal layer (pl) of the cup. The marginal layer (ml) of the cup is continuous posteriorly with the epidermis of the ampullae (see Fig. 4). The papilla everts in the direction of the arrow. The lumen of the vesicle is continuous with the hemocoel of the trunk (h). Bar = 25 μm.

Tubular Invaginated Papillae

Deeply invaginated tubular papillae in *Euherdmania claviformis* (Polyclinidae) and *Pycnoclavella stanleyi* (Polycitoridae) have been described by Trason.[17,18] They evert so rapidly in both species that muscular contractions of the trunk and other details of the process have not been observed. When fully extended the papillae of *E. claviformis* project nearly half the trunk-length in front of the larva. Cells of the innermost tips of the papillae form adhesive glands; these are terminal in the everted papilla and consequently turn inside-out upon eversion. Trason suggested that hydrostatic forces generated by musculature in the trunk could be involved. Retraction has not been described.

Goblet-Shaped Papillae

The larvae of many species in the family Poyclinidae have laterally compressed, goblet-shaped papillae borne on long, thin, tubular stalks.[21,22,42] I have examined the fine structure of the papillae of *Aplidium constellatum.* They contain axial columnar cells with long microvilli and paraxial collocytes with apical secretory granules. Anchor cells are located at the dorsal and ventral margins of each papilla, but myoepithelial cells have not yet been identified. Prior to eversion the bases of the collocytes and axial cells are extensively interdigitated. At the onset of metamorphosis they triple in length within about three seconds (Figs. 6A, 6B), and the interdigitation disappears.

The mechanism of eversion has not been determined. I am investigating the possibility that contraction of the papillar margins compresses the glandular and axial cells and causes them to elongate.

When these papillae retract, dense arrays of microfilaments become aligned in the basal cytoplasm of the cells forming the stalk. Retraction is inhibited by CCB.

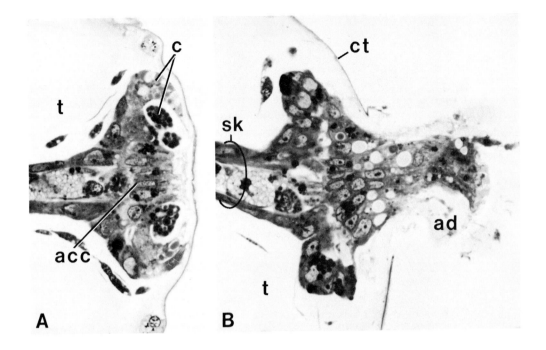

Fig. 6. *Aplidium constellatum*. Sagittal sections of an uneverted (A) and an everted (B) papilla. These goblet-shaped papillae, surrounded by tunic, are borne on long tubular stalks (sk). The parietal layer is composed of axial columnar cells (acc) and collocytes (c). Anchor cells (not shown) are located in the rim of the papilla. Myoepithelial cells have not yet been identified in this species. The axial columnar cells and the collocytes elongate two or threefold upon papillary eversion and rupture the cuticle (ct). Adhesive material (ad) is released by the collocytes. Bar = 25 μm.

Saucer-Shaped Papillae

Ecteinascidia turbinata (Perophoridae) found in Bermuda, Florida and throughout the Caribbean has large saucer-shaped papillae, each with a prominent axial protrusion and a hyaline cap (Fig. 7A). I have not found a method to stimulate eversion; therefore it is necessary to rely on spontaneous settlement. These papillae have myoepithelial cells and processes that resemble those of anchor cells in the marginal layer. Glandular cells extend along the parietal layer and over the axial protrusion. When the papilla everts, the glandular cells are forced outward along the surface of the axial protrusion and form a thick collar (Fig. 7B). During this excursion they exteriorize their granules. A cinemicrographic analysis would be necessary to fully understand this process; from a study of sections I infer that the myoepithelial cells pull down the margins of the saucer to the base of the axial protrusion. The columnar cells of the axial protrusion shorten when the papillae retract (Fig. 7C).

Apapillate Larvae

Larvae of species in the family Molgulidae lack papillae but most of them have an antero-ventral chin-like epidermal protrusion, an ampullar rudiment. The entire surface of the trunk of *Molgula citrina* and *M. manhattensis* becomes sticky just before metamorphosis and larvae may adhere in many orientations.[9] I have examined the structure of the trunk epidermis and settling behavior of the larva of *M. occidentalis*. Most of the epidermal cells contain densely packed, strongly metachromatic, apical secretory granules, but they are especially abundant in the glandular protrusion. When these larvae settle they invariably adhere by the glandular protrusion. When removed from a surface they reattach in the same way. The other epidermal cells probably secrete an adhesive during or following the outgrowth of ampullae from the lateral walls of the trunk.

Discussion

The major types of papillae found in the ascidians are categorized in Table 2 along with some specific examples. Three types, not discussed here, are included to indicate the range of diversity. There are some obvious differences of structure within each type and it may become necessary to modify this classification as more information is obtained.

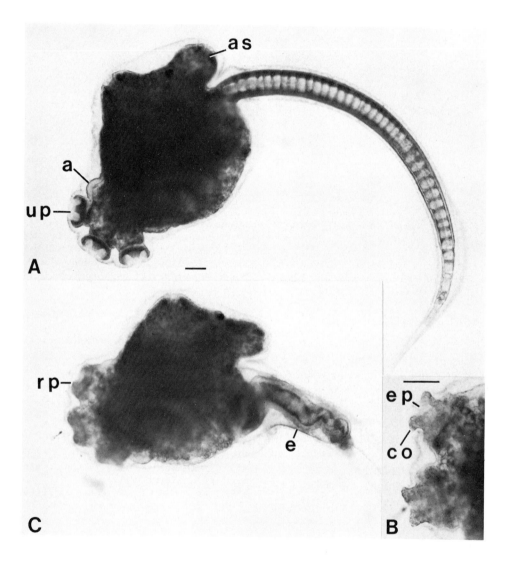

Fig. 7. *Ecteinascidia turbinata*. Larva and early stages of metamorphosis. A. The three uneverted papillae (p) are saucer-shaped and aligned in the sagittal plane. Behind the papillae, one of four thick epidermal buds—rudiments of the ampullae—is visible (a). A long row of notochordal cells, separated by lenticular pockets of matrix form the axis of the tail. The branchial (bs) and atrial siphons (as) covered by the cuticle (ct) are prominent. B. When the papilla everts (ep) the collocytes flow forward and produce a thick collar (co). C. The axial complex is forced into the trunk by contraction of the caudal epidermis (e). Shortly after eversion the papillae are retracted (rp). Bar = 0.1 mm.

TABLE 2

Types of Papillae with Specific Examples

Non-Everting Glandular Papillae	Everting Glandular Papillae	Non-Glandular Papillae
Simple Coniform papillae *Ciona intestinalis* (Cionidae) *Ascidia callosa* (Ascidiidae) *Corella willmeriana* (Corellidae) *Styela partita* (Styelidae) *Boltenia villosa* (Pyuridae) *Halocynthia roretzi* (Pyuridae)	Tubular invaginated papillae[17,18] *Euherdmania claviformis* (Polyclinidae) *Pycnoclavella stanleyi* (Polycitoridae) Scyphate papillae, each with an axial protrusion *Distaplia occidentalis* (Polycitoridae) *Hypsistozoa fasmeriana* (Polycitoridae) *Cystodytes lobatus* (Polycitoridae) *Eudistoma olivaceum* (Polycitoridae) *Diplosoma macdonaldi* (Didemnidae) *Trididemnum savignii* (Didemnidae)	Coniform ganglionated sensory papillae; with holdfast mechanism[20,45] *Botryllus schlosseri* (Styelidae) *Botryllus niger* (Styelidae) *Dendrodoa grossularia* (Styelidae) *Symplegma viride* (Styelidae)
Coniform ganglionated papillae[43] *Distomus variolosus* (Styelidae) Ridged elliptical papillae[4] *Eudistoma fantasiana* (Polycitoridae) *Eudistoma digitatum* (Polycitoridae)	Scyphate papillae, each with an axial vesicle *Eudistoma ritteri* (Polycitoridae) *Eudistoma molle* (Polycitoridae) *Pseudodistoma aurea* (Polyclinidae) Saucer-shaped papillae *Ecteinascidia turbinata* (Perophoridae) *Perophora viridis* (Perophoridae) Goblet-shaped papillae[21,42] *Aplidium constellatum* (Polyclinidae) *Polyclinum indicum* (Polyclinidae)	

Berrill[3] regarded the papillae of aplousobranchs and perophorids as "very much alike"; in fact there are striking histological differences in everting glandular papillae. Papillae of this type have been loosely referred to as "suckers,"[3,17,56] but none is known that actually functions like muscular suckers.

Three discrete types of papillae are found in the genus *Eudistoma*, but the taxonomic significance of these organs has not been evaluated.

In addition to determining how different types of papillae evert, an effort should be made to determine if they have sensory functions of significance in the selection of sites for settlement (see **OTHER PROBLEMS**, below).

MECHANISMS OF TAIL RESORPTION IN DIFFERENT TAXA

Six different forms of tail resorption can be recognized in the ascidians. Three involve fundamentally different mechanisms and three entail secondary, but significant variations (Table 3, Fig. 10).

Form 1

The following mode of tail resorption, found in *Distaplia occidentalis*,[26] is typical of most species in the families Polyclinidae, Polycitoridae and Didemnidae. Within a few seconds after attachment, contraction of the caudal musculature is inhibited. About two minutes later the tail begins to shorten. The notochordal cells round up, and small gaps appear between adjacent cellular processes; the notochord loses its turgor within about 15 seconds owing to the loss of matrix.

TABLE 3

Classification of Tail Resorption in Ascidians with Specific Examples

I. Complete Axial Complex Forced into a Helix or Folded within Body Cavity	II. Complete Axial Complex Not Forced into a Helix or Folded within Body Cavity

A. Epidermis with independent contractile properties. Dense arrays of microfilaments become aligned in apical cytoplasm of contracting cells. Notochordal cells separate causing notochord to lose rigidity. Rate of shortening rapid.
    Type 1 tail resorption:
      Aplidium constellatum (Polycinidae)
      Distaplia occidentalis (Polycitoridae)
      Diplosoma macdonaldi (Didemnidae)
	A. Notochordal cells probably contractile. Prominent aligned arrays of microfilaments in base of notochordal cells. Muscle cells dissociate.
	1. Notochordal cells shorten, dissociate and move out of sheath into body cavity at onset of tail resorption.
	    Type 4 tail resorption:
	      Boltenia villosa (Pyuridae)
	      Styela gibbsii (Styelidae)

B. Epidermis probably contractile; aligned microfilaments in apical cytoplasm of epidermal cells usually less prominent than in Type 1. Rate of tail resorption slow.
    Type 2 tail resorption:
      Ecteinascidia turbinata (Perophoridae)
      Ascidia callosa (Ascidiidae)
      Corella willmeriana (Corellidae)
      Ciona intestinalis (Cionidae)
	2. Notochordal cells shorten, but sheath is not ruptured as in type 4 tail resorption.
	    Type 5 tail resorption:
	      Molgula manhattensis (Molgulidae)
	      Molgula occidentalis (Molgulidae)

C. Epidermis probably contractile; aligned microfilaments prominent in basal cytoplasm. Axial complex tends to dissociate rapidly.
    Type 3 tail resorption:
      Botryllus schlosseri (Styelidae)
	B. Notochord probably not contractile. Part of the axial complex remains in trunk; part is cast off by constriction of epidermis near base.
	    Type 6 tail resorption:
	      Polycitor mutabilis[55] (Polycitoridae)

Simultaneously, the caudal epidermis separates from the axial complex, detaches from its basal lamina and begins to shorten and thicken. A subepidermal fluid-filled space is formed. The axial complex is forced into the body cavity and formed into a right-hand helix; in some species it becomes folded instead of coiled (*Aplidium, Eudistoma*). Within about 7 minutes at 20°C (11 minutes at 14°C) the caudal epidermis is reduced to a coniform mass, about 125 μm in length. During the next 10 minutes the epidermis of the posterior part of the trunk begins to contract causing the loosely coiled axial complex to become compressed. At the same time the caudal epidermal mass begins to invaginate into the body cavity. Invagination is completed within about 90 minutes. The axial complex is gradually phagocytized by cells in the hemocoel. The invaginated epidermal cells are not phagocytized, but the mass gradually diminishes in size and disappears about 45 hours after settlement. I infer that these cells reextend and become part of the body wall of the oozooid.

The force responsible for propelling the axial complex into the trunk in *Distaplia occidentalis*, as well as in *Diplosoma macdonaldi, Aplidium constellatum* and *Hypsistozoa fasmeriana* is generated by the caudal epidermis.[13,26,46,47] The contractile properties of this tissue and its essential role in tail resorption can be demonstrated in each of these species by tearing the tail from the trunk with forceps immediately after it begins to shorten. The epidermis of the isolated tail rapidly contracts over the surface of the axial complex and forms a compact mass at the tip of the tail. The notochord loses its turgor as in normal tail resorption, but the axial complex does not coil, fold or shorten. During similar experiments before the onset of metamorphosis, the caudal epidermis does not contract and the notochord remains turgid. Other experiments in which the tail is cut off at various distances from the base or damaged on one side during tail resorption (causing the tail to bend toward the opposite side) are consistent with the conclusion that the epidermis is contractile. After complete separation from the other tissues of the tail, the epidermis rapidly contracts into a compact mass (Fig. 8).

Since metamorphosis can be easily induced, the caudal epidermis must be competent to contract at any time during the larval life.

Changes in the fine structure of the epidermal cells of *A. constellatum* and *D. occidentalis* have been described in detail.[26,47] In the larva the caudal epidermal cells are squamous and ovoid, but during contraction they become compressed apically and extended basally. They rapidly transform into flask-shaped cells (Fig. 9). Before contraction begins the caudal epidermal cells contain disordered arrays of cytoplasmic filaments (microfilaments). Some of these filaments (50-70 Å in diameter) attach to the plasmalemma in the apical contact zone, where adjacent cells are joined by zonulae occludentes.

In early stages of contraction (fixed 30 to 60 seconds after the onset of tail resorption) many filaments are found aligned parallel to the axis of contraction in the apex of each caudal epidermal cell. As the epidermal cells shorten, this layer of filaments increases in thickness, and the apical plasmalemmata become highly folded. The alignment of the arrays of filaments, parallel to the axis of contraction, persists through the final stages of tail resorption.

During contraction the majority of filaments are localized in the area of greatest shortening at the apex of each caudal epidermal cell. There is an abrupt zone of transition at the junction between the caudal epidermal cells and the epidermal cells of the trunk. Neither filament alignment nor contraction occurs in cells immediately anterior to the zone of transition during the period of rapid tail resorption; a few minutes later, however, these cells also begin to contract, but they have basally aligned filaments.

Cytochalasin B (>0.25 μg/ml) arrests contraction in about 90 seconds and during this period disorganizes the central and subterminal arrays of filaments in the caudal epidermal cells.[26] The

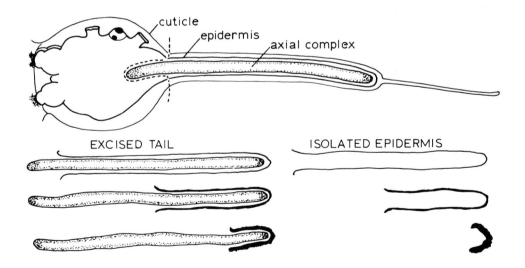

Fig. 8. Contractile properties of the caudal epidermis (*Distaplia, Diplosoma, Hypsistozoa, Aplidium*). When the tail is excised along the dotted line immediately after the beginning of tail resorption, the epidermis contracts forming a compact mass at the tip of the axial complex. The isolated epidermis is also contractile. The axial complex loses turgor, but does not shorten, fold or coil independently. The epidermis does not contract if the tail is excised before metamorphosis.

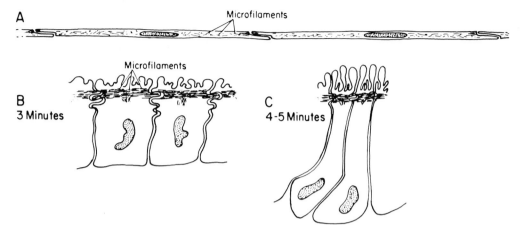

Fig. 9. Diagrammatic representation of contraction of the caudal epidermis during tail resorption in the aplousobranchs (Form 1). In the larva (A) the epidermal cells are squamous and microfilaments are randomly oriented. During contraction (B, C) the cells become shorter and thicker. The apical plasmalemmata become folded, and the microfilaments become aligned principally in the axis of shortening, at the level of the zonulae occludentes.

terminal filaments attached to the plasmalemma in the apical contact zone are not disorganized by CCB. After washing in sea water the central and subterminal filaments become realigned and contraction resumes in 4 to 10 minutes, depending on the length of treatment.

Hydrostatic pressure (6000 to 7000 psi at 22°C) also reversibly inhibits the contraction of the caudal epidermis of *D. occidentalis.* When metamorphosing specimens are fixed under pressure or fixed within about 90 seconds after release from pressure, the aligned apical filaments in the epidermal cells, characteristic of controls, are greatly reduced in number. The central filaments are generally disoriented with reference to the axis of the tail (axis of contraction), and masses of them form one or several subspherical aggregates just below the apex. When specimens are fixed immediately after recovery (after the tail resumes shortening, 7 to 9 minutes after release from pressure), most of the filaments are again organized in apical arrays parallel to the axis of shortening, and the spherical aggregates are reduced in size.

The preceding observations are consistent with the hypothesis that the arrays of apical cytoplasmic filaments identified in the epidermal cells constitute a significant part of the contractile apparatus. The procedure of binding heavy meromyosin (HMM) to actin in glycerol extracted cells has been exceedingly valuable in identifying actin in sections by electron microscopy.[48,49] I have used this method and have found that the microfilaments in the apex of the contractile epidermal cells of *D. occidentalis* bind HMM to form typical "decorated" filaments characteristic of actin.

Myosin has not been demonstrated in the caudal epidermis, but because it is so widespread in other non-muscular cells it is reasonable to suggest that it is present and to try to find ways of testing this hypothesis. Several models have been proposed to explain how contractile proteins might function in non-muscular cells;[59] these are beyond the scope of the present discussion.

Form 2

Tail resorption in species of the families Cionidae, Ascidiidae, Corellidae and Perophioridae is slower than in the aplousobranchs (20 to 60 minutes), but the axial complex coils or folds in a similar way (Figs. 7B, 10). Cytoplasmic filaments become aligned in the species of the caudal epidermal cells during tail resorption in *Ciona intestinalis*,[51] *Perophora viridis*,[52] *Ascidia callosa*[52] and *Ecteinascidia turbinata*, but are usually less prominent than in aplousobranchs. Extensive folds develop at the apices of the epidermal cells of *E. turbinata.* CCB reversibly blocks the shortening of the tail, but at twice the dosages required for aplousobranchs.[52] The epidermis has not been separated from the axial complex to determine if it can shorten independently in this group of ascidians, although ultrastructural studies strongly suggest that the epidermis is contractile. The mechanism that brings about collapse of the notochord has not been investigated.

Form 3

In *Botryllus schlosseri* (Botryllidae) the caudal epidermal cells contain arrays of aligned filaments and folds of the plasmalemmata develop as they shorten, but these occur at the bases of the cells instead of at the apices as in aplousobranchs.[52,53] CCB reversibly inhibits tail resorption, but only at ten or twenty times the effective concentrations used for the aplousobranchs.[52] The muscle and notochordal cells dissociate rapidly once the tail starts to shorten; in this respect there is a resemblance to Form 4 (*Styela* and *Boltenia*) (Fig. 10). I infer that the epidermis provides the driving force for tail resorption, but this inference is based only on ulstructural studies.

Form 4

Tail resorption in *Boltenia villosa* begins with an abrupt rupture of the anterior end of the notochordal sheath and an outward flow of the notochordal cells and the matrix into the body cavity of the trunk[8] (Fig. 10). The tip of the tail moves forward at an average rate of 50 μm/minute. Compression is pronounced at first in the proximal part of the tail, and then progresses distally. The tail thickens and continues to shorten, but the rate of forward movement of the tip gradually slows down. Within 10 minutes most of the notochordal cells have passed into the body cavity where they undergo continuous writhing movements. Within 15 to 20 minutes the sheath of the notochord becomes an empty folded sac; the muscle cells become folded and dissociated from each other; the nerve cord and endodermal strand become highly compressed. The epidermis gradually thickens during tail resorption, and remains in close contact with muscle cells; no subepidermal space is formed.

The notochord does not simply collapse when the sheath is ruptured because if a larval tail is excised near the base, the distal fragment can swim for several days without shortening.[8]

Ultrastructural studies have shown that during early tail resorption the proximal notochordal cells develop conspicuous folds in their basal surfaces, oriented orthogonally to the axis of the tail—like baffles—in complimentary folds of the notochordal sheath. The development of these folds is consistently associated with the formation of a thick layer of aligned cytoplasmic filaments beneath the plasmalemma. These changes gradually extend to the more distal cells. The folds of the plasmalemma and arrays of filaments rapidly disappear as the notochordal cells emerge from the sheath and round up.[13]

During the resorption of the tail the epidermal cells do not develop conspicuous apical or basal folds and cytoplasmic filaments do not become aligned in the cytoplasm as in aplousobranchs and phlebobranchs.

The resorption of the tail can be reversibly blocked with (>5 μg/ml) CCB.[52] The drug disorganizes the filaments of the notochordal cells; when it is removed the filaments become realigned.

These observations are consistent with the following hypothesis: The notochord shortens by active contraction of the notochordal cells; contraction begins anteriorly and spreads to the distal parts of the notochord. As it shortens it drags along the muscle cells, nerve cord and epidermis.[13]

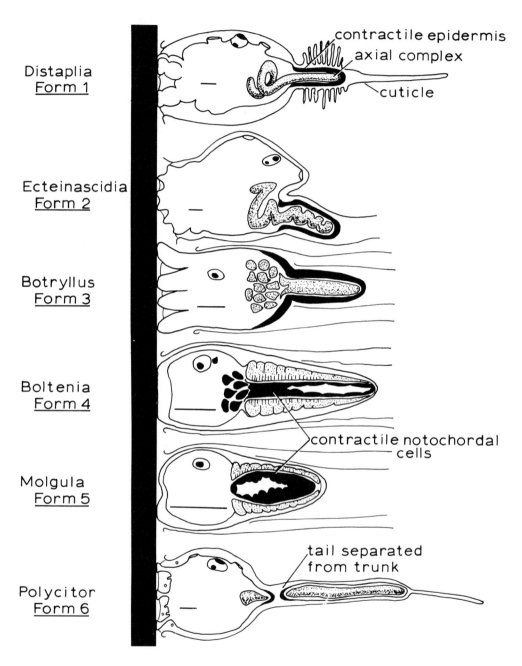

Distaplia
Form 1

Ecteinascidia
Form 2

Botryllus
Form 3

Boltenia
Form 4

Molgula
Form 5

Polycitor
Form 6

contractile epidermis
axial complex
cuticle

contractile notochordal
cells

tail separated
from trunk

Fig. 10. Diagrammatic representation of six forms of tail resorption in ascidians. The solid black parts represent contractile tissue. The stippled parts represent the axial complex (Forms 1, 2, 3, 6) or the muscle cells (Forms 4, 5). Microfilaments become aligned in the apices of the epidermal cells in Forms 1 and 2. In Form 3 they are in the basal cytoplasm. In Forms 4 and 5 the microfilaments become aligned in the basal cytoplasm of the notochordal cells. Form 6 is based on the work of Oka.[55] Bars = 0.1 mm.

The pattern of tail resorption described in *Boltenia villosa* is also characteristic of *Styela gibbsii, Styela partita, Pyura haustor*[8] and *Halocynthia roretzi.*[40] The mechanisms involved in the rupture of the notochordal sheath and the dissociation of the notochordal and muscle cells are not clear. The rupture could be simply mechanical, or it might be caused by hydrolytic enzymes produced by mesenchyme cells (microblasts) around the tip of the notochord. Ishikawa and Numakunai[54] have found acid phosphatase activity in the microblasts of *Halocynthia roretzi* larvae. They infer that putative lysosomes in these cells could have a role in the rupture of the sheath.

It has not been possible to separate the parts of the tail in order to conclusively demonstrate contraction of the notochord. Even though the notochord is apparently the driving force in tail resorption, it should not be inferred that the epidermal cells have no contractile functions in metamorphosis; indeed, the papillae are retractile, the ampullae have contractile properties, and the epidermis may be involved in rotation of the trunk.

Form 5

During tail resorption in *Molgula manhattensis* and *M. occidentalis* the axial complex and epidermis are reduced to a cone-shaped mass. The notochord shortens into an ovoid configuration, at first with a prominent matrix-filled lumen; then the matrix is forced out, leaving a compact mass of cells. Simultaneously the muscle cells are compressed and folded. In contrast to Form 4, the tail shortens without the rupture of the notochordal sheath and without dissociation of the notochordal cells (Fig. 10). Finally the sheath opens anteriorly and the notochordal cells enter the hemocoel. Microfilaments are aligned in the basal cytoplasm of the notochordal cells during tail resorption. The epidermal cells do not contain detectable arrays of aligned filaments.

Tail resorption in the molgulids is relatively slow (20 to 30 minutes), and the thickness of the layer of aligned filaments is less than that found in *B. villosa*. Thirty to forty times the effective concentration of CCB used with aplousobranchs is required to block tail resorption.[52] I favor the hypothesis that the notochord is contractile in molgulids, but it needs to be tested further.

Form 6

Oka[55] has described changes in the tail of *Polycitor mutabilis* (Polycitoridae) that are distinctly different from those in other aplousobranchs (Fig. 10). "When involution begins, the tail is shortened and, at the same time, coiled up in an irregular way, especially at the tip. Then a constriction appears at the base of the tail. Finally the axial organs break in two portions. Those lying before the constriction are resorbed in the trunk, whilst those lying behind are thrown off with the test."

These observations have been neither confirmed nor challenged. Over 80 years ago Salensky[15] claimed that the tail of *Distaplia magnilarva* is shed during metamorphosis. He did not observe larvae at the moment of settling, but failed to find remants of the tail characteristic of postlarval stages he studied in other species. For purposes of classification I have assumed that epidermal contraction would be involved in *P. mutabilis* and *D. magnilarva* (Fig. 10). It must be admitted that this mechanism is wasteful of potential nutrients, although most species can metamorphose and begin feeding normally after the tail is excised. I expect that the observations of Oka and Salensky will be disproved, but if they are correct the details of the mechanism would be interesting to compare with other species.

Conclusions

In the aplousobranchs the caudal epidermis contracts and pushes the axial complex into the trunk. Electron microscopic evidence supports the same interpretation for phlebobranchs and botryllids, but doubt will remain until the caudal epidermises of representative species can be shown to contract independently.

In pyurids, some styelids, and molgulids, electron microscopic evidence is compatible with the hypothesis that contractions of the notochordal cells shorten the tail. Microsurgical experiments should be done to determine if the axial complex can shorten when the epidermis is damaged or removed.

In aplousobranchs the notochord loses turgor when the notochordal cells pull apart, but contraction does not effect shortening of the axial complex. Contraction of the notochordal cells appears to be widespread, but it is manifested in different patterns of tail resorption (Forms 1, 4 and 5).

THE CUTICULAR MOLT

The outer cuticular layer of tunic, the layer that forms the fins of the tail, is split open in the region of the papillae upon settlement and usually is cast off soon after the completion of tail resorption. It retains the superficial configuration of the larva and may be regarded as a molt. Molts of *Boltenia villosa* often contain a dozen or more amoeboid cells.[8] In polyclinids, polycitorids and didemnids, the caudal part of the cuticle is compressed and pleated during tail resorption, but springs outward as soon as the epidermis is reduced to a short cone. The ultimate removal of the molt from the trunk may be aided by water currents, but in compound ascidians other factors often are involved. Within a few seconds after the papillae evert, the oozooid of *Distaplia occidentalis* begins to contract bands of smooth muscle in the body wall. This distorts and compresses the trunk. After 20 to 30 minutes these movements begin to force fluid out of the siphons that helps to dislodge the molt. The oozooid of *Diplosoma macdonaldi* elongates four ampullae radially and forces off the molt.

Removal of the molt is one step in uncovering the siphons; the inner cuticular layer, matrix and cells that occlude the siphons must be removed before feeding begins. The mechanisms involved in this last step have not been investigated.

EMIGRATION OF BLOOD CELLS AND PIGMENT CELLS

Eight morphologically distinct types of blood cells can be identified in the hemocoel of the larvae of *Aplidium constellatum* (Polyclinidae). One type of granulocyte (microgranular basophil) is especially common near the epidermis, but the basal lamina separates them from the plasmalemmata of the epidermal cells. Immediately after the beginning of tail resorption, about 2 minutes after settlement, these cells become activated and begin emigrating across the epidermis into the tunic.[56] The area near the base of the papillae is the most favorable place to observe the phenomenon, but it may occur throughout the trunk. The massive emigration of granulocytes is completed within 10 minutes, although many granulocytes remain in the hemocoel and may emigrate at a later time. From observations of emigrating blood cells in *Clavelina lepadiformis* (Polycitoridae) and living *Salpa democratica* (Thaliacea) Seeliger[57] concluded that the pathway was through the epidermal cells.

Electron microscopic observations are compatible with a transcellular pathway.[56] According to this hypothesis an emigrating granulocyte enters the basal part of an epidermal cell and becomes incorporated into a vacuole; the vacuole and the blood cell move to the apex of the epidermal cell; then the vacuole fuses with the apical plasmalemma releasing the blood cell into the matrix of the tunic. This pathway facilitates emigration without rupturing intercellular junctions.

I have observed emigration of granulocytes across the epidermis in *Eudistoma ritteri* (Polycitoridae); emigration begins about two minutes after papillary eversion, coinciding with the beginning of tail resorption.

Kowalevsky[2] described the emigration of mesenchyme cells into the tunic in metamorphosing *Phallusia mammilata* (Ascidiidae). In the early stages of metamorphosis of *Dendrodoa grossularia*

(Styelidae) the brick-red color of the trunk and tail changes within a few minutes to an orange color. At the same time, according to Berrill,[3] mesenchyme cells migrate through the epidermis into the tunic.

The larva of *Diplosoma macdonaldi* (Didemnidae) has about 100 white pigment cells within the tunic, outside the epidermis. Prior to metamorphosis these cells are clustered in dense patches close to the visceral mass of both the oozooid and the first blastozooid. During tail resorption they begin to move rapidly between large vacuolated cells to the surface of the tunic. This activity lightens the color of the organism when viewed by reflected light (Fig. 11).

In *D. occidentalis* granulocytes begin emigrating through the epidermis during metamorphosis and can be seen wandering between the vacuolated cells of the tunic shortly after tail resorption.

In *Boltenia villosa* cells within the tunic begin to emigrate posteriorly between the cuticular layers of tunic and by the end of tail resorption accumulate in a small cluster at the tip of the resorbed tail; some of these are lost with the cuticular molt.

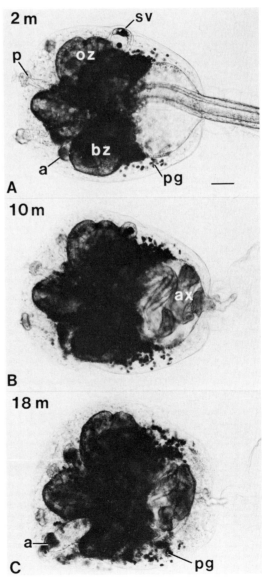

The obvious temporal coupling of blood and pigment cell emigration with the onset of metamorphosis in six families of ascidians provides strong evidence that chemical messengers are implicated in the coordination of some cellular activities.

The role of the emigrating blood cells in the aplousobranchs might be to modify or remodel the tunic in a way that facilitates its expansion or the rotation and expansion of the trunk, but this problem has not been investigated.

ROTATION OF THE TRUNK

At the moment of settlement the branchial siphon of *Distaplia* is directed orthogonally to the longitudinal axis of the larval trunk. During metamorphosis the branchial siphon and other organs rotate 80 to 90° (Fig. 12), as is typical of many species. Rotation usually brings the siphons around so they point more or less away from the substratum.

Fig. 11. *Diplosoma macdonaldi.* Early metamorphosis, 2 to 18 minutes (A-C). The tail is forced into the trunk by contraction of the caudal epidermis. Shortly after the papillae are everted, as shown here (p), they are retracted. The sensory vesicle (sv) with the dorsal ocellus and ventral otocyst retracts rapidly into the oozooid (oz). The blastozooid (bz) is a prominent structure in this species. The pigment cells (pg) (white by reflected light) rapidly move out into the tunic. One of four amupllae is beginning to extend radially at 18 minutes (a). Bar = 0.1 mm.

The time required for reorientation varies among species: *Aplidium constellatum*[58] (18° to 22°C), 2 hours; *Distaplia occidentalis* (13° to 15°C), 45 hours; *Eudistoma ritteri*[19] (14° to 15°C), 30 hours; *Euherdmania claviformis*[17] (17° to 18°C), 48 hours; *Pycnoclavella stanleyi*[18] (13°C), 48 hours; and *Boltenia villosa* (13° to 15°C), 16 hours.

Scott[58] attributes rotation in *A. constellatum* to rhythmic contraction of smooth muscle fibers in the body wall. Trason,[18] in her study of *P. stanleyi* states: "As is usual in ascidian metamorphosis, differential growth results in the reorientation of the internal organs such that

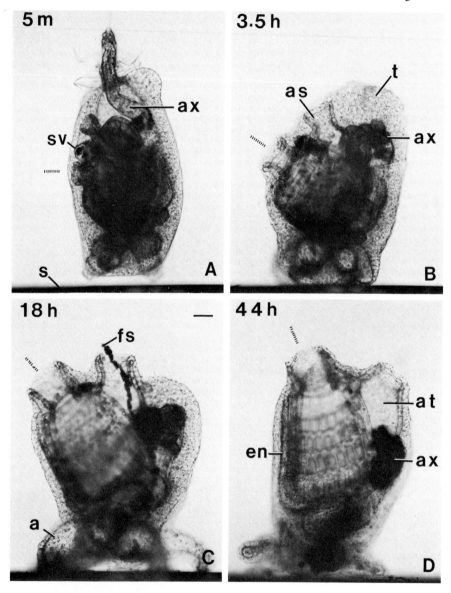

Fig. 12. *Distaplia occidentalis*. Attachment and rotation (5 minutes to 44 hours). A. The axial complex (ax) is coiled within the trunk by contraction of the caudal epidermis. The axis of the branchial siphon (dotted line) is parallel to the substratum (s) (orthogonal to the axis of the tail). B. Rotation has progressed to about 30°, and the axial complex is compressed into a tight coil. The oozooid has pulled closer to the substratum. The outer layer of cuticle has been lost and the siphons are open (as). C. The four ampullae have expanded. A fecal strand (fs) is emerging from the atrial siphon. D. Rotation is complete. The endostyle (en), stigmata and atrium (at) are visible. Bar = 0.1 mm.

they come to lie in a plane 90° from that which they occupied in the tadpole." Levine[19] claims rotation in *Eudistoma ritteri* is caused by differential rates of expansion taking place anteriorly as compared to posteriorly.

None of these hypotheses can be disproven by simple observations or histological analyses, but microsurgical techniques, the use of specific metabolic inhibitors of cell division or contraction, coupled with cinemicrographic and electron microscopic techniques might eliminate some of the suggested mechanisms and help to focus on suitable approaches to the problem.

THE AMPULLAE: ANCILLARY ROLE IN ATTACHMENT

The ampullae are epidermal structures with hemocoelic lumina. In many species they are conspicuous in the larva; in others they develop from thickened parts of the epidermis or unmodified epidermis after the onset of metamorphosis (Table 4). When visible in the larva, ampullae may be digitate, capitate, bulbous or clavate. Larvae of the genus *Cystodytes* are unique in having a single, circumpapillary, ring-shaped ampulla attached by four tubes to the trunk.[4] In *A. constellatum* and other polyclinids, ampulla-like structures develop from the epidermis, but pinch off to form isolated vesicles in the tunic about the trunk.[21]

Shortly after the beginning of metamorphosis the ampullae of *Distaplia occidentalis* expand and form four distended processes that extend over the substratum (Figs. 12, 13). In *Diplosoma macdonaldi* four digitate ampullae push out radially from the oozooid spreading the tunic (Fig. 11). In this species and in *Eudistoma ritteri* [19] the ampullae slowly expand, retract (reciprocate) and shift around in the tunic. The test vesicles of *A. constellatum* move outward in the tunic by an unknown mechanism, secrete an adhesive, and within 8 to 10 minutes, according to Scott,[58] can no longer be identified. In *Polyclinum indicum* the test vesicles persist for a week or more.

TABLE 4

Ampullae and Ampullar Rudiments and Metamorphosis

Species	Larval Structures	Fate Shortly After Settlement
Aplidium constellatum[21]	60 test vesicles	disintegrate in test[55]
Euherdmania claviformis[17]	unmodified trunk epidermis	anterior epidermis expands to form terminal vesicle
Distaplia occidentalis	3 bulbous vesicles behind papillae	4 distended processes form feet
Hypsistozoa fasmeriana[6]	9 to 14 short bulbous vesicles	elongate 10- to 15-fold; processes may branch
Eudistoma ritteri[19]	6 cup-shaped processes	variable number of reciprocating processes
Cystodytes lobatus	1 ring-shaped ampulla	fate unknown
Pycnoclavella stanleyi[18]	8 folds of epidermis	irregular, swollen vascular processes
Diplosoma macdonaldi	4 short digitate processes	4 digitate, reciprocating processes
Ciona intestinalis[2]	unmodified trunk epidermis	single long stalk extends from trunk
Ecteinascidia turbinata	4 thick epidermal buds near papillae	4 tubular processes
Metandrocarpa taylori[24]	24 digitate processes	24 radially arranged digitate processes
Boltenia villosa	unmodified trunk epidermis	8 reciprocating radial processes
Pyura pachydermitina[60]	unmodified trunk epidermis	4 processes extend from trunk
Molgula occidentalis	epidermis glandular, anterior chin-like process	1 long anterior process; 7 lateral ampullae; peristaltic waves generated

then disintegrate.[23] The tunic in *Aplidium* rapidly expands without the aid of ampullae. The possibility that emigrating blood cells secrete material that increases the osmolality of the tunic matrix causing it to swell has not been tested.

Following the early stages of metamorphosis, the tunic of many species becomes sticky; reattachment may occur if specimens are dislodged within several hours of settlement. The tips of the ampullae of *D. macdonaldi* and *Molgula occidentalis,* species that can reattach, contain secretory granules which may contain adhesives, but it is not clear how adhesive substances pass through the cuticular layer of tunic.

Early post-larval stages of *D. occidentalis* and *B. villosa,* once removed from the substratum, tend to float around in culture dishes without reattaching. Their ampullae do not contain conspicuous secretory granules (Fig. 13)

Within a few hours following the onset of metamorphosis, the anteroventral protrusion of *Molgula occidentalis* extends anteriorly to form a tube, two or three times the length of the trunk. Several hours later seven more ampullae grow out from the sides of the trunk. These tubes undergo rhythmic peristaltic contractions for several days and are then withdrawn.

Most species eventually withdraw their ampullae, but the adults of many polystyelids retain vascular ampullae. The fine structure and the mechanism of peristalsis of the vascular ampullae of *Botryllus schlosseri* colonies have been investigated by De Santo and Dudley,[59] who have demonstrated circumferentially oriented microfilaments in the basal cytoplasm of ampullar epithelial cells. During contraction the plasmalemma adjacent to the filaments folds in the same way it does in the cells of the epidermis of aplousobranchs and the notochordal cells of pyurids during tail resorption. Isolated ampullae continue to undergo cycles of contraction and relaxation. This investigation was the first to demonstrate a role of microfilaments in a cyclical mechanism. Further work should be done to determine if contraction can be inhibited with CCB and if the filaments are actin.

It cannot be assumed that all ampullae resemble those in *B. schlosseri.* A circumferential arrangement of cytoplasmic filaments would not seem to provide a mechanism for the elongation, reciprocation and retraction of ampullae.

The ampullae are secondary organs of attachment in most species and are therefore significant in metamorphosis; since they have thin walls and are connected with the hemocoel, they undoubtedly participate at least facultatively in external respiration.

RETRACTION OF THE CEREBRAL VESICLE: PHAGOCYTOSIS OF THE VISCERAL GANGLION, SENSORY ORGANS AND AXIAL COMPLEX

Early work on the fate of the sensory vesicle and larval nervous system has been summarized by Elwyn:[11]

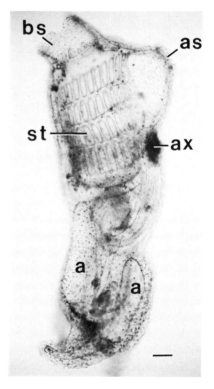

Fig. 13. *Distaplia occidentalis.* Oozooid about 70 hours after settlement. This specimen was removed from the substratum. Phagocytes have engulfed and carried away most of the axial complex. The ampullae have elongated considerably compared with the 44-hour stage (Fig. 11D). Four rows of stigmata (st) are visible; bs, branchial basket; as, atrial siphon; ax, axial complex. Bar = 0.1 mm.

Thus at some stage of Ascidian development two nervous systems, larval and future adult, are present side by side, or one above the other as in *Botrylloides*. The larval nervous system composed of sensory vesicle, visceral ganglion and its caudal continuation completely disintegrates and is absorbed during metamorphosis. The adult is represented by the neurohypophysis which later by cellular proliferation from its walls, furnishes both the adult ganglion and the entire neural gland.

This statement is consistent with the more recent studies of Trason,[18] Levine[19] and Scott.[58]

In *Diplosoma macdonaldi* the sensory vesicle is rapidly compressed and retracted, beginning 3 to 4 minutes after the onset of metamorphosis (Fig. 11). A similar process occurs in *Distaplia occidentalis*, but is less spectacular. This event is reversibly inhibited by CCB. Changes in the fine structure of the epidermal cells have not been investigated.

Phagocytes have ready access to the visceral ganglion since it lies in the hemocoelic compartment; they begin to invade the visceral ganglion of *D. occidentalis* within one hour after settlement. The fate of the otolith and ocellar pigment can often be followed in postlarval development; in *Pycnoclavella stanleyi*[18] (a species without an otolith) the ocellar pigment enters the wall of the gut and is expelled with the feces. Levine[19] described a similar phenomenon in *Eudistoma ritteri*.

There have been no ultrastructural studies of metamorphic changes in the nervous system. It cannot be assumed that the larval nervous tissues merely become senescent and undergo autolysis, because the disintegration of these tissues is normally coupled with the other events of metamorphosis, which may be induced either early or late in the life of a larva. Some larvae of *D. occidentalis* kept in undisturbed glass dishes for 12 to 24 hours often undergo partial metamorphosis without resorbing their tails, papillae or cerebral vesicles; they begin feeding normally, and their ampullae may grow out; their tails may spontaneously twitch, but they do not swim (Fig. 14). In these cases the disintegration of nervous, sensory, locomotor and adhesive organs has been decoupled from rotation, expansion of the trunk and activation of the feeding mechanism. Zhinkin[61] made similar observations in his studies of *Botryllus schlosseri*. He also observed one larva that was feeding normally and able to swim at the same time. A detailed comparison of these terata with normal specimens might have etiologic values of significance in understanding metamorphosis.

After the tail is resorbed, the cells of the axial complex are phagocytized. In *D. occidentalis* the coiled axial complex is slowly dissociated into separate cells beginning at the anterior end. These cells are gradually engulfed by phagocytes over a period of 72 hours and dispersed throughout the body (Fig. 13). Schiaffino et al.[53] have described phagocytosis and the formation of heterolysosomes containing caudal muscle cells in *B. schlosseri*. They have also observed

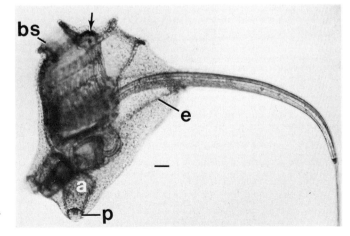

Fig. 14. *Distaplia occidentalis*. Partial metamorphosis. The trunk has rotated, the siphons are open, and the oozooid is able to feed, but the papillae have not everted or retracted, the sensory vesicle (arrow) is unchanged, and the tail has not been resorbed. Other specimens of this type elongate their ampullae and retract their papillae. In all of these cases the normal coupling of metamorphic events is disrupted. Bar = 0.1 mm.

autolysosomes with acid phosphatase activity in the dissociated muscle cells (late in tail resorption), but it should not be inferred that autolysosome formation is induced at the onset of metamorphosis because the larval muscle cells could also contain autolysosomes.

Following tail resorption in *Boltenia villosa* 36 bright orange, separate muscle cells can be counted. They are engulfed by phagocytes and become distributed in a crescent-shaped pattern in the body cavity; they remain identifiable for at least six weeks after metamorphosis.

Nervous tissues, muscle, notochordal cells and the notochordal sheath of the larva are recognized by phagocytes. The problem of how they do this has not been investigated.

OTHER PROBLEMS

How is metamorphosis triggered? More than two dozen reagents (dyes, ions of heavy metals, amino acids, iodine, strychnine, DMSO, ACH), extracts of larvae and some adult tissues, and conditioned sea water (sea water in which larvae have metamorphosed) have been found to promote or accelerate the onset of metamorphosis of various species.[62,63] Copper ions are effective,[64] but the significance of copper in normal metamorphosis is doubtful. Whittaker[65] has shown that when copper ions are reduced to very low levels (low enough to inhibit melanin synthesis) with phenylthiourea, a copper chelating agent, metamorphosis is not inhibited.

The fact that several species (*Distaplia occidentalis*, *Diplosoma macdonaldi*, *Eudistoma ritteri*) can be stimulated to metamorphose by pinching the tail strongly supports the hypothesis that extrinsic factors are not required for metamorphosis, even though they may set off the mechanism.

Crisp and Ghoboshy[65] have convincingly shown that *Diplosoma listerianum* larvae prefer to settle on shaded substrates or dark surfaces. They infer that larvae are capable of accurate orientation and navigation towards a dark stripe just before impact on a surface, or have some limited ability to move over a surface and locate a darker area after making contact. Under favorable lighting conditions these larvae settle much sooner than under unfavorable conditions. I have found that larvae of *Diplosoma macdonaldi* and *Distaplia occidentalis* behave in the same way.

The larvae of *D. macdonaldi* display another example of substrate preference that does not depend on lighting conditions. When larvae are placed in Polycon containers (Tenite polyethylene 18B0; made by Richards Mft. Co. Van Nuys, California) and examined under a dissecting microscope, they can be seen to swim around for a few seconds, but the first time they hit the wall of the container directly with their papillae they instantly settle and begin to metamorphose. This occurs about 60% of the time. About 40% of the larvae require several contacts. Direct contact is required, for if the plastic in an envelope of bolting silk is placed in a culture dish, it does not induce settlement. Newly cut surfaces of plastic are just as effective as the molded surfaces.

These experiments convince me that the larval nervous system and sensory organs are much more important in selecting sites for settlement than previously suspected. We are in need of more detailed studies of the structure of these organs.

Coordination of the Events of Metamorphosis

In the aplousobranchs one may observe within a period of 15 minutes, contractions of the epidermal cells of the papillae, tail, posterior trunk, sensory vesicle and ampullae; the notochordal cells contract, and the blood cells begin rapid locomotion. What coordinates these events? What signals the blood cells to start moving? How are the epithelial cells stimulated to contract in the correct sequence? These are especially difficult problems to study and we have no answers. Large numbers of larvae can be induced to metamorphose simultaneously with DMSO. Extracts of these specimens might contain a coordinating factor that could be detected by observing the effects of fractions on blood cells or isolated larval tails.

ACKNOWLEDGMENTS

The preparation of this paper was supported in part by National Science Foundation Research Grant BMS 7507689 and by The Graduate School Research Fund of the University of Washington. I wish to thank Professor Robert L. Fernald for three decades of enthusiastic support of the field of invertebrate embryology and especially for his critical comments made during the development of my interests in ascidian metamorphosis. Mr. David Borg has provided excellent technical assistance in the preparation of sections and micrographs.

Part of this research was carried out at the Fort Pierce Bureau of the Smithsonian Institution and the Harbor Branch Foundation, Inc. in Fort Pierce, Florida. I am grateful for the use of their facilities and for the assistance offered me by Dr. Mary Rice.

REFERENCES

1. Monniot, C. and Monniot, F. (1972) Arch. Zool. exp. gen., 113, 311-367.
2. Kowalevsky, A. (1866) Mem. Acad. Sci. St. Petersbourg., 8, 1-19.
3. Berrill, N. J. (1950) The Tunicata, Ray Society, London, pp. 1-354.
4. Millar, R. H. (1971) Adv. Mar. Biol., 9, 1-100.
5. Berrill, N. J. (1931) Phil. Trans. Roy. Soc. B, 219, 281-346.
6. Brewin, B. I. (1959) Quart. J. Micro. Sci., 100, 575-589.
7. Grave, C. (1944) J. Morph., 75, 173-191.
8. Cloney, R. A. (1961) Am. Zool., 1, 67-87.
9. Grave, C. (1926) J. Morph., 42, 453-467.
10. Willey, A. (1894) Quart. J. Micro. Sci., 35, 295-316.
11. Elwyn, A. (1937) Bull. Neur. Inst. N.Y., 6, 163-177.
12. Eakin, R. M. and Kuda, A. (1971) Z. Zellforsch., 112, 287-312.
13. Cloney, R. A. (1969) Z. Zellforsch., 100, 31-53.
14. Cloney, R. A. (1964) Acta Embryo. et Morphol. Exp., 7, 11-130.
15. Salensky, W. (1893) Morph. Jahrb., 20, 449-542.
16. Ivanova-Kazas, O. M. (1967) Cahiers de Biol. Marine, 8, 21-62.
17. Trason, W. B. (1957) J. Morph., 100, 509-546.
18. Trason, W. B. (1963) Univ. Calif. Publs. Zool., 65, 283-326.
19. Levine, E. P. (1962) J. Morph., 111, 105-138.
20. Grave, C. (1934) Carneg. Inst. Wash. Publ., 435, 143-156.
21. Grave, C. (1921) J. Morph., 36, 71-101.
22. Scott, F. M. (1946) Biol. Bull., 91, 66-88.
23. Sebastian, V. O. (1942) J. Madras Univ., 14, 251-278.
24. Abbott, D. P. (1955) J. Morph., 97, 569-594.
25. Grave, C. and Woodbridge, H. (1924) J. Morph., 39, 207-247.
26. Cloney, R. A. (1972) Z. Zellforsch., 132, 167-192.
27. Cavey, M. J. and Cloney, R. A. (1976) Cell Tiss. Res., 174, 289-313.
28. Garstang, S. L. and Garstang, W. (1928) Quart. J. Micro. Sci., 72, 1-49.
29. Grave, C. and McCosh, G. K. (1935) Wash. Univ. Stud. Scient., 11, 89-117.
30. Cavey, M. H. and Cloney, R. A. (1972) J. Morph., 138, 349-374.
31. Schiaffino, S., Nunzi, M. G. and Burighel, P. (1976) Tissue and Cell, 8, 101-110.
32. Tannenbaum, A. S. and Rosenbluth, J. (1972) Experientia, 28, 1210-1212.
33. Berrill, N. J. (1948) Quart. J. Micro. Sci., 89, 253-289.
34. Salensky, W. (1893) Morph. Jahrb., 20, 48-74.
35. Barnes, S. N. (1971) Z. Zellforsch., 117, 1-116.
36. Zimmer, R. L. and Woollacott, R. M. (1977) in Biology of Bryozoans, Woollacott, R. M. and Zimmer, R. L. eds., Academic Press, New York, pp. 91-142.
37. Whitten, J. (1968) in Metamorphosis, Edkin, W. and Gilbert, L. I. eds., pp. 43-105.
38. Dent, J. N. (1968) in Metamorphosis, Edkin, W. and Gilbert, L. I. eds. pp. 271-311.
39. Cloney, R. A. (1977) Cell Tiss. Res., 183, 423-444.
40. Numakunai, T., Ishkawa, M. and Hirai, E. (1965) Bull. Mar. Biol. Stat. Asamushi, 12, 161-171.
41. Lane, D.J.W. (1973) Marine Biology, 12, 47-58.
42. Sebastian, V. O. (1954) J. Wash. Acad. Sci., 44, 18-23.
43. Berrill, N. J. (1948) J. Morph., 82, 355-363.
44. Kott, P. (1957) Aust. J. Mar. Freshwater Res., 8, 64-110.

45. Kasas, O. M. (1940) Bull. Acad. Sci. U.S.S.R. Biol., pp. 862-883.
46. Cloney, R. A. (1963) Biol. Bull., 124, 241-253.
47. Cloney, R. A. (1966) J. Ultra. Res., 14, 300-328.
48. Huxley, H. E. (1963) J. Mol. Biol., 7, 281-308.
49. Ishikawa, H., Bischoff, R. and Holtzer, H. (1969) J. Cell. Biol., 43, 312-328.
50. Schroeder, T. E. (1975) in Molecules and Cell Movement, Inoue, S. and Stevens, R. E. eds. Raven Press, N.Y., pp. 305-334.
51. Wessells, N. K., Spooner, B. S., Ash, J. F., Bradley, M. O., Luduena, M. A., Taylor, E. L., Wrenn, J. T. and Yamada, K. M. (1971) Science, 171, 135-143.
52. Lash, J. W., Cloney, R. A. and Minor, R. R. (1973) Biol. Bull., 145, 360-372.
53. Schiaffino, S., Burighel, P. and Nunzi, M. G. (1974) Cell Tiss. Res., 153, 293-305.
54. Ishikawa, M. and Numakunai, T. (1972) Dev. Growth and Diff., 14, 75-83.
55. Oka, H. (1943) Ann. Zool., Jap., 22, 54-58.
56. Cloney, R. A. and Grimm, L. (1970) Z. Zellforsch., 107, 157-173.
57. Seeliger, O. (1893) Z. Wiss. Zool., 56, 488-505.
58. Scott, F. M. (1952) Biol. Bull., 103, 226-241.
59. DeSanto, R. S. and Dudley, P. L (1969) J. Ultra. Res., 28, 259-274.
60. Anderson, D. T., White, B. M. and Egan, E. A. (1976) Proc. Linn. Soc. New South Wales, 100, 205-217.
61. Zhinkin, L. (1939) C.R. Acad. Sci. Moscou, (N.S.), 24, 620-622.
62. Grave, C. (1935) Carneg. Inst. Wash. Publ., 435, 143-156.
63. Lynch, W. F. (1961) Am. Zool., 1, 59-66.
64. Glaser, O. and Anslow, G. A. (1949) J. Exp. Zool., 111, 117-140.
65. Whittaker, J. R. (1964) Nature, 202, 1024-1025.
66. Crisp, D. J. and Ghobashy, A.F.A.A. (1971) Fourth European Marine Biology Symposium, Crisp, D.J. ed., Cambridge, London, pp. 443-465.

PERSPECTIVES: SETTLEMENT AND METAMORPHOSIS OF MARINE INVERTEBRATE LARVAE

Fu-Shiang Chia

Department of Zoology, University of Alberta, Edmonton, Canada T6G 2E9

The activities of most marine invertebrate larvae consist of basically three phases: active moving, settlement, and metamorphosis. The latter two phases transform the larva into a juvenile with an entirely different mode of life. Settlement, including attachment in some cases, refers to general behavioral and habitat changes, whereas metamorphosis denotes morphological and physiological changes. Settlement and metamorphosis usually occur in a relatively short period of time and neither is reversible.

Many invertebrate larvae settle first and then metamorphose, others metamorphose first and then settle, and in others the two processes take place simultaneously.

In this brief communication, I present a model, illustrating the processes and possible mechanisms of settlement and metamorphosis (Fig. 1).

To begin with, the young larva undergoes either an active feeding-differentiation-growth period in planktotrophic development or a differentiation period alone in lecithotrophic development. This is essentially a preparation period for eventual settlement and metamorphosis. Larval behavior during this period is related either to feeding, locomotion or avoidance of predators; the larva is not capable of responding to specific environmental cues which may lead to settlement and metamorphosis.

A larva is said to be "competent" when it has entered a physiological state in which it is capable of metamorphosis when given proper environmental conditions and this state is often marked by a set of recognizable morphological characteristics. The time required for a young larva to become competent is species-specific; it may take from a few hours in some species to many days or months in others. This newly acquired ability of competent larvae to respond to environmental cues usually coincides with the development of certain sensory receptors. There is a good deal of evidence presented in the literature, much of it reviewed in chapters in this symposium, which documents the state of competence of many larval forms. A variety of environmental cues, including some purified metamorphic-inducing substances, also has been reported.

Information on larval sensory receptors is very limited at the present time (see Bonar's review in this symposium). In a recent study of the larval behavior of *Rostanga pulchra*, a nudibranch, we have found that the strong negative phototactic behavior correlates with the development of the larval eyes and the ability to respond to the metamorphic-inducing substrate corresponds to the time of development of rhinophores (presumably bearing chemoreceptors). In this particular case the metamorphic-inducing substrate is an encrusting sponge, *Ophlitaspongia pennata,* an inhabitant of overhangs or crevices of rocky shores. The negative phototactic behavior prior to settlement clearly increases the possibilities of encountering the sponge which is the prey species of *Rostanga pulchra.*

It is perhaps useful to discuss, at this point, the relationships between the environmental cues and invertebrate larvae in general. On the basis of published information, some of which is reported or reviewed in this symposium, two generalizations can be drawn:

1. Specialist species are more dependent on environmental cues for induction of settlement and metamorphosis than are generalist species. The inducing factors are likely to be found in prey species, or host species in cases of symbiotic associations.

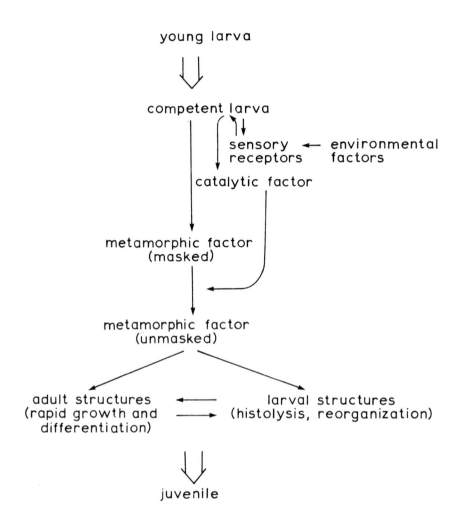

Fig. 1. General events and possible mechanisms of settlement and metamorphosis in marine invertebrate larvae.

2. Gregarious and sedentary species are more dependent on environmental stimuli for induction of settlement and metamorphosis than are non-gregarious and mobile species. The inducing factors are to be found in other individuals of the same speeies.

The above generalizations should provide some guidelines for predicting whether or not a given species is likely to be dependent on environmental stimuli for induction at the time of settlement and metamorphosis and in the selection of species for the study of specific problems. One should choose a specialist, or a gregarious and sedentary species if one is to study the induction of settlement and metamorphosis. Furthermore, one is likely to find that the sensory receptors are better developed in the competent larvae of such species. On the other hand, the generalizations also help one to choose species which are likely to be less dependent on environmental cues for settlement and metamorphosis. In order to have an in-depth understanding of settlement and metamorphosis, comparative studies of a large number of species within a broad spectrum of environmental dependency are necessary. Such studies will help not only to confirm or reject the present generalizations but also will generate further hypotheses.

When applying these hypotheses, it should be kept in mind that many marine invertebrates undergo considerable changes in requirements from juveniles to adults. These include the changes of both food and habitats. In these situations, the requirements of the juvenile are more important than those of the adult when considering induction of settlement and metamorphosis. The actual inducing factors may or may not have anything to do with the adult environment but they are essential to the juveniles.

Returning to Figure 1, it may be noted that a number of steps are proposed in the transformation of a competent larva into a juvenile, steps which account for the processes of settlement and metamorphosis. All these steps, except the changes in larval or adult structures, are speculative. I reasoned here that once a larva becomes competent, it must have acquired a chemical substance which I have called a metamorphic factor. The primary function of this factor is to promote the process of metamorphosis by conditioning the tissues on the one hand and initiating the chain reactions of metamorphosis on the other. This factor, as I have reasoned, is masked, and it cannot exert its effect unless it is unmasked. This belief is based on the fact that many competent larvae can delay metamorphosis for a considerable length of time. The mask itself may undergo a natural decay so that the "aged" larvae will metamorphose even though the proper environmental cues are not available. The rate of decay is species-specific; hence we find great variations as to how long a larva can delay its metamorphosis. There are extreme cases in which the larvae can delay metamorphosis indefinitely and eventually die as larvae. The mask can be easily destroyed by another factor which is produced by the larva when it is stimulated by the proper environmental cues and I have called this the catalytic factor.

Searching for a metamorphic or catalytic factor will be a trying process. The first set of target organs in the larvae for such studies will be in the endocrine and nervous systems. So far, few endocrine organs have been found in marine invertebrate larvae. However, the failure to identify endocrine systems may be due largely to the lack of systematic effort among invertebrate biologists. Moreover, we are pre-conditioned about the morphology of vertebrate endocrine systems and may not be able to recognize an invertebrate larval endocrine system if we see one. There is certainly no lack of glandular cells in the larval tissue and we have very little knowledge about the function of many types of the glands. It is possible that some of the glands are endocrine in nature. Neurosecretions of the larval ganglia should be closely examined, especially in the competent larvae. Nerve impulse alone may be sufficient to function as a catalytic factor for initiating the metamorphic changes. A wide range of approaches, concentrating on neuroendocrine systems, should be employed.

Once the metamorphic factor is unmasked, it can exert its effect in three ways (Fig. 1). The first is to affect the larval structures directly, causing them either to break down or to reorganize. The changes in larval structures may in turn set off a rapid process of differentiation and growth of adult structures. The second way in which the metamorphic factor may act is the reverse of the first: that is, it stimulates formation of adult structures and loss of larval structures follows. The third, and most likely way, is to affect the larval and adult structures simultaneously. The end result of these changes is the transformation of larva to juvenile. Numerous reports, including most chapters in this symposium, have provided descriptions of morphological changes in both larval and adult structures, either at the tissue or cellular lvel. Some reports (e.g., Cloney's in this symposium) have addressed the mechanisms of the morphological changes during metamorphosis.

In spite of all the information which has been gained on the morphological and structural changes of metamorphosis, a systematic effort is still needed before we can make a meaningful synthesis or prediction as to what sorts of larval structures are likely to be destroyed and what sorts are likely to be conserved. The ability to delay metamorphosis may provide us with a good system for studying the genetic programming of the destined changes of the larval structures. In addition we need more work on the mechanisms of the structural changes.

In closing, it must be recorded that some of the ideas in this communication have resulted from discussions with several of my colleagues and students. The writing of this paper was supported by a grant from the National Research Council of Canada.

285

INDEX